W9-ABN-182

Non-Standard and Improperly Posed Problems

This is volume 194 in
MATHEMATICS IN SCIENCE AND ENGINEERING
Edited by William F. Ames, *Georgia Institute of Technology*

A list of recent titles in this series appears at the end of this volume.

Non-Standard and Improperly Posed Problems

Karen A. Ames

DEPARTMENT OF MATHEMATICS
UNIVERSITY OF ALABAMA AT HUNTSVILLE, ALABAMA, USA

and

Brian Straughan

DEPARTMENT OF MATHEMATICS
UNIVERSITY OF GLASGOW, GLASGOW, UK

ACADEMIC PRESS
Harcourt Brace & Company, Publishers
San Diego London Boston New York
Sydney Tokyo Toronto

This book is printed on acid-free paper.

ACADEMIC PRESS, INC.
525 B Street, Suite 1900, San Diego, CA 92101-4495, USA
http://www.apnet.com

ACADEMIC PRESS LIMITED
24–28 Oval Road, London NW1 7DX, UK
http://www.hbuk.co.uk/ap/

ISBN 0-12-056745-8

A catalogue record for this book is available from the British Library

Printed and bound in Great Britain by the University Press, Cambridge

97 98 99 00 01 02 EB 9 8 7 6 5 4 3 2 1

Contents

Preface

We both owe an immense amount of gratitude to Professor Lawrence E. Payne of Cornell University, whose influence on both the subject and us personally has been profound. It gives us great pleasure to have the opportunity to thank him.

We should like to thank Professor William F. Ames of the Georgia Institute of Technology, for encouraging us to publish this volume in the present series. For pertinent comments on this and related work, special thanks to Professor James N. Flavin of University College Galway, Professoressa Franca Franchi of the University of Bologna, and to Professor Salvatore Rionero of the University of Napoli, Federico II. The work of B.S. was partly completed with the help of a Max Planck Forschungspreis, of the Alexander von Humboldt Foundation and the Max Planck Institute. We both wish to thank NATO for their support through grant number CRG 910630.

This book is intended to be a resource for mathematicians, scientists, and engineers interested in the area of improperly posed problems for partial differential equations. Although there has been a significant amount of research on such problems in the last fifteen years, little of this work has appeared in any unified form. The aim of this monograph is to survey some of the recent literature on ill posed problems and to examine new directions and applications in the field. We are particularly interested in filling a substantial gap in the mathematical literature by collecting an assortment of results for non-standard problems in partial differential equations that have received attention recently in connection with a number of physical applications.

Since it would not be possible to thoroughly cover the broad spectrum of improperly posed problems, we shall focus on the following topics: continuous dependence of solutions on spatial and initial time geometry; continuous dependence on modeling questions; problems in which the data are prescribed on only part of the domain boundary; the first two belong to the area often known as structural stability. We do not treat in detail other classes of improperly posed problems such as inverse problems or Dirichlet problems for partial differential equations. Our goal is to discuss the progress made with relatively new classes of ill posed problems.

After an introductory chapter that includes a description of the methodology typically used to study ill posed problems, the book is divided into five chapters. These address, respectively, the following subjects: continuous dependence on initial - time geometry and on spatial geometry, continuous dependence on modeling backward in time, continuous dependence on modeling forward in time, non-standard or non-characteristic problems such as those where the data are only prescribed on a part of the boundary, and other relevant improperly posed problems. A bibliography is included in order to provide researchers with an up to date compendium of the mathematical literature on the aforementioned topics. If any relevant literature is omitted this is entirely unintentional.

Chapter 5 and sections 1.6, 3.1, 3.2, 3.3, 3.4, 6.1, 6.5, and 6.6 were written by K.A. Ames, while chapters 2 and 4 and sections 1.4, 6.2, 6.3, and 6.4 were written by B. Straughan. The remaining sections were jointly written.

K.A. Ames
Associate Professor of Mathematics
University of Alabama at Huntsville

B. Straughan
Simson Professor of Mathematics
University of Glasgow

K.A. Ames & B. Straughan

Non-Standard and Improperly Posed Problems

1

Introduction

1.1 Introductory Remarks

The study of improperly posed problems for partial differential equations has received considerable recent attention in response to the realization that a number of physical phenomena lead to these kinds of mathematical models. In this book, we intend to describe some of the topics confronted in the investigation of several classes of ill-posed and non-standard problems.

A boundary or initial value problem for a partial differential equation is understood to be well posed in the sense of Hadamard if it possesses a unique solution that depends continuously on the prescribed data. This definition is made precise by indicating the space to which the solution must belong and the measure in which continuous dependence is desired. In general we shall refer to a boundary and/or initial value problem for a partial differential equation, or a system of partial differential equations as simply a problem. A problem is called *improperly posed* (or *ill-posed*) if it fails to have a global solution, or if it fails to have a unique solution, or if the solution does not depend continuously on the data. In the subsequent chapters we shall discuss the three major questions of existence, uniqueness, and stabilization of solutions to such problems.

Considerable progress has been made in the analysis of improperly posed problems for partial differential equations. The monograph by Payne (1975) is a primary source on various topics in this field and includes a substantial bibliography of the work done prior to 1975. In the last twenty years, research on ill-posed problems has intensified and moved in so many diverse directions that it would be virtually impossible to give a comprehensive, up-to-date account of all the work that has appeared on these kinds of problems. It is no surprise then that the few existing books on improperly posed problems deal with very specific classes of problems. In addition to Payne's volume, there are the books of Bloom (1981), Flavin & Rionero (1995), Lattes & Lions (1969), Lavrentiev (1967), Lavrentiev, Romanov & Sisatskii (1986), Morozov (1984), Straughan (1982), and Tikhonov & Arsenin (1977). These books deal with various aspects of improperly posed problems although not exclusively so.

Improperly posed problems occur in many mundane areas. We mention attempts to calculate the age of the Earth by extrapolation backward in time, notably work of Lord Kelvin. Similarly, there are studies of extrapolating the behaviour of the Earth's magnetic field to previous times; these are both examples of improperly posed problems. Another important area involves oil recovery from an underground source. Geophysicists take measurements using seismic waves and need to understand the state of the oil encapsulated in the rocks, often well below ground. There are many improperly posed problems in oil reservoir engineering, as may be seen in the accounts of Ewing (1984) and Ewing *et al.* (1987). Also important in oil reservoir recovery theory is the question posed when a less viscous fluid breaks through a more viscous one. This is important when oil is attempted to be withdrawn from an underground reservoir by pumping water into the reservoir. The mathematical theory for this is closely connected with the Hele-Shaw problem where a less viscous fluid is driven against a more viscous one. This theory gives rise to an improperly posed problem and is studied by Di Benedetto & Friedman (1984), where other relevant references may be found. Other familiar occurrences of ill posed problems are in glacier physics, cf. Hutter (1984), Payne & Straughan (1990b), where only the surface of the glacier is available for measurement. Climatology is another area which has led to several novel improperly posed problems, cf. Diaz & Lions (1993). Chorin (1985) studies numerically a solidification/melting problem in three - dimensions. He shows how this is an ill-posed problem and how care must be taken. Joseph & Saut (1990) give several examples of problems in continuum mechanics where the equations change type and thus lead to improperly posed problems, and Joseph *et al.* (1985) do likewise, concentrating on flows of viscoelastic fluids. In the important field of multi - component mixtures of fluids, which have many practical applications, including cloud physics, upper atmospheric physics, and other areas, several practical improperly posed problems arise; a study of waves and related questions in these theories may be found in e.g. Muller & Villaggio (1976), Morro & Straughan (1990a,b), and Morro *et al.* (1993). Further instances of improperly posed problems in thermal convection studies in fluid flow in porous media are given in e.g. McKay (1992), Morro & Straughan (1992), and Richardson (1993). A uniqueness theorem in an improperly posed problem for a piezoelectric material is given by Morro & Straughan (1991). We also mention Carasso's (1994) important paper which describes a technique for solving a class of ill posed problems. He illustrates his technique by showing how one can sharpen photographs of brain imagery. This is a highly practical study of an improperly posed problem which has application in the field of medicine.

It is worth making the connection that an improperly posed problem may have with studies of human brain models. Hawking (1993) remarks that the human brain consists of about 10^{26} particles. It is thus very sensitive to the initial conditions and use of a brain model will be difficult to predict the behaviour of a human. Finally, we note that related phenomena is involved with fragile economies. Here a small change in a parameter can lead to stock market chaos, currency fluctuations, and several other highly changeable situations.

Mathematicians who investigated ill-posed problems for partial differential equations before 1950 dealt primarily with questions of uniqueness for linear equations since such problems typically have at most one solution. However, it should be pointed out that examples of nonunique continuation for linear elliptic and parabolic equations have been constructed by Plis (1960,1963) and Miller (1973,1974). In fact, the literature of the 1950's and early 1960's is rich with studies which contain examples of equations, often nonlinear, for which uniqueness fails.

While it is typical that improperly posed problems will have at most one solution, it is often the case that solutions will not exist for arbitrary data and if they do, such solutions will not depend continuously on the data. Throughout this book, the term "existence" refers to global existence. For some of the problems considered, local existence can be deduced from the Cauchy-Kowalevski theorem.

Global existence of solutions to ill-posed problems is difficult to discuss with much generality. It is frequently the case that very strong regularity and compatibility conditions must be imposed on the data in order for a solution to exist. Since errors can be made in measuring data, such conditions cannot always be guaranteed in practice. Attempts to deal with the existence question have often followed the strategy of modifying either the concept of a solution or the mathematical model. In many physical situations, it is sufficient to define the "solution" as a domain function that belongs to an appropriate class and best approximates the prescribed data. Alternatively, changing the underlying model has led to the development of techniques to generate approximate solutions, including Tikhonov regularization (Tikhonov, 1963) and the quasireversibility method (Lattès & Lions, 1969). The primary aim of these methods is the numerical solution of ill-posed initial-boundary value problems. The calculation of approximate solutions may be as close as we come to establishing the existence of a "solution" to such problems. One issue we encounter in attempting to compute solutions of these problems is the question of stability. Not only do we need to derive stability estimates but we also need to develop numerical methods that effectively incorporate the prescribed bounds usually required to stabilize ill-posed problems. The work of Bellomo & Preziosi (1995), Cannon (1964a,b,c,1984), Cannon & Douglas (1967,1968), Cannon & Ewing (1976), Cannon & Knightly (1969), Cannon & Miller (1965), Carasso (1976,1977,1982,1987,1993,1994),Carasso & Stone (1975), Chorin (1985),

Demmel & Kagstrom (1988), Douglas (1960), Elden (1987,1988,1990), Engl & Groetsch (1987),Ewing (1975,1984), Ewing & Falk (1979), Ewing *et al.* (1987), Falk & Monk (1986), Han (1982), Hansen (1992), Ingham *et al.* (1991), Klibanov & Rakesh (1992), Klibanov & Santosa (1991), Lin & Ewing (1992), Miller (1973b), Monk (1986) deals with these difficulties.

The subject of continuous dependence on the data has received considerable attention since 1960 with the result that there exist a number of methods for deriving continuous dependence inequalities. It is now well established that for many improperly posed problems, if the class of admissible solutions is suitably restricted, then solutions in this class will depend continuously on the data in the appropriate norms. In many physically interesting problems, stability, in the modified sense of Hölder (John 1955,1960), is guaranteed when a uniform bound for the solution or some functional of the solution is prescribed. Since such a bound need not be sharp, it is reasonable to anticipate that it can be obtained from the physics of the problem via observation and measurement. We emphasize that this type of continuous dependence should not be confused with asymptotic stability or with the usual notion of continuous dependence for well posed problems.

Our discussion of continuous dependence in the context of improperly posed problems will focus on three important sources of errors that arise in the derivation and analysis of a mathematical model for a physical process:

(1) errors in measuring data (initial or boundary values, coefficients in the equations or boundary operators, parameters),

(2) errors in characterizing the spatial geometry of the underlying domain and the initial-time geometry, and

(3) errors in formulating the model equation(s).

In connection with the subject of continuous dependence on the model itself we point out that this is a topic attracting the attention of many people. Bellomo & Preziosi (1994), pp. 85-86, and Drazin & Reid (1981), pp. 407-415, also devote attention to this topic and present several examples. Bellomo & Preziosi (1994) call continuous dependence on the model itself *structural stability*, or *total stability*. They point out that into this category one should include changes to the equations themselves, such as the addition of dissipative terms. Also important are questions of continuous dependence on the coefficients such as the body force. Bellomo & Preziosi (1994) write that structural stability is more dificult to deal with than stability of equilibria with respect to perturbations of the initial state. They also state that structural stability is hard to verify and to obtain in practice. We devote much attention to questions of continuous dependence on modeling, i.e. structural stability, in the course of this book.

In order to recover continuous dependence in ill posed problems, we must constrain the solution in some mathematically and physically realizable manner. It would be optimal if this constraint could simultaneously stabilize the problem against all possible errors that might arise as we set up the model. However, it is usually the case that the appropriate constraint is difficult to determine. Moreover, such a restriction will turn an otherwise linear problem into a nonlinear one and thus caution must be exercised in any attempt to study the effects of various errors separately and to then superpose these effects.

Since this book is dedicated to studying novel classes of improperly posed problems we now list some of the relevant literature in the appropriate sub-fields. The lists we give are not meant to be exhaustive, but give a guide as to where further references may be found. In the field of continuous dependence on the initial - time geometry we cite the work of Ames & Payne (1994b,1995a), Ames & Straughan (1995), Franchi & Straughan (1995), Knops & Payne (1969,1988), Payne (1987a,b,1989,1992), Payne & Straughan (1990c,1996a), and Song (1988), while in the field of continuous dependence on the spatial geometry we refer to Crooke & Payne (1984), Lin & Payne (1994), Payne (1993a), Payne & Straughan (1990a,1996b), and Persens (1986). The subject of continuous dependence on modeling, or structural stability, is treated in Adelson (1973,1974), Ames (1982a,b,1992), Ames *et al.* (1987), Ames & Payne (1992,1994a,c,1995b), Ames & Straughan (1992), Bellomo & Preziosi (1995), Bennett (1986,1991), Chirita & Rionero (1991), Flavin & Rionero (1995,1996), Franchi (1984, 1985,1995), Franchi & Straughan (1993a,b,1994a,b,1996), Knops & Payne (1969, 1988), Lin & Payne (1994), McKay (1992), Morro & Straughan (1992), Mulone *et al.* (1992), Payne (1967,1971,1975,1987a,b,1993b), Payne & Sather (1967a,b), Payne & Straughan (1989a,b,1990a,b), Richardson (1993), Rionero & Chirita (1987,1989), and Song (1988). The question of continuous dependence in improperly posed problems when data are given on non-characteristic surfaces, including the so-called "sideways" problems, or data are given on other non-standard surfaces, is investigated in Ames (1992), Ames & Isakov (1991), Bell (1981a,1981b), Cannon & Knightly (1969), Klibanov & Rakesh (1992), Lavrentiev *et al.* (1983), Levine & Payne (1985), Manselli & Miller (1980), Monk (1986), Payne (1985), Payne & Straughan (1990b), Talenti (1978), and Talenti & Vessella (1982). Finally, several other relevant classes of interesting ill-posed and connected problems, including those involving spatial decay in a cylinder, and quasi-reversibility methods, may be found in Bellomo & Preziosi (1995), Cannon (1984), Elden (1987,1988,1990), Engl & Groetsch (1987), Flavin (1989,1991,1992,1993,1995,1996a,1996b), Flavin & Rionero (1993, 1995), Franchi & Straughan (1994b), Kazemi & Klibanov (1993), Klibanov & Santosa (1991), Knabner & Vessella (1987a,b,1988), Lattès & Lions (1969),

Lavrentiev *et al.* (1983), Levine (1983), Levine & Vessella (1985a,b), Lin & Payne (1993,1994), Miller (1973a,b,1974,1975), and Showalter (1975).

1.2 Notation and Some Important Inequalities

We introduce here some of the notation that will be used throughout this book as well as some important inequalities that will be referred to frequently in the course of the exposition.

Many of the problems in this book will be defined on a bounded region Ω of N dimensional Euclidean space $\mathbf{R}^N$. The boundary of Ω will be denoted by Γ or $\partial\Omega$ and the exterior of Ω will usually be denoted by Ω^*. The spatial variable is denoted by $\mathbf{x} = (x_1, \dots, x_N)$ while the time variable is denoted by t. On the boundary of Ω, $\mathbf{n} = (n_1, \dots, n_N)$ will represent the outward pointing unit normal and the normal derivative will be given by $\partial u/\partial n$. We shall employ a subscript comma to denote partial differentiation with respect to the spatial variable and adopt the summation convention so that repeated indices in an expression are summed from 1 to N.

Throughout this book, the symbol $L^p(\Omega)$, $p \geq 1$, will refer to the Banach space consisting of all measurable functions on Ω that have finite norm

$$\|u\|_p = \left(\int_\Omega |u|^p dx \right)^{1/p}.$$

The norm in $L^2(\Omega)$ will usually be abbreviated to $\|\cdot\|$. For the most part, we have used integral notation but have occasionally employed the inner product notation $(\cdot, \cdot)$, where the inner product is understood to be that on the Hilbert space $L^2(\Omega)$.

The following four common inequalities will be used throughout the book:

(i) the arithmetic-geometric mean inequality (sometimes abbreviated to the AG inequality), with weight $\alpha > 0$,

$$2uv \leq \alpha u^2 + \frac{1}{\alpha} v^2,$$

or its integrated form

$$2 \int_\Omega uv \, dx \leq \alpha \int_\Omega u^2 \, dx + \frac{1}{\alpha} \int_\Omega v^2 \, dx,$$

(ii) the Cauchy-Schwarz inequality (or simply Schwarz's inequality),

$$\int_\Omega uv\,dx \le \left(\int_\Omega u^2 dx\right)^{1/2}\left(\int_\Omega v^2 dx\right)^{1/2},$$

(iii) Hölder's inequality,

$$\int_\Omega uv\,dx \le \left(\int_\Omega u^p dx\right)^{1/p}\left(\int_\Omega v^q dx\right)^{1/q},$$

where

$$\frac{1}{p}+\frac{1}{q}=1, \qquad p,q>1,$$

and

(iv) Young's inequality,

$$\int_\Omega uv\,dx \le \frac{1}{p}\int_\Omega u^p dx + \frac{1}{q}\int_\Omega v^q dx,$$

with

$$\frac{1}{p}+\frac{1}{q}=1, \qquad p,q>1,$$

or the pointwise form

$$uv \le \frac{u^p}{p}+\frac{v^q}{q}.$$

We shall occasionally want to formulate a problem in an abstract setting, in which case we will deal primarily with a real Hilbert space, denoted usually by H. In such spaces, we define a scalar product (u, v) for an arbitrary pair of elements u and v. As the norm of an element u we take $\|u\| = (u,u)^{1/2}$. A subset D of H is called *dense* if the closure of D is H. An operator A, defined on some set $\mathcal{D}(A)$, assigns to each element $u \in \mathcal{D}(A)$ a certain element $v \in H$. This will usually be written $v = Au$ or $v = A(u)$. If the equality

$$A(\lambda u_1 + \mu u_2) = \lambda A u_1 + \mu A u_2,$$

holds, we say that A is *linear*. If, in addition, there exists a constant c such that

$$\|Au\| \le c\|u\|, \qquad \forall u \in \mathcal{D}(A), \tag{1.2.1}$$

then A is called a *bounded* operator. The adjoint of an operator A is denoted by A^* and is defined by the relation

$$(Au, v) = (u, A^* v). \tag{1.2.2}$$

When A is a bounded linear operator it can be shown that $\mathcal{D}(A) = H$ and there exists a unique operator A^* which satisfies (1.2.2). If, however, A does not satisfy (1.2.1) it is then unbounded and a stronger condition has to be placed on $\mathcal{D}(A)$ as we discuss below. An operator A is symmetric if

$$(Au, v) = (u, Av), \qquad \forall u, v \in \mathcal{D}(A),$$

and skew-symmetric if, instead,

$$(Au, v) = -(u, Av) \qquad \forall u, v \in \mathcal{D}(A).$$

It is a well known result that the domain of an unbounded operator A must be dense in H. However, for completeness, we now show that in general it is impossible to define the adjoint operator uniquely unless $\mathcal{D}(A)$ is dense in H.

Let $V = \mathcal{D}(A)$ and let V be dense in H. Suppose that for $y \in H$, there exists $z \in H$, such that

$$(Ax, y) = (x, z) \qquad \forall x \in V.$$

Denote by D^* the set of all such elements y and define the operator $A^* : D^* \to H$ with domain of definition $\mathcal{D}(A^*) = D^*$, by

$$A^* y = z.$$

Then A^* is the adjoint operator to A. The vector z is uniquely defined by the vector y, for, suppose there exists $z' \in V$ such that

$$(Ax, y) = (x, z') \qquad \forall x \in V.$$

Then,

$$(x, z - z') = 0 \qquad \forall x \in V.$$

But, since V is dense in H, this argument may be extended to show

$$(h, z - z') = 0 \qquad \forall h \in H. \tag{1.2.3}$$

(The extension goes as follows. Fix a, b in H. Suppose $a_n \to a$ as $n \to \infty$, with $a_n \in V$, i.e.

$$\|a_n - a\| \to 0 \quad \text{as} \quad n \to \infty.$$

Suppose also $(a_n, b) = 0$ for all n. Then,

$$\begin{aligned} |(a, b)| &= |(a, b) - (a_n, b)| \\ &= |(a - a_n, b)| \\ &\leq \|a - a_n\| \, \|b\|. \end{aligned}$$

Hence

$$(a, b) = 0.)$$

From (1.2.3) it follows that

$$(z - z', z - z') = 0.$$

So

$$z = z'.$$

Hölder Stability

In dealing with improperly posed problems we frequently have recourse to use the concept of Hölder continuous dependence on the data. This concept was explained lucidly by John (1960) whose notation we now employ.

Consider a problem with a set, U, of solutions and a set, F, of data. Let A be the mapping from the set of data to the set of solutions. In order for the difference of two data terms or solutions to be meaningful, we assume that the solutions to the problem are defined on a subset of U, say R, and the data is defined on a subset of F, say G, such that R and G are normed linear spaces with $\|\cdot\|_R$ and $\|\cdot\|_G$ as the norms on R and G, respectively. For any subset S, of R, with norm $\|\cdot\|_S$, if $u_1, u_2 \in U$, and $f_1, f_2 \in F$, such that $u_1 = Af_1$, $u_2 = Af_2$, we say the mapping A is *Hölder continuous* at f_1, if and only if,

$$\sup_{u_2 \in S} \|u_1 - u_2\|_S < M\epsilon^\alpha \qquad \text{whenever} \qquad \|f_1 - f_2\|_G < \epsilon,$$

where M, α are positive constants depending on U and S.

This definition has obvious applications in physical problems, in which case we let

$$R = \{u(t) \in U \big| t \in [0, T)\},$$

where $[0, T)$ refers to an interval of time,

$$S = \{u(t) \in U \big| t \in [0, T_1)\}, \qquad T_1 \leq T,$$

and

$$G = \{f \in F \big| \text{ such that } \exists\, u \in U \text{ with } u(0) = f\}.$$

We may then reformulate the above definition in order to discuss the continuous dependence of the solution on the initial data of the problem. The extension to other forms of data is clear.

The solution $u_1(\cdot, t)$ is *Hölder stable* on the interval $t \in [0, T)$, if and only if, given $\epsilon(> 0)$, then

$$\sup_{0 \leq t \leq T_1 < T} \|u_1(\cdot, t) - u_2(\cdot, t)\|_t < C\epsilon^\alpha,$$

whenever

$$\|u_1(\cdot,0) - u_2(\cdot,0)\|_0 < \epsilon,$$

$\forall u_2(\cdot,0) \in F$, where $\alpha \in (0,1]$, $\|\cdot\|_t$ and $\|\cdot\|_0$ are norms defined on the solution at time t and initially, respectively, and C is a positive constant independent of ϵ. We could equally well have said the solution depends Hölder continuously on the initial data for $t \in [0,T)$.

The point about the above definition is that Hölder stability is defined over compact subintervals of a finite time interval.

1.3 Examples of Non-Uniqueness and Breakdown of Continuous Dependence

Mathematicians who investigated improperly posed problems for partial differential equations before 1950 dealt primarily with questions of uniqueness for linear equations, since such problems typically have at most one solution. The literature of the 1950's and early 1960's is rich with uniqueness studies, some of which involve nonlinear equations. Serrin (1963), for example, established uniqueness for classical solutions to the Navier-Stokes equations on a bounded spatial domain backward in time using weighted energy arguments based on work of Lees & Protter (1961). A number of uniqueness studies for various problems associated with the Navier-Stokes equations backward in time have appeared since, cf. Ames & Payne (1993), Bardos & Tartar (1973), Calmelet-Eluhu & Crooke (1990), Crooke (1973), Galdi & Straughan (1988), Ghidaglia (1986), Knops & Payne (1968), Payne (1971), Straughan (1982), and the references therein.

To illustrate the point that uniqueness and continuous dependence are not trivial, we present some simple examples which show the solution to a boundary and/or initial value problem for a partial differential equation may not be unique, or even if it is, it need not depend continuously on the data, at least in the sense of Lyapunov.

Non-Uniqueness for a Nonlinear Wave Equation

Consider the boundary-initial value problem:

$$\begin{aligned} \frac{\partial^2 u}{\partial t^2} &= \Delta u + u^\varepsilon, \qquad 0 < \varepsilon < 1, \\ u(\mathbf{x},0) &= \frac{\partial u}{\partial t}(\mathbf{x},0) = 0, \end{aligned} \tag{1.3.1}$$

where the spatial domain is $\mathbf{R}^3$, and we are restricting attention to the possibility of a positive (real) solution. Clearly, one solution to (1.3.1) is

$u \equiv 0$. However, inspection shows another solution is

$$u = \left[\frac{(1-\varepsilon)^2}{2(1+\varepsilon)}\right]^{1/(1-\varepsilon)} t^{2/(1-\varepsilon)}.$$

The Initial Value Problem for the Backward Heat Equation

Suppose we wish to know the behaviour of the temperature at a past time of a rod, length l, whose ends are kept at a fixed temperature. If we specify the initial conditions as a sine wave, then the temperature of the rod is effectively governed by the following problem:

$$\frac{\partial^2 u}{\partial x^2} + \frac{\partial u}{\partial t} = 0, \qquad t \in (0,T), \quad x \in (0,l),$$

subject to the boundary data

$$u(0,t) = u(l,t) = 0,$$

and the initial data

$$u(x,0) = \frac{C}{m} \sin\left(\frac{m\pi x}{l}\right).$$

The solution to this problem, which may be shown to be unique, is

$$u(x,t) = \frac{C}{m} \exp\left(\frac{m^2\pi^2 t}{l^2}\right) \sin\left(\frac{m\pi x}{l}\right).$$

We note that as $m \to \infty$, $|u(x,0)| \to 0$, but $|u(x,t)| \to \infty$. Thus, the data becomes arbitrarily small, but the solution becomes unbounded in time. Hence we have a problem in which the solution does not depend continuously on the initial data in the sense of Lyapunov. (Continuous dependence may, however, be recovered if one adopts the notion of continuous dependence in the sense of Hölder. For details, see pages 16 - 18. In particular, the two paragraphs following (1.5.20) are especially relevant at this point.)

Non-Uniqueness for a Nonlinear Diffusion Equation

Payne (1975) presented an example which may be regarded as non - uniqueness, or decay to zero in finite time. His argument is the following. Consider the initial-boundary value problem:

$$\frac{\partial u}{\partial t} = \Delta u - u|u|^{-\alpha}, \qquad \text{in} \quad \Omega \times (0,T), \tag{1.3.2}$$

$$u = 0, \qquad \text{on} \quad \Gamma \times [0,T), \tag{1.3.3}$$

$$u(\mathbf{x},0) = f(\mathbf{x}), \qquad \mathbf{x} \in \Omega. \tag{1.3.4}$$

Here Ω is a bounded domain in $\mathbf{R}^n$ with smooth boundary Γ, Δ is the Laplace operator, $f(\mathbf{x})$ is a prescribed function, and the fixed exponent α satisfies $0 < \alpha < 1$. Equation (1.3.2) might be regarded as a model for heat conduction in a body with a sublinear heat sink. We now use Payne's (1975) argument to show the problem (1.3.2) - (1.3.4) exhibits nonuniqueness of solutions backward in time.

Define the functional

$$\Phi_p(t) = \int_\Omega u^{2p} dx, \tag{1.3.5}$$

where p is a positive integer. Differentiating (1.3.5), we obtain

$$\frac{d\Phi_p}{dt} = 2p \int_\Omega u^{2p-1} \frac{\partial u}{\partial t} dx.$$

Substitution of the differential equation and integration by parts leads to the expression

$$\frac{d\Phi_p}{dt} = -2p(2p-1) \int_\Omega u^{2p-2} u_{,i} u_{,i} dx - 2p \int_\Omega u^{2p} |u|^{-\alpha} dx.$$

Thus,

$$\frac{d\Phi_p}{dt} \le -2p \int_\Omega u^{2p} |u|^{-\alpha} dx.$$

If we now let

$$v(t) = \max_\Omega |u(\mathbf{x}, t)|, \tag{1.3.6}$$

it follows that

$$\frac{d\Phi_p}{dt} \le -2p v^{-\alpha} \Phi_p.$$

Integrating this inequality from 0 to t, we obtain

$$\Phi_p(t) \le \Phi_p(0) \exp\Big(-2p \int_0^t \big[v(s)\big]^{-\alpha} ds\Big). \tag{1.3.7}$$

Let us now raise both sides of (1.3.7) to the $1/2p$ power and then let $p \to \infty$. We find that

$$v(t) \le v(0) \exp\Big(-\int_0^t \big[v(s)\big]^{-\alpha} ds\Big). \tag{1.3.8}$$

The asumption that $v(t) > 0$ allows us to write (1.3.8) in the form

$$\frac{1}{\big[v(0)\big]^\alpha} \le \frac{1}{\big[v(t)\big]^\alpha} \exp\Big(-\alpha \int_0^t \big[v(s)\big]^{-\alpha} ds\Big),$$

and an integration from 0 to t results in the inequality

$$\frac{t}{\left[v(0)\right]^{\alpha}} \leq \frac{1}{\alpha}\left[1 - \exp\Big(-\alpha\int_0^t \left[v(s)\right]^{-\alpha} ds\Big)\right] \leq \frac{1}{\alpha}.$$

Consequently, we conclude that

$$t \leq \frac{1}{\alpha}\left[v(0)\right]^{\alpha}. \tag{1.3.9}$$

Observing that (1.3.9) cannot hold for all time, we see that if a solution to (1.3.2) - (1.3.4) exists, it must decay to zero in a finite time T satisfying the upper bound

$$T \leq \frac{1}{\alpha}\left[v(0)\right]^{\alpha}.$$

The above exhibits a non-trivial example for which we fail to have a unique solution backward in time.

Almost Uniqueness

In connection with the subject matter of this section we draw attention to the interesting paper of Flavin (1993). In this paper Professor Flavin observes that Rellich (1933) and Courant & Hilbert (1962) show that there are at most *two* solutions to the Dirichlet problem for the Monge-Ampère equation,

$$\frac{\partial^2\psi}{\partial x^2}\frac{\partial^2\psi}{\partial y^2} - \left(\frac{\partial^2\psi}{\partial x\partial y}\right)^2 = f(x,y), \qquad (x,y)\in\Omega,$$

$$\psi = g(x,y), \qquad \text{on } \Gamma,$$

for $f > 0$; here Ω is a bounded domain of $\mathbf{R}^2$ with boundary Γ, and $\psi \in C^2(\Omega)\cup C(\bar{\Omega})$. Flavin (1993) refers to the phenomenon of having at most *two* solutions as *almost uniqueness*.

Flavin (1993) extends this idea to the situation where ψ is a stream function for a fluid flow, which may be steady or unsteady. The function f is then

$$f = \frac{1}{2\rho}\Delta p,$$

with ρ the constant density and p the pressure. Flavin's (1993) paper is concerned with the equation arising from axisymmetric flow,

$$\frac{\partial}{\partial r}\left(\frac{1}{r}\frac{\partial\psi}{\partial r}\right)\frac{\partial^2\psi}{\partial z^2} - \frac{1}{r}\left(\frac{\partial^2\psi}{\partial r\partial z}\right)^2 + \frac{1}{2}\frac{\partial}{\partial r}\left(\frac{1}{r^2}\left[\frac{\partial\psi}{\partial z}\right]^2\right) = \Pi(r,z), \tag{1.3.10}$$

in cylindrical polar coordinates r, z. The function ψ is the stream function, such that the velocity field $\mathbf{u} = (u, v)$ satisfies

$$(u, v) = \Big(\frac{1}{r}\frac{\partial\psi}{\partial z}, -\frac{1}{r}\frac{\partial\psi}{\partial r}\Big),$$

and

$$\Pi(r, z) = \frac{r}{2\rho}\,\Delta p. \tag{1.3.11}$$

Flavin's (1993) paper, in fact, shows that if $\Pi > 0$ then the Dirichlet problem for equation (1.3.10) has at most two solutions.

We conclude this section by noting that many other counterexamples to uniqueness, existence and stability occur in applied mathematics such as in geophysics, Ramm (1995). The nonexistence proof of Kutev & Tomi (1995) is also very interesting. Finally, section 1.8 of the book of Flavin & Rionero (1995), contains a variety of interesting examples of non-uniqueness which typically occur in continuum mechanics. In fact, within the field of continuum mechanics, they include counterexamples in the Navier-Stokes theory for viscous, incompressible fluids, in viscoelasticity, in dynamic linear elasticity, and in linear elastostatics. Flavin & Rionero (1995) (theorem 1.2, p. 10,) also show that lack of uniqueness immediately implies lack of continuous dependence on the initial data. In connection with the Navier-Stokes theory for the behaviour of a viscous, incompressible fluid, we draw the reader's attention to the book of Doering & Gibbon (1995) which includes a very readable account of properties of solutions to these equations.

1.4 Four Classical Improperly Posed Problems

There are some improperly posed problems which are cornerstones of the subject. To help with the understanding of later work we now present four such problems. One of these has already been encountered in the one - dimensional situation in section 1.3, namely the initial - value problem for the backward heat equation. We now record it in greater generality.

The Initial - Boundary Value Problem for the Backward Heat Equation

Let Ω be a domain in $\mathbf{R}^n$ with boundary Γ. This problem seeks the solution to the equation

$$\frac{\partial u}{\partial t} = -\Delta u, \qquad \text{in} \quad \Omega \times (0, T), \tag{1.4.1}$$

and the boundary and initial conditions are

$$\begin{aligned} u(\mathbf{x},t) &= u_B(\mathbf{x},t), \qquad \mathbf{x} \in \Gamma,\ t \in [0,T], \\ u(\mathbf{x},t) &= u_0(\mathbf{x}), \qquad \mathbf{x} \in \Omega. \end{aligned} \tag{1.4.2}$$

We describe this problem in further detail in section 1.5.

The "Sideways" Problem for the Heat Equation

Here the whole of the boundary Γ is not accessible in order to be able to make data measurements and consequently the data are not given on all of Γ. To compensate for this it is usual to give u and its normal derivative $\partial u/\partial n$ on a part of Γ, Γ_1 say. In one space dimension such a problem is

$$\frac{\partial u}{\partial t} = \frac{\partial^2 u}{\partial x^2}, \qquad x > 0, \tag{1.4.3}$$

with

$$u(0,t) = g(t), \qquad \frac{\partial u}{\partial x}(0,t) = h(t), \tag{1.4.4}$$

where g and h are given functions.

This problem is discussed in more detail in section 4.1.

The Cauchy Problem for the Laplace Equation

For simplicity we consider this problem on a strip in $\mathbf{R}^2$. Let

$$\Omega = \{x \in (0,L)\} \times \{y > 0\}.$$

Then consider

$$\Delta u = f(x,y), \qquad (x,y) \in \Omega, \tag{1.4.5}$$

with

$$u(0,y) = f_1(y), \qquad u(L,y) = f_2(y). \tag{1.4.6}$$

Rather than prescribe values for u at two fixed y points as in the Dirichlet problem for the Laplace equation, we here give "initial" conditions as follows,

$$u(x,0) = g_1(x), \qquad \text{and} \qquad \frac{\partial u}{\partial y}(x,0) = g_2(x). \tag{1.4.7}$$

This has been known to be an improperly posed problem for a long time and is dealt with in greater detail in section 4.3.

The Dirichlet Problem for the Wave Equation

In general in two dimensions this problem investigates the solution to the equation

$$\frac{\partial^2 u}{\partial t^2} = \frac{\partial^2 u}{\partial x^2}, \qquad (x,t) \in \Omega, \tag{1.4.8}$$

where $\Omega = (a,b) \times (c,d)$. The solution u is given on Γ, i.e.

$$\begin{aligned} u(a,t) &= u_1(t), \quad u(b,t) = u_2(t), \\ u(x,c) &= v_1(x), \quad u(x,d) = v_2(x). \end{aligned} \tag{1.4.9}$$

This is another classical improperly posed problem. In general, even existence and uniqueness may fail. In this book we do not deal with this problem, but nevertheless it is a historically important one and ought to be mentioned. The existence question is considered by Fox & Pucci (1958) for

$$\Omega = (0,\pi) \times (0,\alpha\pi).$$

They investigate

$$\begin{aligned} &\frac{\partial^2 u}{\partial t^2} = \frac{\partial^2 u}{\partial x^2}, \qquad \text{in } \Omega, \\ &u(0,t) = u(\pi,t) = 0, \qquad t \in [0,\alpha\pi], \\ &u(x,0) = \phi(x), \quad u(x,\alpha\pi) = \psi(x), \qquad x \in [0,\pi]. \end{aligned}$$

They consider the existence question carefully. A very interesting paper on the stability question of a solution to the Dirichlet problem for the wave equation is that of Papi Frosali (1979). Existence and continuous dependence on data is considered for an abstraction of problem (1.4.8), (1.4.9) by Levine & Vessella (1985) who treat the problem in Hilbert space,

$$\begin{aligned} &\frac{d^2 u}{dt^2} + Au = 0, \\ &u(0) = 0, \quad u(T) = u_0, \end{aligned}$$

where A is a densely defined, positive, self adjoint linear operator with compact inverse.

1.5 Methods of Analysis

Researchers have adapted a variety of techniques in order to resolve existence, uniqueness, or continuous dependence questions for improperly posed problems. These methods of analysis typically rely heavily on the judicious use of inequalities. In this book we do not describe all the methods presently used in the literature. However, an elementary exposition of those most frequently employed is now included.

Logarithmic Convexity

One method that was first successfully applied to ill-posed problems by Pucci (1955), John (1955,1960), Lavrentiev (1956), and Payne (1960), is logarithmic convexity, a technique that makes use of second order differential inequalities. A detailed treatment of this method in a variety of settings can be found, for example, in Agmon (1966).

A sufficient condition for a function $f(t)$ to be a convex function of t on the interval $[t_1, t_2]$ is that $f''(t) \geq 0$, $\forall t \in [t_1, t_2]$. If $f'' \geq 0$, then f' is increasing and thus

$$\int_{t_1}^{t} f'(s)ds \leq f'(t)(t - t_1),$$

as well as,

$$f'(t)(t_2 - t) \leq \int_{t}^{t_2} f'(s)ds,$$

for $t \in [t_1, t_2]$. Eliminating $f'(t)$ from these two inequalities, we obtain the result

$$f(t) \leq \left(\frac{t_2 - t}{t_2 - t_1}\right) f(t_1) + \left(\frac{t - t_1}{t_2 - t_1}\right) f(t_2). \tag{1.5.1}$$

If $f(t) = \ln F(t)$, we say that the function $F(t)$ is a *logarithmically convex* function of t. We can therefore show that $\ln F(t)$ is convex on the interval $[t_1, t_2]$ by demonstrating that

$$\frac{d^2}{dt^2} \ln F(t) \geq 0, \tag{1.5.2}$$

or, provided $F(t) > 0$ $\forall t \in [t_1, t_2]$, that $F(t)$ satisfies the inequality

$$FF'' - (F')^2 \geq 0. \tag{1.5.3}$$

Thus, when we interpret (1.5.1) in terms of $F(t)$ we find that $F(t)$ satisfies

$$F(t) \leq \big[F(t_1)\big]^{(t_2 - t)/(t_2 - t_1)} \big[F(t_2)\big]^{(t - t_1)/(t_2 - t_1)}, \tag{1.5.4}$$

$\forall t \in [t_1, t_2]$. This inequality proves useful in establishing stability estimates for improperly posed problems.

The logarithmic convexity method involves constructing a non-negative, differentiable functional, $F(t)$, defined on solutions of the problem under consideration, such that $F(t)$ satisfies (1.5.3). In addition, $F(t) = 0$ if and only if the solution is identically zero on the time interval in question. We note that it is often the case that rather than (1.5.3), we can only establish the weaker inequalities,

$$FF'' - (F')^2 \geq -kF^2, \tag{1.5.5}$$

or

$$FF'' - (F')^2 \geq -k_1 FF' - k_2 F^2, \tag{1.5.6}$$

where k, k_1, k_2, are positive constants. Fortunately, both of these inequalities can be transformed into ones similar to (1.5.2) for $F > 0$ on $[t_1, t_2]$. Inequality (1.5.5) can be written as

$$\frac{d^2}{dt^2}\left[\ln\left(Fe^{kt^2/2}\right)\right] \geq 0, \tag{1.5.7}$$

while setting

$$\sigma = e^{-k_1 t}, \tag{1.5.8}$$

allows us to rewrite (1.5.6) in the form

$$\frac{d^2}{d\sigma^2}\left(\ln\left[F(\sigma)\sigma^{-k_2/k_1^2}\right]\right) \geq 0, \tag{1.5.9}$$

where the notation $F(\sigma)$ is used even though the functional form of F may be changed. It follows from (1.5.7), (1.5.9), respectively, and the result (1.5.1), that

$$F(t) \leq \exp\left[\frac{1}{2}k(t-t_1)(t_2-t)\right]\left[F(t_1)\right]^{(t_2-t)/(t_2-t_1)} \times \left[F(t_2)\right]^{(t-t_1)/(t_2-t_1)}, \tag{1.5.10}$$

and

$$F(\sigma)\sigma^{-k_2/k_1^2} \leq \left[F(\sigma_1)\sigma_1^{-k_2/k_1^2}\right]^{(\sigma_2-\sigma)/(\sigma_2-\sigma_1)} \times \left[F(\sigma_2)\sigma_2^{-k_2/k_1^2}\right]^{(\sigma-\sigma_1)/(\sigma_2-\sigma_1)}, \tag{1.5.11}$$

for $t \in [t_1, t_2]$, and $\sigma \in [\sigma_1, \sigma_2]$, where $\sigma_1 = e^{-k_1 t_1}$, $\sigma_2 = e^{-k_1 t_2}$.

Useful information concerning solutions of ill posed problems for partial differential equations can be obtained from these convexity inequalities. To illustrate how the method is applied and the type of results that can be deduced, let us consider a Cauchy problem for the backward heat equation, namely

$$\begin{aligned} \frac{\partial u}{\partial t} + \Delta u &= 0, && \text{in} \quad \Omega \times (0, T), \\ u &= 0, && \text{on} \quad \Gamma \times [0, T), \\ u(\mathbf{x}, 0) &= f(\mathbf{x}), && \mathbf{x} \in \Omega, \end{aligned} \tag{1.5.12}$$

where Ω is a bounded domain in $\mathbf{R}^3$ with a smooth boundary Γ, and Δ is the Laplace operator. We select the functional

$$F(t) = \|u(t)\|^2, \tag{1.5.13}$$

where $\|\cdot\|$ denotes the $L^2(\Omega)$ norm, and we now show that $F(t)$ satisfies inequality (1.5.3).

Differentiating (1.5.13), we find that

$$F'(t) = 2\int_\Omega uu_t dx, \tag{1.5.14}$$

or, after substituting from the differential equation and integrating by parts, we see that

$$\begin{aligned} F'(t) =& 2\int_\Omega u(-\Delta u)dx \\ =& 2\int_\Omega |\nabla u|^2 dx, \end{aligned} \tag{1.5.15}$$

where ∇ is the gradient operator. A second differentiation and integration by parts now leads to

$$\begin{aligned} F''(t) =& 4\int_\Omega \nabla u \cdot \nabla u_t dx \\ =& -4\int_\Omega u_t \Delta u dx, \end{aligned}$$

and consequently, on replacing Δu by $-u_t$ using $(1.5.12)_1$ it follows that

$$F''(t) = 4\|u_t\|^2. \tag{1.5.16}$$

If we now form the expression $FF'' - (F')^2$, using (1.5.13), (1.5.14) and (1.5.16), we derive

$$FF'' - (F')^2 = 4\|u\|^2\|u_t\|^2 - 4\Big(\int_\Omega uu_t dx\Big)^2 \geq 0, \tag{1.5.17}$$

where the non-negativity of the right hand side is assured by the Cauchy-Schwarz inequality. Thus inequality (1.5.3) has been established. The convexity inequality (1.5.4) then becomes

$$\|u(t)\|^2 \leq \|f\|^{2(1-t/T)}\|u(T)\|^{2t/T}, \tag{1.5.18}$$

for $t \in [0, T)$.

Properties of solutions to the boundary-initial value problem (1.5.12) can be obtained from (1.5.17) and (1.5.18). We first observe that if $F(t)$ satisfies (1.5.17) and if F vanishes at some point $t_1 \in [0, T]$, then a continuity argument shows that $F(t)$ must be identically zero for all $t \in [0, T]$. In view of the properties of the functional F, it follows that the solution of the linear Cauchy problem (1.5.12) is unique.

If we interpret u as the difference between two solutions corresponding to Cauchy data whose difference is f, then inequality (1.5.18) looks like a stability inequality on the finite time interval $[0, T)$. However, since $\|f\|$ being small does not necessarily guarantee that the product

$$\|f\|^{2(1-t/T)}\|u(T)\|^{2t/T}$$

will remain small for $t \in [0, T)$, we need to restrict the class of permissible perturbations in order to infer continuous dependence results. Inequality (1.5.18) clearly indicates a class of solutions in which we may establish Hölder stability, namely the class of functions which satisfy the bound

$$\|u(T)\|^2 \leq M^2, \tag{1.5.19}$$

for some prescribed constant M. For such functions the convexity inequality becomes

$$\|u(t)\|^2 \leq \|f\|^{2(1-t/T)} M^{2t/T}, \tag{1.5.20}$$

from which we obtain the result that solutions depend Hölder continuously on the Cauchy data in $L^2(\Omega)$, on compact subintervals of $[0, T)$.

It should be noted that such a stability result is of practical value only if a constant M can be computed from the physical data of the system under consideration. As observed on page 4 because such *a priori* bounds need not be sharp, this computation is frequently possible. For example, if u represented the temperature in a particular problem then an upper bound for the temperature may usually be derived.

From the preceding analysis we observe the fact that the imposition of some sort of *a priori* bounds leads to the recovery of continuous dependence in this modified sense. Such a stability result is typical of the kind that can be obtained for ill posed problems and should not be confused with our usual notion of continuous dependence.

Additional information can be obtained from the fact that $\ln F(t)$ is a convex function of t. Integration of the inequality $FF'' - (F')^2 \geq 0$, leads also to the lower bound

$$F(t) \geq F(0) \exp\left[\frac{tF'(0)}{F(0)}\right]. \tag{1.5.21}$$

For the model problem (1.5.12) we then deduce

$$\|u(t)\|^2 \geq \|f\|^2 \exp\left[\frac{2t\|\nabla f\|^2}{\|f\|^2}\right]. \tag{1.5.22}$$

Consequently, if $\|u(t)\|$ is defined on $[0, \infty)$, it must grow exponentially. Inequalities of the form (1.5.21) or (1.5.22) thus provide growth estimates for solutions.

The Method of Lagrange Identities

Another method that has been used to establish uniqueness and continuous dependence results for improperly posed problems governed by linear equations is the Lagrange identity method. Brun (1965a,b,1967,1969) employed this method to study problems in elastodynamics, viscoelasticity, and thermoelasticity. Novel recent extensions which employ Lagrange identities include the work of Rionero & Chirita (1987) and Chirita & Rionero (1991) which extends the method to unbounded spatial domains, that of Knops & Payne (1988) which obtains particularly sharp results on continuous dependence on the initial - time geometry in linear elastodynamics, Franchi & Straughan (1993d) who use Lagrange identities to study modeling questions in theories of heat propagation which allow for finite wavespeeds, and Ames & Payne (1995c) who utilize the technique in their study of continuous dependence on the initial - time geometry in thermoelasticity. The technique entails the introduction of a function that satisfies the adjoint of the equation under consideration and integration of the identity

$$0 = \int_0^t \{(v, Lu) - (u, L^*v)\} d\eta. \tag{1.5.23}$$

Here L is a linear operator on an appropriately defined space with inner product $(,)$, L^* denotes the adjoint of L, and the functions u and v satisfy $Lu = 0$ and $L^*v = 0$, respectively. A judicious use of inequalities leads to uniqueness and stability results. Let us return to the model problem (1.5.12) in order to demonstrate how the Lagrange identity method is implemented.

Suppose $v(\mathbf{x}, t)$ is any solution of the forward heat equation that vanishes on $\Gamma \times [0, T)$. Then (1.5.23) becomes,

$$0 = \int_0^t \int_\Omega \Big[v(\frac{\partial u}{\partial \eta} + \Delta u) - u(-\frac{\partial v}{\partial \eta} + \Delta v) \Big] dx d\eta, \tag{1.5.24}$$

or, after integrations with respect to η and $\mathbf{x}$,

$$\int_\Omega v(\mathbf{x}, t) u(\mathbf{x}, t) dx = \int_\Omega v(\mathbf{x}, 0) f(\mathbf{x}) dx. \tag{1.5.25}$$

Let us now assume that $0 \leq t \leq 2t < T$, and choose

$$v(\mathbf{x}, \eta) = u(\mathbf{x}, 2t - \eta). \tag{1.5.26}$$

Then we obtain from (1.5.25) the equation

$$\|u(t)\|^2 = \int_\Omega u(\mathbf{x}, 2t) f(\mathbf{x}) dx,$$

or, as a consequence of Schwarz's inequality, the result

$$\|u(t)\|^2 \le \|u(2t)\| \, \|f\|, \qquad t \in [0, T/2]. \tag{1.5.27}$$

To see how inequality (1.5.27) yields continuous dependence results analogous to (1.5.20), we rewrite (1.5.27) as

$$F(t) \le \big[F(2t)\big]^{1/2} \big[F(0)\big]^{1/2}. \tag{1.5.28}$$

Replacing t by $2t$ in this inequality, we have

$$F(2t) \le \big[F(4t)\big]^{1/2} \big[F(0)\big]^{1/2}.$$

It then follows that

$$F(t) \le \big[F(0)\big]^{1/2} \Big\{ \big[F(4t)\big]^{1/2} \big[F(0)\big]^{1/2} \Big\}^{1/2} = \big[F(4t)\big]^{1/4} \big[F(0)\big]^{3/4}.$$

Observing that (1.5.28) yields, for positive integer n, the inequality

$$F(nt) \le \big[F(2nt)\big]^{1/2} \big[F(0)\big]^{1/2},$$

we find that continual resubstitution of this result leads to

$$F(t) \le \big[F(2nt)\big]^{1/2n} \big[F(0)\big]^{1-1/2n}. \tag{1.5.29}$$

If we now take $2nt = T$, then (1.5.29) implies

$$F(t) \le \big[F(0)\big]^{1-t/T} \big[F(T)\big]^{t/T}, \tag{1.5.30}$$

which is the same type of estimate as (1.5.20) for $t = T/2n$, provided we again restrict the class of solutions to that whose L^2 norm at $t = T$ satisfies a uniform bound. The translation $t \to t - T/2$ in (1.5.28) permits us to generate the desired continuous dependence result for $t = (n+1)T/2n$, integer $n \ge 1$. The results at these time levels can in turn be used to establish Hölder continuity at intermediary levels (e.g. $T/3$). Hence, the Lagrange identity method leads to the inequality (1.5.30) for all t in the interval $0 \le t < T$.

We observe briefly that instead of using Schwarz's inequality, we could have used Hölder's inequality to obtain, instead of (1.5.27), L^p results. For example, since Hölder's inequality gives

$$\|u(t)\|^2 \le \|u(2t)\|_p \|f\|_q \qquad \frac{1}{p} + \frac{1}{q} = 1,$$

it follows that if we restrict u to the set of sufficiently smooth functions whose L^p norm is bounded at T by the constant M_p, then

$$\|u(t)\|^2 \le M_p \|f\|_q ,$$

for $0 \le t < T/2$.

Thus, the Lagrange identity method is not restricted to Hilbert space as it would appear the logarithmic convexity method is.

The Weighted Energy Method

Let us now turn to a discussion of the weighted energy method which has also been used successfully on a number of improperly posed problems for partial differential equations and operator equations to establish uniqueness theorems and continuous dependence inequalities. The thrust for this method was largely provided by the fundamental work of Protter (1953,1954,1960,1961,1962,1963), see also Lees and Protter (1961), Murray & Protter (1973), Payne (1985), and Protter (1974). Until rather recently, the usual result obtained via a weighted energy method was a weak logarithmic continuous dependence one. However, a modification of the method will in many cases actually lead to Hölder continuous dependence of solutions on the data, as is shown in Payne (1985), and Ames *et al.* (1987).

To illustrate the weighted energy method, we consider the initial value problem for the operator equation:

$$\frac{d^2u}{dt^2} = Pu, \qquad t \in (0,T),\ T < \infty, \tag{1.5.31}$$

$$u(0) = u_0, \quad \frac{du}{dt}(0) = v_0, \tag{1.5.32}$$

where the linear operator P maps a dense subdomain D of a real Hilbert space H into H and, along with D and H, is independent of t. We further assume that P is a skew symmetric operator, i.e.

$$(u, Pv) = -(v, Pu),$$

which is possibly unbounded, and that strong solutions $u \in C^2([0,T); D)$ of (1.5.31), (1.5.32) exist.

We set $u = e^{\lambda t}v$, for some $\lambda > 0$ to be prescribed. Then v satisfies the equation

$$Lv \equiv v_{tt} + 2\lambda v_t + \lambda^2 v - Pv = 0. \tag{1.5.33}$$

We next form the expression

$$\int_0^t (Lv, 2\lambda v_\eta - Pv)d\eta = 0, \tag{1.5.34}$$

which upon integration and manipulation leads to

$$\int_0^t \|2\lambda v_\eta - Pv\|^2 d\eta + \left[\lambda\|v_\eta\|^2 + \lambda^3\|v\|^2 - (v_\eta, Pv)\right]\Big|_0^t = 0,$$

after using the skew symmetry of the operator P to observe that $(P\phi, \phi) = 0$, for all functions $\phi \in D$. Dropping the non-negative integral, we have the inequality

$$\left[\lambda\|v_\eta\|^2 + \lambda^3\|v\|^2 - (v_\eta, Pv)\right]\Big|_0^t \leq 0. \tag{1.5.35}$$

If we now assume $\lambda > 1$, reintroduce u, and use the skew symmetry of P, we find that

$$\lambda\|u_t\|^2 + 2\lambda^3\|u\|^2 - 2\lambda^2(u_t, u) - (u_t, Pu) \leq \lambda^3 Q_0^2 e^{2\lambda t}, \tag{1.5.36}$$

where Q_0 is a data term, specifically,

$$Q_0^2 = \alpha\|u_0\|^2 + \beta\|v_0\|^2 + \gamma\|Pu_0\|^2,$$

for positive constants α, β, γ. Substituting the differential equation $Pu = u_{tt}$ in (1.5.36) and then integrating the result from 0 to t leads us to the inequality

$$\lambda\int_0^t \left[\|u_\eta\|^2 + 2\lambda^2\|u\|^2\right]d\eta - \frac{1}{2}\|u_t\|^2 - \lambda^2\|u\|^2 \leq \frac{1}{2}\lambda^2 Q_0^2 e^{2\lambda t}. \tag{1.5.37}$$

Let us define

$$G(t) = \frac{1}{2}\int_0^t \left[\|u_\eta\|^2 + 2\lambda^2\|u\|^2\right]d\eta, \tag{1.5.38}$$

and observe that (1.5.37) can be written as

$$2\lambda G - \frac{dG}{dt} \leq \frac{1}{2}\lambda^2 Q_0^2 e^{2\lambda t}, \tag{1.5.39}$$

or,

$$\frac{d}{dt}\Big[-Ge^{-2\lambda t}\Big] \le \frac{1}{2}\lambda^2 Q_0^2. \qquad (1.5.40)$$

Integrating (1.5.40) from t to T, we have

$$G(t) \le G(T)e^{-2\lambda(T-t)} + \frac{1}{2}\lambda^2 Q_0^2 (T-t)e^{2\lambda t}. \qquad (1.5.41)$$

If we now assume that u belongs to the constraint set

$$\mathcal{M} = \{u \in C^2([0,T);D)\Big|\lambda^{-2}G(T) \le N^2\}, \qquad (1.5.42)$$

where N is a prescribed constant and choose, for Q_0^2 sufficiently small,

$$\lambda = \frac{1}{2T}\ln\left(\frac{N^2}{Q_0^2}\right), \qquad (1.5.43)$$

we obtain the result

$$G(t) \le \lambda^2 N^{2t/T} Q_0^{2(1-t/T)}\big[1 + \frac{1}{2}(T-t)\big], \qquad (1.5.44)$$

for $0 \le t < T$. Thus, recalling the defintion of $G(t)$, we are lead to the inequality

$$\int_0^t \|u\|^2 d\eta \le N^{2t/T} Q_0^{2(1-t/T)}\big[1 + \frac{1}{2}(T-t)\big]. \qquad (1.5.45)$$

Consequently, if we interpret u as the difference between two solutions u_1 and u_2 corresponding to different initial data, then inequality (1.5.45) implies that $u \in \mathcal{M}$ depends Hölder continuously on the Cauchy data on compact sub-intervals of $[0, T)$.

The method outlined here is better suited to use on equations with non-symmetric operators and on non-standard Cauchy problems than either logarithmic convexity or Lagrange identity arguments. For such problems, the weighted energy method commences with the introduction of

$$u = e^{\lambda\phi}v,$$

where the function ϕ, which may depend on any or all of the independent variables of the problem, is appropriately chosen. (In our previous example, we took $\phi = t$.)

The Weighted Energy Method for Unbounded Spatial Domain Problems

Let u satisfy the heat equation forward in time, i.e.

$$\frac{\partial u}{\partial t} = \Delta u, \qquad \text{in } \Omega \times (0, T), \tag{1.5.46}$$

where Ω is a region exterior to the closure of a bounded region Ω_0, $\Omega_0 \subset \mathbf{R}^3$. (We can easily allow $\Omega_0 \subset \mathbf{R}^N$.) On the boundary of Ω_0, i.e. the inner boundary of Ω, we assume

$$u = 0. \tag{1.5.47}$$

We are interested in the uniqueness question for (1.5.46) and so let u be the difference of two solutions each of which satisfies (1.5.47) together with the *same* initial data. For conditions at infinity, we assume

$$|u|, \ \left|\frac{\partial u}{\partial r}\right| \leq e^{kr^2}, \tag{1.5.48}$$

for some $k > 0$. (We could easily put a constant, Q say, on the right of (1.5.48), i.e. write it as Qe^{kr^2}.)

Let (,) denote the inner product on $L^2(\Omega)$. To allow the solution to grow like (1.5.48) we introduce a weight

$$g = \exp\left(-f(t)r^\gamma\right), \tag{1.5.49}$$

where $f(t) > 0$ is a function to be chosen and $\gamma > 0$ is a constant we select later. Note that

$$\frac{\partial g}{\partial t} = -f' r^\gamma g, \qquad \frac{\partial g}{\partial x_i} = -f\gamma r^{\gamma-1}\frac{x_i}{r}\, g\,.$$

Multiply (1.5.46) by gu and integrate over Ω, i.e.

$$(gu, u_t) = (gu, \Delta u). \tag{1.5.50}$$

We integrate by parts, assuming that on the right we replace the domain Ω by Ω_R which is $\Omega \cap B_R$, B_R being the ball of radius R, centered at the origin. Then the right hand side leads to

$$\int_{\Omega_R} gu\Delta u dV = -\int_{\Omega_R} g|\nabla u|^2 dV - \int_{\Omega_R} \nabla g\, u \nabla u dV + \oint_{\Gamma_R} gu\frac{\partial u}{\partial r}\, dS. \tag{1.5.51}$$

Assume now that $\gamma \geq 2$ and

$$f(t) > 2k, \tag{1.5.52}$$

for suitable t. Then we allow $R \to \infty$ in (1.5.51) and from (1.5.50) we find

$$\frac{d}{dt}\frac{1}{2}(gu, u) + \frac{1}{2}f'(r^\gamma gu, u) = - \int_\Omega g|\nabla u|^2 dV + \gamma f\Big(r^{\gamma-1}\frac{\mathbf{x}}{r}gu, \nabla u\Big).$$

We use the arithmetic - geometric mean inequality in the last term on the right as

$$\gamma f\Big(r^{\gamma-1}\frac{\mathbf{x}}{r}gu, \nabla u\Big) \leq \int_\Omega g|\nabla u|^2 dV + \frac{\gamma^2 f^2}{4}\int_\Omega g r^{2(\gamma-1)} u^2 dV.$$

Combining the above two expressions we thus obtain

$$\frac{d}{dt}\frac{1}{2}(gu, u) \leq \int_\Omega gu^2\Big[-\frac{1}{2}f' r^\gamma + \frac{\gamma^2 f^2}{4} r^{2(\gamma-1)}\Big] dV. \tag{1.5.53}$$

We wish to allow the solution to grow at infinity as rapidly as possible and thus in order for the right hand side of (1.5.53) to have a possibility of being non-positive, the "best case" is found if we select

$$\gamma = 2(\gamma - 1),$$

i.e. $\gamma = 2$. Then, to further ensure that the right hand side of (1.5.53) is non-positive we need

$$-\frac{1}{2}f' + f^2 \leq 0.$$

So by integration we see that

$$f \geq \frac{1}{A - 2t},$$

for a suitable positive constant A. Require now that

$$A > 2T. \tag{1.5.54}$$

Thus (1.5.52) and (1.5.54) are conditions we must satisfy in selecting f.

We may select

$$f = \frac{1}{A - 2t},$$

for $2T < A < 1/2k$, e.g. choose

$$T = \frac{1}{8k}, \qquad A = \frac{3}{8k}\,.$$

We then find from (1.5.53) that $u \equiv 0$ on $\Omega \times [0,T]$, for $T = 1/8k$. This argument may be extended to establish uniqueness on an arbitrarily large interval $[0,T]$ since we apply it to $[T,2T]$, and so on. The number k is fixed and once $u = 0$ in $[0,T]$ we simply rescale to $[T,2T]$, etc.

The growth condition in (1.5.48) is the sharpest one can use. Protter & Weinberger (1967), p. 181, show that uniqueness holds if

$$|u| \le Ae^{cr^2}. \tag{1.5.55}$$

They also conclude that if growth more rapid than (1.5.55) is allowed then a counterexample to uniqueness can be exhibited.

We might include that the method just described together with that of the subsection to follow is extended to a system which arises in a theory for the simultaneous diffusion of heat and moisture by Flavin & Straughan (1997). Further generalisations of the method just presented may be found in Russo (1987), who treats the uniqueness question for the system

$$\frac{\partial u_i}{\partial t} = \frac{\partial}{\partial x_j}\left(A_{ijhk}\frac{\partial u_h}{\partial x_k}\right) + b_i.$$

The spatial region chosen by Russo (1987) is unbounded and the growth conditions are very sharp. The boundary conditions satisfied on the boundary of the region (not the boundary at infinity) are of mixed type, i.e.

$$u_i = \bar{u}_i,$$

on a part of the boundary, while

$$n_j A_{ijhk}\frac{\partial u_h}{\partial x_k} = s_i,$$

on the rest of the boundary. The functions $\bar{u}_i$ and s_i are prescribed.

The Weighted Energy Method in Conjunction with the Lagrange Identity Method

Let now Ω be an unbounded exterior domain in $\mathbf{R}^3$ with smooth interior boundary $\partial\Omega_0$. We consider the heat equation backward in time, namely,

$$\frac{\partial u}{\partial t} + \Delta u = 0, \qquad \text{in} \quad \Omega \times (0,T), \tag{1.5.56}$$

together with zero boundary data

$$u = 0 \qquad \text{on} \quad \partial\Omega_0,$$

and zero initial data. Since (1.5.56) is a linear equation, a study of this equation together with these boundary and initial conditions is equivalent to the uniqueness question for a solution to (1.5.56).

The combination of the weighted energy method plus the Lagrange identity technique was first used by Galdi *et al.* (1986) in the context of linear elastodynamics.

The conditions at infinity we consider are that u satisfy

$$|u|,\ |\nabla u| = o(e^{\alpha R}), \qquad \text{as } R \to \infty, \tag{1.5.57}$$

for a suitable $\alpha > 0$. In addition, we require

$$\lim_{\alpha\to 0} \alpha^2 \int_0^{2t} \int_\Omega e^{-\alpha r} |\nabla u|^2 dV\, ds = 0. \tag{1.5.58}$$

Conditions (1.5.57) and (1.5.58) are, in fact, those of Galdi *et al.* (1986).

Galdi *et al.* (1986) show that (1.5.58) is not trivial and is satisfied, for example, provided

$$|\nabla u| = O(R^{-1/2-\epsilon}), \qquad \text{as } R \to \infty,$$

for some $\epsilon > 0$.

With (,) again being the inner product on $L^2(\Omega)$ we introduce the weight g by

$$g = e^{-\alpha r},$$

where $r^2 = x_i x_i$, and $\alpha > 0$ is to be selected.

Let a superposed dot denote differentiation with respect to the "t" argument, i.e.

$$\dot{u}(s) = \frac{\partial u}{\partial s}(\mathbf{x}, s), \qquad \dot{u}(2t - s) = \frac{\partial u}{\partial (2t - s)}(\mathbf{x}, 2t - s).$$

We then evaluate (1.5.56) at $2t - s$ and multiply by $gu(s)$, integrate over Ω_R, integrate by parts and let $R \to \infty$ to find (using (1.5.57)),

$$\int_0^t \Big(g\dot{u}(2t-s), u(s)\Big) ds = \int_0^t \Big(g\nabla u(2t-s), \nabla u(s)\Big) ds + \int_0^t \Big(\nabla g \nabla u(2t-s), u(s)\Big) ds. \tag{1.5.59}$$

A similar procedure with s and $2t-s$ reversed yields

$$\int_0^t \Big(g\dot u(s), u(2t-s)\Big)ds = \int_0^t \Big(g\nabla u(s), \nabla u(2t-s)\Big)ds + \int_0^t \Big(\nabla g \nabla u(s), u(2t-s)\Big)ds. \quad (1.5.60)$$

Observe that

$$\dot u(2t-s) = -\frac{d}{ds}\,u(2t-s),$$

then subtract (1.5.60) from (1.5.59) and we obtain

$$-\int_0^t \frac{d}{ds}\Big(gu(2t-s), u(s)\Big)ds = \int_0^t \Big(\nabla g\, u(s), \nabla u(2t-s)\Big)ds - \int_0^t \Big(\nabla g \nabla u(s), u(2t-s)\Big)ds.$$

Hence, since $u(\mathbf{x},0)\equiv 0$,

$$\big(gu(t), u(t)\big) = \int_0^t \Big(\nabla g\, u(2t-s), \nabla u(s)\Big)ds - \int_0^t \Big(\nabla g\, u(s), \nabla u(2t-s)\Big)ds. \quad (1.5.61)$$

Note that

$$\nabla g = -\alpha\,\frac{\mathbf{x}}{r}\,g,$$

let

$$F(t) = \big(gu(t), u(t)\big),$$

and by using the arithmetic - geometric mean inequality on the right of (1.5.61) we may see that

$$\begin{aligned}\big(gu(t), u(t)\big) \le& \frac{1}{2}\int_0^t \Big(gu(2t-s), u(2t-s)\Big)ds \\ &+ \frac{1}{2}\alpha^2 \int_0^t \Big(g\nabla u(s), \nabla u(s)\Big)ds + \frac{1}{2}\int_0^t \Big(gu(s), u(s)\Big)ds \\ &+ \frac{1}{2}\alpha^2 \int_0^t \Big(g\nabla u(2t-s), \nabla u(2t-s)\Big)ds \\ =& \frac{1}{2}\int_0^{2t} F(s)ds + \alpha^2 Q(t), \end{aligned} \quad (1.5.62)$$

where $Q(t)$ has been defined by

$$Q(t) = \frac{1}{2}\int_0^{2t} \Big(g\nabla u(s), \nabla u(s)\Big) ds.$$

We next integrate (1.5.62) to find, since $Q(t)$ is increasing,

$$(1-t)\int_0^{2t} F(s)ds \le 2t\alpha^2 Q(t). \tag{1.5.63}$$

Suppose now $t < 1$, with t fixed. Then for $R > 0$ arbitrary, from (1.5.63) we see that

$$(1-t)e^{-\alpha R}\int_0^{2t}\int_{\Omega_R} u^2(\mathbf{x}, s)dV\, ds \le 2t\alpha^2 Q(t). \tag{1.5.64}$$

We now let $\alpha \to 0$ and then thanks to (1.5.58), we see that

$$0 \le \int_0^{2t}\int_{\Omega_R} u^2 dV\, ds \le 0. \tag{1.5.65}$$

Hence $u \equiv 0$ on $\Omega_R \times (0, 2t)$. The number R is arbitrary and we conclude by a bootstrap argument that $u \equiv 0$ on $\Omega \times (0, T)$.

The above is an illustration of how the Lagrange identity method may be combined with the weighted energy one to yield uniqueness in an improperly posed problem without requiring severe decay restrictions at infinity when the spatial domain is unbounded. In fact, the combination of these techniques leads to many other results including continuous dependence, and similar results for half space regions or regions with other non-compact boundaries. Details may be found in Galdi *et al.* (1986).

1.6 Existence of Solutions and Outline of the Book

The last section of the first chapter contains some remarks concerning attempts to deal with the question of existence of solutions to improperly posed problems. One strategy that has been followed is to modify the underlying mathematical model and then develop techniques to approximate solutions to the problem under consideration. The primary aim of these methods which include Tikhonov regularisation (Tikhonov & Arsenin, 1977) and quasi-reversibility (Lattès & Lions, 1969) is the numerical solution of ill posed problems. Since existence theorems for such problems are generally not available, approximate solutions may be as close as we can come to establishing the existence of a "solution".

The method of quasi-reversibility was introduced by Lattes & Lions (1969) and later improved by Miller (1973b) who developed a stabilized version of the technique. Application of the method to operator equations have been successfully made by Ewing (1975) and Showalter (1975) as well as Miller. More recently, quasi-reversibility has been used to obtain approximate solutions to the Cauchy problem for Laplace's equation (see Klibanov & Santosa (1993)).

The basic idea of the quasi-reversibility method is to perturb the operator in the differential equation in such a way that the perturbed problem is properly posed. The usual procedure is to add a differential operator multiplied by a small coefficient that formally tends to zero. One then also needs to include in the perturbed problem appropriate boundary and /or initial conditions that degenerate as the small parameter vanishes. Using the solution of this perturbed problem as a guide, an approximate solution of the original ill-posed problem is constructed, usually by numerical procedures. It should be pointed out that the perturbation used in the method of quasi-reversibility is not unique. Physical motivation or computational simplicity serve as guidelines in choosing the appropriate perturbation.

Approximate solutions have been obtained numerically using the regularization methods of Tikhonov (1963) and those of Nashed and Wahba (1972). A variety of other numerical procedures have been used to compute approximate solutions to ill-posed problems. For a more complete description and application of these methods, see the book edited by Engl & Groetsch (1987) as well as the work of Tikhonov & Arsenin (1977). More details of numerical work in improperly posed problems and results for practical problems may be found in the articles of Bellomo & Preziosi (1995), Cannon (1964a,b,c,1984), Cannon & Douglas (1967,1968), Cannon & Ewing (1976), Cannon & Knightly (1969), Cannon & Miller (1965), Carasso (1976,1977,1982,1987,1993,1994), Carasso & Stone (1975), Chorin (1985), Demmel & Kagstrom (1988), Elden (1987,1988,1990), Engl & Groetsch (1987), Ewing (1975,1984), Ewing & Falk (1979), Ewing *et al.* (1987), Falk & Monk (1986), Han (1982), Hansen (1992), Ingham *et al.* (1991), Klibanov & Rakesh (1992), Klibanov & Santosa (1993), Lin & Ewing (1992), Monk (1986), and Seidman & Elden (1990).

The remainder of the book is divided into five chapters, each of which addresses some aspect of the major topics of existence, uniqueness and continuous dependence for improperly posed problems. Chapter 2 will deal with continuous dependence of solutions on changes in the geometry. This is followed by chapter 3 which treats continuous dependence of solutions on changes of the model under study, for problems which extrapolate backward in time. Chapter 4 deals with continuous dependence of the solution on changes of the model itself, for problems going forward in time. The penultimate chapter then explores some recent results for non - standard and non - characteristic problems. For example, we devote sections of this chapter to problems in which the data are prescribed on only part of the domain boundary. Other improperly posed problems are the focus of the final chapter, including Alfred Carasso's important recent work on overcoming Hölder continuity and its highly practical application to image restoration.

2

Continuous Dependence on the Geometry

2.1 Continuous Dependence on the Initial - Time Geometry for the Heat Equation

The study of continuous dependence on the initial - time geometry was initiated by Knops & Payne (1969) who studied this problem in the context of linear elasticity. This paper used the method of logarithmic convexity and actually investigates various aspects of continuous dependence for improperly posed problems in linear elastodynamics in addition to the initial - time geometry one. The results of Knops & Payne (1969) have recently been markedly improved in Knops & Payne (1988) where the main technique employed is the Lagrange identity method. In three influential papers dealing with various aspects of modeling and continuous dependence on the initial - time geometry and on the spatial geometry in improperly posed problems Payne (1987a,b,1989) emphasized how important initial - time geometry problems are in general, and in particular, in improperly posed problems for partial differential equations. Payne's papers have certainly been highly influential on the ensuing research, see e.g. Ames & Payne (1994b,1995a), Ames & Straughan (1995), Franchi & Straughan (1995), Payne & Straughan (1990c,1996a), and Song (1988).

Study of Errors in the Initial - Time Geometry for the Forward Heat Equation

The topic of this subsection was first addressed by Payne (1987b). The idea is to compare the solution to the heat equation when the initial data are given at time $t = 0$ to the solution to the equivalent problem but with the "initial" data given over a "curve" $t = \epsilon f(\mathbf{x})$. In reality, data is rarely measured at the same instant of time and thus it is important to know if the variation of where the data is actually measured has a pronounced effect on the solution. The data curves in one space dimension are sketched schematically in figure 1.

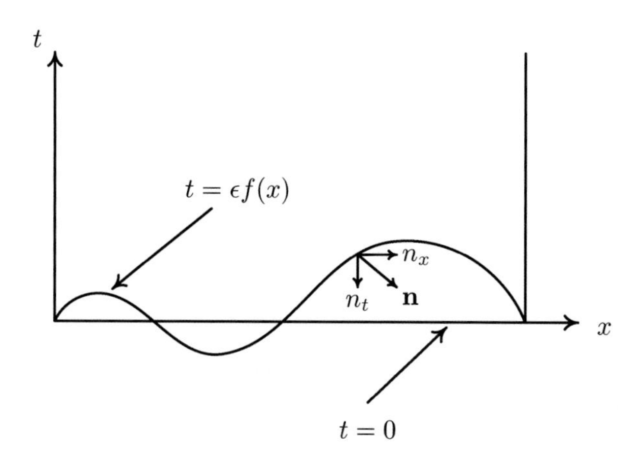

Figure 1. Configuration for a Study of the Initial - Time Geometry Error.

Let now Ω be a bounded domain in $\mathbf{R}^n$ with boundary Γ. Let Σ denote the surface $t = \epsilon f(\mathbf{x})$ and suppose

$$|f(\mathbf{x})| < 1 \tag{2.1.1}$$

and for $\delta > 0$,

$$n_t < -\delta, \tag{2.1.2}$$

where n_t is the component of the unit normal to Σ in the $t-$direction; n_i will denote the component in the spatial direction, cf. figure 1.

Suppose u solves the boundary - initial value problem:

$$\begin{aligned}
&\frac{\partial u}{\partial t} = \Delta u, \qquad \text{in} \quad \Omega \times (0,T),\\
&u = 0, \qquad \text{on} \quad \Gamma \times [0,T),\\
&u(\mathbf{x},0) = g(\mathbf{x}), \qquad \mathbf{x} \in \Omega,
\end{aligned} \tag{2.1.3}$$

and let v solve the following boundary - initial value problem:

$$\begin{aligned}
&\frac{\partial v}{\partial t} = \Delta v, \qquad \mathbf{x} \in \Omega, \quad \epsilon f(\mathbf{x}) < t < T,\\
&v = 0, \qquad \mathbf{x} \in \Gamma, \quad \epsilon f(\mathbf{x}) \le t < T,\\
&v\big(\mathbf{x}, \epsilon f(\mathbf{x})\big) = g(\mathbf{x}).
\end{aligned} \tag{2.1.4}$$

Observe that v is the solution to the problem where the data are measured on $t = \epsilon f(\mathbf{x})$. The object is to determine whether v differs appreciably from u in some sense, since it is the u–problem which is usually studied in practice.

Payne (1987b) tackles this question by introducing

$$w = u - v, \tag{2.1.5}$$

and then on $\Omega \times \big(\epsilon f(\mathbf{x}), T\big)$ w satisfies the equation

$$\frac{\partial w}{\partial t} = \Delta w, \tag{2.1.6}$$

and the boundary conditions

$$w = 0, \qquad \mathbf{x} \in \Gamma. \tag{2.1.7}$$

Multiplication of (2.1.6) by w and integration over Ω with an integration by parts shows

$$\frac{d}{dt}\frac{1}{2}\|w\|^2 = -\int_\Omega \frac{\partial w}{\partial x_i}\frac{\partial w}{\partial x_i}dx,$$

where $\|\cdot\|$ denotes the norm on $L^2(\Omega)$, and then use of Poincaré's inequality gives

$$\frac{d}{dt}\frac{1}{2}\|w\|^2 \leq -\lambda_1\|w\|^2.$$

This is integrated from ϵ to t to find

$$\|w(t)\|^2 \leq \|w(\epsilon)\|^2 \exp\big[-2\lambda_1(t-\epsilon)\big]. \tag{2.1.8}$$

The object now is to bound $\|w(\epsilon)\|$ in terms of known data and ϵ and thus find an estimate of how v compares with u as a function of ϵ, and in so doing we describe Payne's (1987b) method.

If $\tilde{u}$ denotes u but extended backward of $t = 0$ by $g(\mathbf{x})$ and $\tilde{v}$ likewise denotes the backward extension of v beyond $t = \epsilon f(\mathbf{x})$ by $g(\mathbf{x})$ then for $\tilde{w} = \tilde{u} - \tilde{v}$,

$$\begin{aligned}\|w(\epsilon)\|^2 &= 2\int_{-\epsilon}^{\epsilon}\int_\Omega \tilde{w}\frac{\partial \tilde{w}}{\partial s}\,dx\,ds,\\ &\leq \frac{8\epsilon}{\pi}\int_{-\epsilon}^{\epsilon}\|\frac{\partial \tilde{w}}{\partial s}\|^2\,ds,\end{aligned} \tag{2.1.9}$$

where the Cauchy - Schwarz inequality and Poincaré's inequality have been employed. To proceed from (2.1.9) Payne (1987b) uses the arithmetic - geometric mean inequality to find

$$\|w(\epsilon)\|^2 \leq \frac{16}{\pi}\,\epsilon\,(J_1 + J_2), \tag{2.1.10}$$

where

$$J_1 = \int_0^\epsilon \|\frac{\partial u}{\partial s}\|^2 ds, \tag{2.1.11}$$

and

$$J_2 = \int_\Omega \int_{\epsilon f(\mathbf{x})}^\epsilon \left(\frac{\partial v}{\partial s}\right)^2 ds\, dx. \tag{2.1.12}$$

The idea is to now bound J_1 and J_2 in terms of data, since (2.1.10) already contains an ϵ on the right hand side.

By using (2.1.3),

$$\begin{aligned}
J_1 =& \int_0^\epsilon \int_\Omega \frac{\partial u}{\partial s} \Delta u\, dx\, ds, \\
=& -\frac{1}{2}\int_0^\epsilon \int_\Omega \frac{\partial}{\partial s}\left(\frac{\partial u}{\partial x_i}\frac{\partial u}{\partial x_i}\right) dx\, ds, \\
=& \frac{1}{2}\int_\Omega g_{,i}g_{,i}\, dx - \frac{1}{2}\int_\Omega u_{,i}u_{,i}\, dx\Big|_{t=\epsilon}, \\
\leq& \frac{1}{2}\int_\Omega g_{,i}g_{,i}\, dx.
\end{aligned} \tag{2.1.13}$$

By a similar procedure employing (2.1.4) but recollecting that v has data assigned over $t = \epsilon f(\mathbf{x})$

$$\begin{aligned}
J_2 =& \int_\Omega \int_{\epsilon f(\mathbf{x})}^\epsilon \frac{\partial v}{\partial s} \Delta v\, ds\, dx, \\
\leq& \int_\Sigma \left(\frac{\partial v}{\partial t}\frac{\partial v}{\partial x_i} n_i - \frac{1}{2}\frac{\partial v}{\partial x_i}\frac{\partial v}{\partial x_i} n_t\right) dS,
\end{aligned} \tag{2.1.14}$$

dS denoting integration over Σ. Next, (2.1.14) must be written in terms of data and so recalling the tangential derivative of v on Σ is

$$\frac{\partial v}{\partial t} n_i - \frac{\partial v}{\partial x_i} n_t\,,$$

Payne (1987b) writes

$$J_2 \leq \int_\Sigma \frac{\partial v}{\partial x_i}\left(\frac{\partial v}{\partial t} n_i - \frac{\partial v}{\partial x_i} n_t\right) dS + \frac{1}{2}\int_\Sigma \frac{\partial v}{\partial x_i}\frac{\partial v}{\partial x_i} n_t\, dS. \tag{2.1.15}$$

Condition (2.1.2) on n_t now allows further deduction from this, namely

$$\begin{aligned}
J_2 \leq& \int_\Sigma \frac{\partial v}{\partial x_i}(n_i g_{,t} - n_t g_{,i}) dS - \frac{1}{2}\delta \int_\Sigma \frac{\partial v}{\partial x_i}\frac{\partial v}{\partial x_i}\, dS, \\
\leq& \frac{1}{2\delta}\int_\Sigma (n_i g_{,t} - n_t g_{,i})(n_i g_{,t} - n_t g_{,i})\, dS.
\end{aligned} \tag{2.1.16}$$

Estimates (2.1.13) and (2.1.16) are now employed in (2.1.8) to derive the a *priori* bound

$$\|u(t) - v(t)\|^2 \leq M^2\epsilon, \tag{2.1.17}$$

where the data term M^2 is given by

$$\begin{aligned} M^2 =& \exp\left[-2\lambda_1(t-\epsilon)\right] \\ & \times \frac{8}{\pi}\left[\int_\Omega g_{,i}g_{,i}\,dx + \frac{1}{\delta}\int_\Sigma (n_i g_{,t} - n_t g_{,i})(n_i g_{,t} - n_t g_{,i})\,dS\right]. \end{aligned}$$

Inequality (2.1.17) establishes continuous dependence on the initial - time geometry and since it is for a forward in time problem it is an *a priori* result. To extend such a result to the improperly posed backward in time problem requires restrictions and the introduction of a constraint set.

Study of Errors in the Initial - Time Geometry for the Backward Heat Equation

Payne (1987a) gave the first analysis of continuous dependence on the initial - time geometry for the backward heat equation. In a recent paper Ames & Payne (1994) have weakened the constraints imposed by Payne's earlier analysis and thus we review the more recent work.

Ames & Payne (1994) argue that in an improperly posed problem one cannot assume existence for arbitrary $g(\mathbf{x})$ and, therefore, they allow the $u-$ and $v-$ problems to have different prescribed data.

They let u be a solution to

$$\begin{aligned} &\frac{\partial u}{\partial t} = -\Delta u, \qquad \text{in} \quad \Omega\times(0,T), \\ &u = 0, \qquad \text{on} \quad \Gamma\times[0,T], \\ &u(\mathbf{x},0) = g_1(\mathbf{x}), \qquad \mathbf{x}\in\Omega, \end{aligned} \tag{2.1.18}$$

while v solves

$$\begin{aligned} &\frac{\partial v}{\partial t} = -\Delta v, \qquad \mathbf{x}\in\Omega, \quad \epsilon f(\mathbf{x}) < t < T, \\ &v = 0, \qquad \mathbf{x}\in\Gamma, \quad \epsilon f(\mathbf{x}) \leq t \leq T, \\ &v\big(\mathbf{x},\epsilon f(\mathbf{x})\big) = g_2(\mathbf{x}), \qquad \mathbf{x}\in\Omega. \end{aligned} \tag{2.1.19}$$

The object is again to produce an estimate of how much v varies from u. The condition (2.1.1) is again required and the prescribed data g_1, g_2 are such that

$$\begin{aligned} &\int_\Omega g_{1,i}g_{1,i}\,dx < \infty, \qquad \int_\Omega g_{2,i}g_{2,i}\,dx < \infty, \\ &\|g_1 - g_2\| \leq K_0\epsilon^{1/2}, \end{aligned} \tag{2.1.20}$$

for a positive constant K_0.

The constraint set of Ames & Payne (1994) is relatively weak and imposes a bound at only one instant of time, T. They suppose

$$\|u(T)\|^2 \le M_1^2, \qquad \|v(T)\|^2 \le M_1^2, \tag{2.1.21}$$

for some known constant M_1. We point out that as described in chapter 1, the imposition of a constraint set in an improperly posed problem is natural, and indeed, necessary.

Since $w = u - v$ satisfies

$$\frac{\partial w}{\partial t} = -\Delta w,$$

for $t \in \big(\epsilon f(\mathbf{x}), T\big)$, with $w = 0$ on Γ, Ames & Payne (1994) use a logarithmic convexity method just as in section 1.5 to show that

$$\|w(t)\| \le \|w(\epsilon)\|^{(T-t)/(T-\epsilon)} \|w(T)\|^{(t-\epsilon)/(T-\epsilon)},$$

and then due to (2.1.21),

$$\|w(t)\| \le (2M_1)^{(t-\epsilon)/(T-\epsilon)} \|w(\epsilon)\|^{(T-t)/(T-\epsilon)}. \tag{2.1.22}$$

It remains to bound $\|w(\epsilon)\|$ as a function of ϵ.

Ames & Payne (1994) begin this with the triangle inequality to observe

$$\|w(\epsilon)\| \le \|u(\epsilon) - g_1\| + \|g_1 - g_2\| + \|v(\epsilon) - g_2\|. \tag{2.1.23}$$

A bound for the middle term on the right follows from (2.1.20). The bound for the first term is easier than that for the third but follows along the same lines. Only that appropriate to the third term is described here.

Since

$$\int_{\epsilon f(\mathbf{x})}^{\epsilon} \frac{\partial v}{\partial s}(x,s)\,ds = v(x,\epsilon) - g_2(x),$$

there follows

$$\begin{aligned} \|v(\epsilon) - g_2\|^2 &= \int_\Omega \big[v(x,\epsilon) - g_2(x)\big] \left[\int_{\epsilon f(\mathbf{x})}^{\epsilon} \frac{\partial v}{\partial s}(x,s)\,ds\right] dx, \\ &= -\int_\Omega \big[v(x,\epsilon) - g_2(x)\big] \left[\int_{\epsilon f(\mathbf{x})}^{\epsilon} \Delta v(x,s)\,ds\right] dx, \\ &= \int_\Omega \big[v_{,i}(x,\epsilon) - g_{2,i}(x)\big] \left[\int_{\epsilon f(\mathbf{x})}^{\epsilon} v_{,i}(x,s)\,ds\right] dx, \end{aligned} \tag{2.1.24}$$

where in the second step the differential equation (2.1.19) has been used, while the last step has been achieved by integration by parts. From this stage the Cauchy - Schwarz inequality is employed to yield

$$\begin{aligned}
\|v(\epsilon) - g_2\|^2 \leq & \int_\Omega \nabla(v - g_2)\Big(\int_{\epsilon f(\mathbf{x})}^{\epsilon} ds\Big)^{1/2}\Big(\int_{\epsilon f(\mathbf{x})}^{\epsilon} |\nabla v|^2 ds\Big)^{1/2} dx,\\
\leq & \Big[\int_\Omega \epsilon(1-f)|\nabla(v-g_2)|^2 dx \int_\Omega \int_{\epsilon f(\mathbf{x})}^{\epsilon} |\nabla v|^2 ds\, dx\Big]^{1/2},
\end{aligned}$$

and then noting $|f| \leq 1$ and using the triangle inequality one may show

$$\begin{aligned}
\|v(\epsilon) - g_2\|^2 \leq & (2\epsilon)^{1/2}\big[\|\nabla v(\epsilon)\| + \|\nabla g_2\|\big]\\
& \times \Big[\int_\Omega \int_{\epsilon f(\mathbf{x})}^{\epsilon} |\nabla v(x,s)|^2 ds\, dx\Big]^{1/2}. \qquad (2.1.25)
\end{aligned}$$

The term $\|\nabla g_2\|$ may be treated as data and so bounds are needed for $\|\nabla v(\epsilon)\|$ and $\int_\Omega \int_{\epsilon f(\mathbf{x})}^{\epsilon} |\nabla v(x,s)|^2 ds\, dx$. To establish the second of these note

$$\begin{aligned}
\int_\Omega \int_{\epsilon f(\mathbf{x})}^{\epsilon} |\nabla v(x,s)|^2 ds\, dx = & \int_\Omega \int_{\epsilon f(\mathbf{x})}^{\epsilon} v_{,i}(v_{,i} - g_{2,i}) ds\, dx\\
& + \int_\Omega \int_{\epsilon f(\mathbf{x})}^{\epsilon} v_{,i} g_{2,i} ds\, dx,\\
\leq & - \int_\Omega \int_{\epsilon f(\mathbf{x})}^{\epsilon} \Delta v\,(v - g_2)\, ds\, dx\\
& + \Big(\int_\Omega \int_{\epsilon f(\mathbf{x})}^{\epsilon} |\nabla v|^2 ds\, dx\Big)^{1/2} \Big(\int_\Omega \int_{\epsilon f(\mathbf{x})}^{\epsilon} |\nabla g_2|^2 ds\, dx\Big)^{1/2},
\end{aligned}$$

where we have integrated by parts and used the Cauchy - Schwarz inequality and then using the differential equation one may find

$$\begin{aligned}
\int_\Omega \int_{\epsilon f(\mathbf{x})}^{\epsilon} |\nabla v(x,s)|^2 ds\, dx \leq & \int_\Omega \int_{\epsilon f(\mathbf{x})}^{\epsilon} \frac{\partial v}{\partial s}(v - g_2) ds\, dx\\
& + \Big(\int_\Omega \int_{\epsilon f(\mathbf{x})}^{\epsilon} |\nabla v|^2 ds\, dx\Big)^{1/2} \Big(\int_\Omega \int_{\epsilon f(\mathbf{x})}^{\epsilon} |\nabla g_2|^2 ds\, dx\Big)^{1/2},\\
\leq & \frac{1}{2}\|v(\epsilon) - g_2\|^2 + (2\epsilon)^{1/2}\|\nabla g_2\| \Big(\int_\Omega \int_{\epsilon f(\mathbf{x})}^{\epsilon} |\nabla v|^2 ds\, dx\Big)^{1/2},
\end{aligned}$$

where the first term has simply been integrated and the fact that $|f| \leq 1$ has been used on the second. This quadratic inequality leads to

$$\Big(\int_\Omega \int_{\epsilon f(\mathbf{x})}^{\epsilon} |\nabla v|^2 ds\, dx\Big)^{1/2} \leq (2\epsilon)^{1/2}\|\nabla g_2\| + \frac{1}{\sqrt{2}}\|v(\epsilon) - g_2\|. \qquad (2.1.26)$$

For the other term in (2.1.25) the analysis begins with

$$\|\nabla v(\epsilon)\|^2 = -\int_\Omega \left(\frac{T-s}{T-\epsilon}\right) |\nabla v(x,s)|^2 \, n_s \, dx,$$

and then using the divergence theorem this gives

$$\begin{aligned}\|\nabla v(\epsilon)\|^2 &= \frac{1}{(T-\epsilon)} \int_\epsilon^T \|\nabla v\|^2 ds - 2\int_\epsilon^T \int_\Omega \left(\frac{T-s}{T-\epsilon}\right) \frac{\partial v}{\partial x_i} \frac{\partial^2 v}{\partial x_i \partial s} \, dx \, ds,\\ &= \frac{1}{(T-\epsilon)} \int_\epsilon^T \|\nabla v\|^2 ds + 2\int_\epsilon^T \int_\Omega \left(\frac{T-s}{T-\epsilon}\right) \Delta v \frac{\partial v}{\partial s} \, dx \, ds.\end{aligned}$$

Integrating by parts, and then using the differential equation (2.1.19), there follows

$$\begin{aligned}\|\nabla v(\epsilon)\|^2 &= \frac{1}{(T-\epsilon)} \int_\epsilon^T \|\nabla v\|^2 ds - 2\int_\epsilon^T \int_\Omega \left(\frac{T-s}{T-\epsilon}\right) \left(\frac{\partial v}{\partial s}\right)^2 dx \, ds,\\ &\leq \frac{1}{(T-\epsilon)} \int_\epsilon^T \|\nabla v\|^2 ds,\\ &= -\frac{1}{(T-\epsilon)} \int_\epsilon^T \int_\Omega v \Delta v \, dx \, ds,\\ &= \frac{1}{(T-\epsilon)} \int_\epsilon^T \int_\Omega v \frac{\partial v}{\partial s} \, dx \, ds,\end{aligned}$$

where further integration by parts and use of (2.1.19) has been made. The right hand side may now be integrated to find

$$\|\nabla v(\epsilon)\|^2 \leq \frac{1}{2(T-\epsilon)} \int_\Omega v^2(x,T) \, dx.$$

Dropping a non-positive term, and then an appeal to (2.1.21) yields

$$\|\nabla v(\epsilon)\|^2 \leq \frac{M_1^2}{2(T-\epsilon)}. \tag{2.1.27}$$

Bounds (2.1.26) and (2.1.27) are next used in (2.1.25) to find

$$\begin{aligned}\|v(\epsilon) - g_2\|^2 \leq &(2\epsilon)^{1/2} \left(\frac{M_1}{2(T-\epsilon)^{1/2}} + \|\nabla g_2\|\right)\\ &\times \left((2\epsilon)^{1/2} \|\nabla g_2\| + \frac{1}{\sqrt{2}} \|v(\epsilon) - g_2\|\right),\end{aligned}$$

and this quadratic inequality is manipulated to give

$$\|v(\epsilon) - g_2\| \leq K_2 \epsilon^{1/2}, \tag{2.1.28}$$

where

$$K_2 = \frac{M_1}{\sqrt{2(T-\epsilon)}} + \|\nabla g_2\| + (2\|\nabla g_2\|)^{1/2}\Big(\frac{M_1}{\big[2(T-\epsilon)\big]^{1/2}} + \|\nabla g_2\|\Big)^{1/2}.$$

Use of (2.1.28), (2.1.20), and the analogous bound to (2.1.28) for $\|u(\epsilon) - g_1\|$ then leads Ames & Payne (1994) to an inequality of form

$$\|u(t) - v(t)\|^2 \le K\epsilon^{(T-t)/(T-\epsilon)}, \qquad t \in [\epsilon, T]. \tag{2.1.29}$$

This estimate establishes continuous dependence on the initial - time geometry as we set out to achieve.

2.2 Continuous Dependence on the Initial - Time Geometry for the Heat Equation on an Exterior Domain

In this section we study the equivalent problem to that analyzed in section 2.1 except now the spatial domain Ω^* is the exterior to the closure of a bounded region $\Omega(\subset \mathbf{R}^n)$ which contains the origin. If we assume integrability of everything in section 2.1 then we could simply use the analysis there. However, Payne & Straughan (1990c) used weighted integrals to allow the solution to possibly even grow at infinity, and so we include an exposition of that work.

The bounded boundary of Ω^*, i.e. the boundary of Ω, is denoted by Γ.

Study of Errors in the Initial - Time Geometry for the Forward Heat Equation on an Exterior Domain

The surface Σ is now that surface $t = \epsilon f(\mathbf{x}), \mathbf{x} \in \Omega^*$, and we suppose

$$\begin{aligned} &|f(\mathbf{x})| < 1, \qquad \forall \mathbf{x} \in \Omega^*, \\ &|\nabla f(\mathbf{x})| < K, \qquad \forall \mathbf{x} \in \Omega^*, \end{aligned} \tag{2.2.1}$$

where K is a known constant.

In this section we compare the solution to the boundary - initial value problem:

$$\begin{aligned} &\frac{\partial u}{\partial t} = \Delta u, \qquad \text{in} \quad \Omega^* \times (0,T), \\ &u = 0, \qquad \text{on} \quad \Gamma \times [0,T), \\ &u(\mathbf{x},0) = g(\mathbf{x}), \qquad \mathbf{x} \in \Omega^*, \end{aligned} \tag{2.2.2}$$

with the solution v to the following boundary - initial value problem:

$$\begin{aligned} &\frac{\partial v}{\partial t} = \Delta v, \qquad \mathbf{x} \in \Omega^*, \quad \epsilon f(\mathbf{x}) < t < T, \\ &v = 0, \qquad \mathbf{x} \in \Gamma, \quad \epsilon f(\mathbf{x}) \leq t < T, \\ &v\big(\mathbf{x}, \epsilon f(\mathbf{x})\big) = g(\mathbf{x}) \qquad \mathbf{x} \in \Omega^*. \end{aligned} \tag{2.2.3}$$

We note again that v is the solution to the problem where the data are measured on $t = \epsilon f(\mathbf{x})$, and the object is to determine whether v differs appreciably from u in some sense, since it is the $u-$problem which is usually studied in practice.

As in section 2.1 we put

$$w = u - v,$$

and then for $t > \epsilon$, we see that w satisfies the equation

$$\frac{\partial w}{\partial t} = \Delta w, \tag{2.2.4}$$

and the boundary conditions

$$w = 0, \qquad \mathbf{x} \in \Gamma. \tag{2.2.5}$$

The starting point of the continuous dependence proof given by Payne & Straughan (1990c) is to multiply (2.2.4) by $e^{-\alpha r}w$, where $\alpha > 0$ is a constant to be selected and $r = \sqrt{x_i x_i}$. The effect of the weight $e^{-\alpha r}$ is to allow the solution and its derivatives to grow at infinity. This variant of the weighted energy method was used by Rionero & Galdi (1976) to study uniqueness for the Navier - Stokes equations. The function $F(t)$ is defined by

$$F(t) = \int_{\Omega^*} e^{-\alpha r} w^2(x,t)\, dx, \tag{2.2.6}$$

and then after multiplication of (2.2.4) by $e^{-\alpha r}w$ and integration over Ω^* we find

$$\begin{aligned} F'(t) =& 2\int_{\Omega^*} e^{-\alpha r} w \frac{\partial w}{\partial t}\, dx, \\ =& 2\int_{\Omega^*} e^{-\alpha r} w \Delta w\, dx, \\ =& 2\alpha \int_{\Omega^*} e^{-\alpha r} \frac{x_i}{r} w \frac{\partial w}{\partial x_i}\, dx - 2\int_{\Omega^*} e^{-\alpha r} \frac{\partial w}{\partial x_i}\frac{\partial w}{\partial x_i}\, dx, \end{aligned} \tag{2.2.7}$$

where we have substituted from (2.2.4) and integrated by parts. The solutions are assumed to grow no faster at infinity than the weight $e^{-\alpha r}$ will dominate and this is why the boundary terms at infinity disappear. The arithmetic - geometric mean inequality is next used on the first term on the right of (2.2.7), noting $x_i/r \leq 1$, to see that

$$\begin{aligned} 2\alpha \int_{\Omega^*} e^{-\alpha r} \frac{x_i}{r} w \frac{\partial w}{\partial x_i} \, dx \leq & \frac{1}{2}\alpha^2 \int_{\Omega^*} e^{-\alpha r} w^2 dx \\ & + 2 \int_{\Omega^*} e^{-\alpha r} \frac{\partial w}{\partial x_i} \frac{\partial w}{\partial x_i} \, dx. \end{aligned} \tag{2.2.8}$$

Upon using (2.2.8) in (2.2.7) we then derive

$$F' \leq \frac{1}{2}\alpha^2 F. \tag{2.2.9}$$

This inequality may be integrated from ϵ to t to obtain

$$F(t) \leq F(\epsilon) \exp\left[\frac{1}{2}\alpha^2 (t - \epsilon)\right]. \tag{2.2.10}$$

Inequality (2.2.10) is the beginning of the continuous dependence estimate on the initial - time geometry and it now remains to estimate $F(\epsilon)$ in terms of known data and a function of ϵ and thus find an estimate of how v compares with u as a functional relationship in ϵ.

Payne & Straughan (1990c) commence their estimate of $F(\epsilon)$ by writing

$$\begin{aligned} v(\mathbf{x}, \epsilon) =& g(\mathbf{x}) + \int_{\epsilon f(\mathbf{x})}^{\epsilon} \frac{\partial v}{\partial s} \, ds, \\ u(\mathbf{x}, \epsilon) =& g(\mathbf{x}) + \int_0^{\epsilon} \frac{\partial u}{\partial s} \, ds, \end{aligned} \tag{2.2.11}$$

and hence note that

$$F(\epsilon) = \int_{\Omega^*} e^{-\alpha r} \left(\int_0^{\epsilon} \frac{\partial u}{\partial s} \, ds - \int_{\epsilon f(\mathbf{x})}^{\epsilon} \frac{\partial v}{\partial s} \, ds \right)^2 dx. \tag{2.2.12}$$

This term may then be estimated with the aid of the Cauchy - Schwarz inequality and the arithmetic - geometric mean inequality observing that $1 - f \leq 2$ to see that

$$F(\epsilon) \leq 3\,\epsilon\,(J_1 + J_2), \tag{2.2.13}$$

where the terms J_1 and J_2 are defined by

$$\begin{aligned} J_1 &= \int_0^{\epsilon} \int_{\Omega^*} e^{-\alpha r} \left(\frac{\partial u}{\partial s}\right)^2 dx \, ds, \\ J_2 &= \int_{\Omega^*} e^{-\alpha r} \int_{\epsilon f(\mathbf{x})}^{\epsilon} \left(\frac{\partial v}{\partial s}\right)^2 ds \, dx. \end{aligned} \tag{2.2.14}$$

We must now show J_1 and J_2 may be bounded by data.

One of the $\partial u/\partial s$ terms in J_1 is substituted from (2.2.2) and then after an integration by parts, we derive

$$\begin{aligned} J_1 =&\alpha \int_0^\epsilon \int_{\Omega^*} e^{-\alpha r} \frac{x_i}{r} \frac{\partial u}{\partial s} \frac{\partial u}{\partial x_i}\, dx\, ds + \frac{1}{2} \int_{\Omega^*} e^{-\alpha r} \frac{\partial g}{\partial x_i} \frac{\partial g}{\partial x_i}\, dx, \\ &- \frac{1}{2} \int_{\Omega^*} e^{-\alpha r} \frac{\partial u}{\partial x_i} \frac{\partial u}{\partial x_i}\, dx \bigg|_{t=\epsilon}. \end{aligned} \tag{2.2.15}$$

Next, note $|x_i/r| \leq 1$, and then from the arithmetic - geometric mean inequality

$$\begin{aligned} \alpha \int_0^\epsilon \int_{\Omega^*} e^{-\alpha r} \frac{x_i}{r} \frac{\partial u}{\partial s} \frac{\partial u}{\partial x_i}\, dx\, ds \leq& \frac{1}{2} \alpha^2 \int_0^\epsilon \int_{\Omega^*} e^{-\alpha r} \frac{\partial u}{\partial x_i} \frac{\partial u}{\partial x_i}\, dx\, ds \\ &+ \frac{1}{2} J_1. \end{aligned} \tag{2.2.16}$$

Inequality (2.2.16) is used in (2.2.15) and the last term of (2.2.15) is discarded to deduce

$$J_1 \leq \alpha^2 \int_0^\epsilon \int_{\Omega^*} e^{-\alpha r} \frac{\partial u}{\partial x_i} \frac{\partial u}{\partial x_i}\, dx\, ds + \int_{\Omega^*} e^{-\alpha r} \frac{\partial g}{\partial x_i} \frac{\partial g}{\partial x_i}\, dx. \tag{2.2.17}$$

Further progress requires an estimate for the first term on the right and to this end note that by integration by parts and (2.2.2),

$$\begin{aligned} \int_0^\epsilon \int_{\Omega^*} e^{-\alpha r} \frac{\partial u}{\partial x_i} \frac{\partial u}{\partial x_i}\, dx\, ds =& - \int_0^\epsilon \int_{\Omega^*} e^{-\alpha r} u \Delta u\, dx\, ds \\ &+ \alpha \int_0^\epsilon \int_{\Omega^*} e^{-\alpha r} \frac{x_i}{r} u \frac{\partial u}{\partial x_i}\, dx\, ds \\ =& - \int_0^\epsilon \int_{\Omega^*} e^{-\alpha r} u \frac{\partial u}{\partial s}\, dx\, ds \\ &+ \alpha \int_0^\epsilon \int_{\Omega^*} e^{-\alpha r} \frac{x_i}{r} u \frac{\partial u}{\partial x_i}\, dx\, ds \\ =& \frac{1}{2} \int_{\Omega^*} e^{-\alpha r} g^2\, dx - \frac{1}{2} \int_{\Omega^*} e^{-\alpha r} u^2(\mathbf{x}, \epsilon)\, dx \\ &+ \alpha \int_0^\epsilon \int_{\Omega^*} e^{-\alpha r} \frac{x_i}{r} u \frac{\partial u}{\partial x_i}\, dx\, ds. \end{aligned}$$

Now again use the fact that $|x_i/r| \leq 1$ and use the arithmetic - geometric mean inequality on the last term. By dropping the second term we obtain

$$\begin{aligned} \int_0^\epsilon \int_{\Omega^*} e^{-\alpha r} \frac{\partial u}{\partial x_i} \frac{\partial u}{\partial x_i}\, dx\, ds \leq& \alpha^2 \int_0^\epsilon \int_{\Omega^*} e^{-\alpha r} u^2\, dx\, ds \\ &+ \int_{\Omega^*} e^{-\alpha r} g^2\, dx. \end{aligned} \tag{2.2.18}$$

The triangle inequality may now be used to show

$$\begin{aligned}\Big(\int_0^\epsilon\int_{\Omega^*} e^{-\alpha r}u^2\,dx\,ds\Big)^{1/2} &\leq \Big(\int_0^\epsilon\int_{\Omega^*} e^{-\alpha r}(u-g)^2\,dx\,ds\Big)^{1/2}\\ &\quad + \epsilon^{1/2}\Big(\int_{\Omega^*} e^{-\alpha r}g^2\,dx\Big)^{1/2},\\ &\leq \frac{2\epsilon}{\pi}\sqrt{J_1}\\ &\quad + \epsilon^{1/2}\Big(\int_{\Omega^*} e^{-\alpha r}g^2\,dx\Big)^{1/2}, \qquad (2.2.19)\end{aligned}$$

where Poincaré's inequality (in t) has been employed. Squaring (2.2.19) and using the arithmetic - geometric mean inequality we find

$$\begin{aligned}\int_0^\epsilon\int_{\Omega^*} e^{-\alpha r}u^2\,dx\,ds &\leq \frac{4\epsilon^2}{\pi^2}(1+\epsilon)J_1\\ &\quad + (1+\epsilon)\int_{\Omega^*} e^{-\alpha r}g^2\,dx. \qquad (2.2.20)\end{aligned}$$

Inequality (2.2.20) is now employed on the right in (2.2.18) and the resulting inequality is then used in (2.2.17) and one finds

$$\begin{aligned}\Big[1-\frac{4\alpha^4\epsilon^2}{\pi^2}(1+\epsilon)\Big]J_1 &\leq \int_{\Omega^*} e^{-\alpha r}\frac{\partial g}{\partial x_i}\frac{\partial g}{\partial x_i}\,dx\\ &\quad + \alpha^2\{1+(1+\epsilon)\alpha^2\}\int_{\Omega^*} e^{-\alpha r}g^2\,dx. \qquad (2.2.21)\end{aligned}$$

If we now require α (which is at our disposal) so small that the coefficient of J_1 is positive, i.e. for some $\delta > 0$ we have

$$\alpha^4\epsilon^2 < \frac{\pi^2(1-\delta)}{4(1+\epsilon)},$$

then (2.2.21) gives the result for J_1 which we seek, namely,

$$J_1 \leq \frac{Q_1}{\delta}, \qquad (2.2.22)$$

where

$$Q_1 = \int_{\Omega^*} e^{-\alpha r}\frac{\partial g}{\partial x_i}\frac{\partial g}{\partial x_i}\,dx + \alpha^2\{1+(1+\epsilon)\alpha^2\}\int_{\Omega^*} e^{-\alpha r}g^2\,dx. \qquad (2.2.23)$$

The bound for J_2 proceeds along similar lines but is more complicated because v is given as g on $t = \epsilon f(\mathbf{x})$, not $t = 0$. However, in that case one may show (Payne & Straughan (1990c)) that provided we restrict α such that

$$\alpha^4\epsilon^2 < \frac{\pi^2(1-\gamma)}{16(1+\epsilon)}, \tag{2.2.24}$$

for some $\gamma > 0$, then

$$J_2 \le \frac{Q_2}{\gamma}, \tag{2.2.25}$$

where

$$\begin{aligned} Q_2 =&2\alpha^2 \int_{\Omega^*} e^{-\alpha r} \frac{\partial f}{\partial x_i}\frac{\partial f}{\partial x_i}\, g^2\, dx \\ &+ \alpha^2\{1+2(1+\epsilon)\alpha^2\} \int_{\Omega^*} e^{-\alpha r} g^2\, dx \\ &+ 2\int_{\Omega^*} e^{-\alpha r} \frac{\partial g}{\partial x_i}\frac{\partial g}{\partial x_i}\, dx. \end{aligned} \tag{2.2.26}$$

Thus, if we now employ (2.2.22) and (2.2.25) in (2.2.13) we obtain a constant K_1 such that

$$F(\epsilon) \le K_1\epsilon,$$

and then this inequality in (2.2.10) gives the continuous dependence estimate

$$\int_{\Omega^*} e^{-\alpha r} w^2\, dx \le \exp\{\frac{1}{2}\alpha^2(t-\epsilon)\}K_1\epsilon. \tag{2.2.27}$$

Payne & Straughan (1990c) show that condition (2.2.24) is actually necessary, and not just artificial, by constructing a specific example.

To use (2.2.27) effectively as an estimate for continuous dependence on the initial - time geometry we note that if B_R denotes the ball center 0, radius R, then

$$\int_{\Omega^*} e^{-\alpha r} w^2\, dx \ge e^{-\alpha R} \int_{\Omega^* \cup B_R} w^2\, dx,$$

and this in (2.2.27) shows

$$\int_{\Omega^* \cup B_R} w^2(x,t)\, dx \le K\epsilon, \tag{2.2.28}$$

where

$$K(t,R) = K_1 e^{\alpha R}\exp\{\frac{1}{2}\alpha^2(t-\epsilon)\}.$$

Inequality (2.2.28) is thus an *a priori* estimate which demonstrates continuous dependence on the initial - time geometry on compact subdomains of Ω^*.

Results (2.2.27) and (2.2.28) establish continuous dependence in an L^2 measure involving w. Payne & Straughan (1990c) also show how continuous dependence may be established in other norms.

Study of Errors in the Initial - Time Geometry for the Backward Heat Equation on an Exterior Domain

Payne & Straughan (1990c) also show how one may establish continuous dependence on the initial - time geometry for the backward heat equation when Ω^* is an exterior domain as in the beginning of this section. To do this they utilize a logarithmic convexity method with a weighted functional, the purpose of the weight again being to dominate terms at infinity and thus allow growth of the solution and its derivatives.

To briefly review this work let u be a solution to the boundary - initial value problem

$$\begin{aligned} &\frac{\partial u}{\partial t} = -\Delta u, \qquad \text{in} \quad \Omega^* \times (0,T), \\ &u = 0, \qquad \text{on} \quad \Gamma \times [0,T], \\ &u(\mathbf{x},0) = g(\mathbf{x}), \qquad \mathbf{x} \in \Omega^*, \end{aligned} \tag{2.2.29}$$

and let v solve the boundary - initial value problem

$$\begin{aligned} &\frac{\partial v}{\partial t} = -\Delta v, \qquad \mathbf{x} \in \Omega^*, \quad \epsilon f(\mathbf{x}) < t < T, \\ &v = 0, \qquad \mathbf{x} \in \Gamma, \quad \epsilon f(\mathbf{x}) \le t \le T, \\ &v\big(\mathbf{x}, \epsilon f(\mathbf{x})\big) = g(\mathbf{x}), \qquad \mathbf{x} \in \Omega^*. \end{aligned} \tag{2.2.30}$$

The idea here is as before to produce an estimate of the variation of v as compared to u. The initial - time geometry function f again satisfies (2.1.1) and $w = u - v$.

The constraint set of Payne & Straughan (1990c) consists of those functions satisfying

$$\begin{aligned} \alpha^2 \bigg(\int_0^T \int_{\Omega^*} e^{-\alpha r} u^2 \, dx \, ds + \int_{\Omega^*} e^{-\alpha r} \int_{\epsilon f}^T v^2 \, ds \, dx \bigg) & \\ + \int_{\Omega^*} e^{-\alpha r} \big[u^2(\mathbf{x},T) + v^2(\mathbf{x},T) \big] dx \le M_1^2, & \end{aligned}$$

and

$$- \int_\Sigma e^{-\alpha r} \frac{\partial v}{\partial x_i} \frac{\partial v}{\partial x_i} n_t \, dS \le M_2^2,$$

for constants M_1, M_2.

The proof of continuous dependence begins with the functional $F(t)$ defined by

$$F(t) = \int_{\Omega^*} e^{-\alpha r} w^2(\mathbf{x},t) \, dx. \tag{2.2.31}$$

Due to the weight $e^{-\alpha r}$ the logarithmic convexity method for the backward heat equation as described in section 1.5 needs modifying. In fact, we find

$$F'(t) = 2\int_{\Omega^*} e^{-\alpha r}\frac{\partial w}{\partial x_i}\frac{\partial w}{\partial x_i}\,dx - \int_{\Omega^*}\Delta(e^{-\alpha r})w^2\,dx, \tag{2.2.32}$$

and

$$\begin{aligned} F''(t) \geq & 4\int_{\Omega^*} e^{-\alpha r}\Big(\frac{\partial w}{\partial t}\Big)^2 dx - 4\alpha^2\int_{\Omega^*} e^{-\alpha r}\frac{\partial w}{\partial x_i}\frac{\partial w}{\partial x_i}\,dx \\ & + \int_{\Omega^*}\Delta^2(e^{-\alpha r})w^2\,dx. \end{aligned} \tag{2.2.33}$$

Upon evaluation of $\Delta(e^{-\alpha r})$ and $\Delta^2(e^{-\alpha r})$ this inequality leads to (for $N \geq 3$; the case $N = 2$ is also dealt with in Payne & Straughan (1990c)),

$$F''(t) \geq 4\int_{\Omega^*} e^{-\alpha r}\Big(\frac{\partial w}{\partial t}\Big)^2 dx - 2\alpha^2 F'(t) - \alpha^4 F(t). \tag{2.2.34}$$

From (2.2.31), (2.2.32) and (2.2.34) one may then show $\exists$ constants k_1, k_2 such that

$$FF'' - (F')^2 \geq -k_1 FF' - k_2 F^2.$$

This is the basic logarithmic convexity inequality and is integrated as in section 1.5 to yield the result

$$F(t) \leq e^{-mt}\big[F(T)e^{mT}\big]^{\lambda}\big[F(\epsilon)e^{m\epsilon}\big]^{1-\lambda}, \tag{2.2.35}$$

where m and λ are defined in terms of k_1, k_2 by

$$m = \frac{k_2}{k_1}, \qquad \lambda = \frac{\sigma_2 - \sigma}{\sigma_2 - \sigma_1},$$

with $\sigma, \sigma_1, \sigma_2$ being given by

$$\sigma = e^{-k_1 t}, \qquad \sigma_1 = e^{-k_1 T}, \qquad \sigma_2 = e^{-k_1 \epsilon}.$$

The term $F(T)$ is bounded using the constraint. The bound for the $F(\epsilon)$ term begins by noting that (2.2.13) is still valid in the backward in time problem. The relevant bounds for J_1 and J_2 are, however, different. For example, to bound J_1 we define $\omega(t)$ by

$$\omega = \frac{T - t}{T - \epsilon}$$

and then note that

$$J_1 \leq \int_0^T \omega \int_{\Omega^*} e^{-\alpha r} \left(\frac{\partial u}{\partial t} \right)^2 dx\, dt. \tag{2.2.36}$$

Upon using the differential equation and integrating by parts and using the arithmetic - geometric mean inequality one may show

$$\int_0^T \omega \int_{\Omega^*} e^{-\alpha r} \left(\frac{\partial u}{\partial t} \right)^2 dx\, dt \leq \left(\frac{1+\alpha^2 T}{T-\epsilon} \right) \times \int_0^T \int_{\Omega^*} e^{-\alpha r} \frac{\partial u}{\partial x_i} \frac{\partial u}{\partial x_i}\, dx\, dt. \tag{2.2.37}$$

Further integrations reveal

$$\int_0^T \int_{\Omega^*} e^{-\alpha r} \frac{\partial u}{\partial x_i} \frac{\partial u}{\partial x_i}\, dx\, dt \leq \alpha^2 \int_0^T \int_{\Omega^*} e^{-\alpha r} u^2\, dx\, dt + \int_{\Omega^*} e^{-\alpha r} u^2(\mathbf{x}, T)\, dx. \tag{2.2.38}$$

Putting (2.2.36) - (2.2.38) together shows that

$$\begin{aligned} J_1 &\leq \left(\frac{1+\alpha^2 T}{T-\epsilon} \right) \left[\alpha^2 \int_0^T \int_{\Omega^*} e^{-\alpha r} u^2\, dx\, dt \right. \\ &\quad \left. + \int_{\Omega^*} e^{-\alpha r} u^2(\mathbf{x}, T)\, dx \right], \\ &\leq \left(\frac{1+\alpha^2 T}{T-\epsilon} \right) M_1^2, \end{aligned} \tag{2.2.39}$$

where in the last step the constraint set has been employed.

Similar, but more tricky calculations involving integrals over the surface Σ lead to an analogous bound for J_2, involving data and the constraint set constants. Further details may be found in Payne & Straughan (1990c).

Once the relative bounds for J_1 and J_2 are used in (2.2.35) along with the constraint, a continuous dependence on the initial - time geometry result for the backward in time problem follows from (2.2.35).

2.3 Continuous Dependence on the Initial - Time Geometry for an Equation from Dynamo Theory

In this section we again investigate the error due to the fact that in an evolution problem not all of the data assigned at $t = t_0$ are actually measured at the same instant in time but rather over some interval of time.

Precisely, we describe results of Ames & Straughan (1995) and focus our attention on the question of continuous dependence on initial - time geometry by deriving stabilizing inequalities for an equation that arises in the description of the behaviour of the toroidal part of the magnetic field in a dynamo problem when the flow is compressible and time - dependent. Details of the relevant dynamo theory can be found in Lortz & Meyer - Spasche (1982). Franchi & Straughan (1995) also studied the same initial - time geometry problem, but when the spatial region is one exterior to a bounded set. The unbounded spatial geometry problem investigated by Franchi & Straughan (1995) is more involved and technically much harder; we here concentrate on the bounded domain case. (Franchi & Straughan (1994) have also recently considered the equation arising from the same dynamo problem with a different goal in mind; their work resulted in spatial decay estimates as well as continuous dependence on the prescribed velocity field for both the forward and backward in time problems and is described in sections 3.8 and 4.4.) First we investigate the forward in time problem associated with the theory and afterward we study the improperly posed backward in time problem, obtaining in both cases, results that demonstrate the solution depends continuously on the initial - time geometry, when the spatial domain is bounded.

The Forward in Time Problem

We begin by comparing the solutions to the following two problems:

$$\frac{\partial H}{\partial t} = \big(\eta(\mathbf{x})H_{,i}\big)_{,i} - \big(v_i(\mathbf{x})H\big)_{,i}, \qquad \text{in } \Omega \times (0,\infty), \tag{2.3.1}$$

$$H = 0, \qquad \text{on } \Gamma \times [0,\infty), \tag{2.3.2}$$

$$H(\mathbf{x},0) = g(\mathbf{x}), \qquad \mathbf{x} \in \Omega; \tag{2.3.3}$$

$$\frac{\partial K}{\partial t} = \big(\eta(\mathbf{x})K_{,i}\big)_{,i} - \big(v_i(\mathbf{x})K\big)_{,i}, \qquad \epsilon f(\mathbf{x}) < t,\ \mathbf{x} \in \Omega, \tag{2.3.4}$$

$$K = 0, \qquad \epsilon f(\mathbf{x}) \le t,\ \mathbf{x} \in \Gamma, \tag{2.3.5}$$

$$K(\mathbf{x}, \epsilon f(\mathbf{x})) = g(\mathbf{x}), \qquad \mathbf{x} \in \Omega. \tag{2.3.6}$$

Here H and K denote the toroidal part of a magnetic field, Ω is a bounded domain in $\mathbf{R}^3$ with boundary Γ, while $\eta(\mathbf{x})$ and $\mathbf{v}(\mathbf{x})$ represent the diffusivity and velocity, respectively, both of which depend on the spatial variables. Standard indicial notation is employed with summation from 1 to 3.

We assume that for $\mathbf{x} \in \Omega$,

$$|f(\mathbf{x})| \leq 1, \tag{2.3.7}$$

$$0 < \eta_0 \leq \eta(\mathbf{x}), \tag{2.3.8}$$

and denote by M and Q the bounds

$$M = \sup_{\Omega} |v_{i,i}|, \qquad Q = \sup_{\Omega} |\mathbf{v}|. \tag{2.3.9}$$

Since the velocity field in this problem is prescribed these bounds are simply the maximum values which the indicated functions achieve. We let Σ denote the surface $t = \epsilon f(\mathbf{x})$ for $\mathbf{x} \in \Omega$ in the following analysis and assume that this surface is sufficiently smooth.

We now set $h = H - K$ and observe that h satisfies the equation

$$\frac{\partial h}{\partial t} = \left(\eta(\mathbf{x}) h_{,i}\right)_{,i} - \left(v_i(\mathbf{x}) h\right)_{,i}, \tag{2.3.10}$$

for $t > \epsilon$. To establish continuous dependence on the initial - time geometry, we intend to derive an explicit inequality of the form

$$\|h(t)\|^2 \leq \epsilon F(t), \tag{2.3.11}$$

where F depends on the initial data and

$$\|h\|^2 = \int_{\Omega} h^2 \, dx.$$

To obtain (2.3.11), we multiply (2.3.10) by h and integrate the result over Ω. After integration by parts, we find that

$$\frac{1}{2} \frac{d}{dt} \|h\|^2 = - \int_{\Omega} \eta |\nabla h|^2 dx + \int_{\Omega} v_i h h_{,i} dx, \tag{2.3.12}$$

$$= - \int_{\Omega} \eta |\nabla h|^2 dx - \frac{1}{2} \int_{\Omega} v_{i,i} h^2 dx, \tag{2.3.13}$$

where we have also made use of the fact that $h = 0$ on the boundary of Ω. In view of (2.3.8), (2.3.9), and Poincaré's inequality, we obtain the inequality

$$\frac{d}{dt} \|h\|^2 \leq -2\eta_0 \lambda_1 \|h\|^2 + M \|h\|^2, \tag{2.3.14}$$

where λ_1 is the first eigenvalue in the membrane problem for Ω. Setting $c = M - 2\eta_0\lambda_1$, it follows from (2.3.14) that

$$\frac{d}{dt}\left(e^{-ct}\|h\|^2\right) \leq 0,$$

and hence, after an integration from ϵ to t,

$$\|h(t)\|^2 \leq e^{c(t-\epsilon)}\|h(\epsilon)\|^2. \tag{2.3.15}$$

We next need to find a bound for $\|h(\epsilon)\|$ in terms of ϵ. To this end we observe that

$$H(\mathbf{x},\epsilon) = g(\mathbf{x}) + \int_0^\epsilon H_{,s}(\mathbf{x},s)ds, \tag{2.3.16}$$

and

$$K(\mathbf{x},\epsilon) = g(\mathbf{x}) + \int_{\epsilon f(\mathbf{x})}^\epsilon K_{,s}(\mathbf{x},s)ds. \tag{2.3.17}$$

Thus,

$$\|h(\epsilon)\|^2 = \int_\Omega\left(\int_0^\epsilon H_{,s}ds - \int_{\epsilon f}^\epsilon K_{,s}ds\right)^2 dx,$$

and after an application of the arithmetic - geometric mean inequality, we obtain

$$\|h(\epsilon)\|^2 \leq 3\int_\Omega\left(\int_0^\epsilon H_{,s}ds\right)^2 dx + \frac{3}{2}\int_\Omega\left(\int_{\epsilon f}^\epsilon K_{,s}ds\right)^2 dx.$$

Using Schwarz's inequality and the fact that $1 - f \leq 2$, we have

$$\begin{aligned} \|h(\epsilon)\|^2 \leq& 3\epsilon\left[\int_0^\epsilon\int_\Omega H_{,s}^2 dxds + \int_\Omega\int_{\epsilon f}^\epsilon K_{,s}^2 dsdx\right], && (2.3.18)\\ =& 3\epsilon[J_1 + J_2], && (2.3.19)\end{aligned}$$

where J_1 and J_2 are defined by

$$J_1 = \int_0^\epsilon\int_\Omega H_{,s}^2 dxds,$$

and

$$J_2 = \int_\Omega\int_{\epsilon f}^\epsilon K_{,s}^2 dsdx.$$

We now need to find bounds for J_1 and J_2.

Substitution of the differential equation and integration by parts leads to

$$J_1 = \int_0^\epsilon \int_\Omega H_{,s}\big[(\eta H_{,i})_{,i} - (v_i H)_{,i}\big]dx\, ds,$$
$$= -\frac{1}{2}\int_0^\epsilon \int_\Omega \eta(H_{,i}H_{,i})_{,s}dx\, ds - \int_0^\epsilon \int_\Omega H_{,s}(v_i H)_{,i}dx\, ds.$$

An application of the arithmetic - geometric mean inequality results in

$$J_1 \le -\frac{1}{2}\int_\Omega \eta H_{,i}H_{,i}\Big|_{t=\epsilon} dx + \frac{1}{2}\int_\Omega \eta g_{,i}g_{,i}dx$$
$$+\frac{1}{2}\int_0^\epsilon \int_\Omega H_{,s}^2 dx\, ds + \frac{1}{2}\int_0^\epsilon \int_\Omega \big[(v_i H)_{,i}\big]^2 dx\, ds.$$

It then follows that

$$J_1 \le -\int_\Omega \eta H_{,i}H_{,i}\Big|_{t=\epsilon} dx + \int_\Omega \eta g_{,i}g_{,i}dx + \int_0^\epsilon \int_\Omega \big[v_{i,i}H + v_i H_{,i}\big]^2 dx\, ds,$$

and in view of (2.3.9),

$$J_1 \le \int_\Omega \eta g_{,i}g_{,i}dx + 2M^2 \int_0^\epsilon \int_\Omega H^2 dx\, ds + 2Q^2 \int_0^\epsilon \int_\Omega H_{,i}H_{,i}dx\, ds. \quad (2.3.20)$$

Since H satisfies (2.3.1) - (2.3.3), it also satisfies an equation like (2.3.13), i.e.

$$\frac{1}{2}\frac{d}{dt}\int_\Omega H^2 dx = -\int_\Omega \eta|\nabla H|^2 dx - \frac{1}{2}\int_\Omega v_{i,i}H^2 dx, \quad (2.3.21)$$

where $|\nabla H|^2 = H_{,i}H_{,i}$.

Integration of (2.3.21) from 0 to ϵ and use of the bounds (2.3.8) and (2.3.9) yield the inequality

$$\frac{1}{2}\int_\Omega H^2(\epsilon)dx + \eta_0 \int_0^\epsilon \int_\Omega |\nabla H|^2 dx\, ds \le \frac{1}{2}\int_\Omega g^2 dx$$
$$+\frac{1}{2}M \int_0^\epsilon \int_\Omega H^2 dx, \quad (2.3.22)$$

from which we can conclude

$$\int_0^\epsilon \int_\Omega |\nabla H|^2 dx\, ds \le \frac{1}{2\eta_0}\int_\Omega g^2 dx + \frac{M}{2\eta_0}\int_0^\epsilon \int_\Omega H^2 dx. \quad (2.3.23)$$

Inserting this result into (2.3.20) we obtain

$$J_1 \le \int_\Omega \eta g_{,i}g_{,i}dx + \frac{Q^2}{\eta_0}\int_\Omega g^2 dx + \left(2M^2 + \frac{MQ^2}{\eta_0}\right)\int_0^\epsilon \int_\Omega H^2 dx\, ds. \quad (2.3.24)$$

If we now integrate (2.3.21) from 0 to t, then instead of (2.3.22) we can derive the inequality

$$\int_\Omega H^2(t)dx \le \int_\Omega g^2 dx + M\int_0^t\int_\Omega H^2 dx\, ds. \tag{2.3.25}$$

We now set $\phi = \int_0^t \int_\Omega H^2 dx\, ds$ so that (2.3.25) becomes

$$\phi' - M\phi \le \int_\Omega g^2 dx.$$

Solving this differential inequality, we find that

$$\phi(t) \le \frac{1}{M}(e^{Mt} - 1)\int_\Omega g^2 dx,$$

from which it follows that

$$\int_0^\epsilon \int_\Omega H^2 dx\, ds \le \frac{1}{M} e^{M\epsilon} \int_\Omega g^2 dx. \tag{2.3.26}$$

Using this in (2.3.24) we conclude that

$$J_1 \le \int_\Omega \eta g_{,i} g_{,i} dx + \left[\frac{Q}{\eta_0} + \left(2 + \frac{Q}{\eta_0}\right)e^{M\epsilon}\right]\int_\Omega g^2 dx, \tag{2.3.27}$$

which is the desired bound for J_1.

To bound J_2 we assume that n_t, the t-component of the unit normal, is negative on Σ and, moreover,

$$n_t < -\delta, \tag{2.3.28}$$

for a positive constant δ. Then upon substituting the equation (2.3.4) and integrating by parts we find that

$$\begin{aligned} J_2 = &-\frac{1}{2}\int_\Omega \int_{\epsilon f}^\epsilon \eta(K_{,i}K_{,i})_{,s} ds\, dx + \int_\Sigma \eta K_{,t} K_{,i} n_i dS \\ &- \int_\Omega \int_{\epsilon f}^\epsilon K_{,s}(v_i K)_{,i} ds\, dx. \end{aligned} \tag{2.3.29}$$

Integrating the first term on the right hand side of (2.3.29) and dropping the non-positive term evaluated at $t = \epsilon$, we obtain

$$\begin{aligned} J_2 \le &-\frac{1}{2}\int_\Sigma \eta K_{,i} K_{,i} n_t dS + \int_\Sigma \eta K_{,t} K_{,i} n_i dS \\ &- \int_\Omega \int_{\epsilon f}^\epsilon K_{,s}(v_i K)_{,i} ds\, dx, \end{aligned} \tag{2.3.30}$$

$$\begin{aligned} \le &\int_\Sigma \eta K_{,i}(n_i K_{,t} - n_t K_{,i}) dS - \frac{1}{2}\delta \int_\Sigma \eta K_{,t} K_{,i} dS \\ &- \int_\Omega \int_{\epsilon f}^\epsilon K_{,s}(v_i K)_{,i} ds\, dx. \end{aligned} \tag{2.3.31}$$

An application of the arithmetic - geometric mean inequality to the last term on the right yields

$$
\begin{aligned}
J_2 \leq & \int_\Sigma \eta K_{,i}(n_i K_{,t} - n_t K_{,i}) dS - \frac{1}{2}\delta \int_\Sigma \eta K_{,i} K_{,i} dS + \frac{1}{2} \int_\Omega \int_{\epsilon f}^{\epsilon} K_{,s}^2 ds\, dx \\
& + M^2 \int_\Omega \int_{\epsilon f}^{\epsilon} K^2 ds\, dx + Q^2 \int_\Omega \int_{\epsilon f}^{\epsilon} K_{,i} K_{,i} ds\, dx, \qquad (2.3.32)
\end{aligned}
$$

from which it follows that

$$
\begin{aligned}
J_2 \leq & 2 \int_\Sigma \eta K_{,i}(n_i K_{,t} - n_t K_{,i}) dS - \delta \int_\Sigma \eta K_{,i} K_{,i} dS \\
& + 2M^2 \int_\Omega \int_{\epsilon f}^{\epsilon} K^2 ds\, dx + \frac{2Q^2}{\eta_0} \int_\Omega \int_{\epsilon f}^{\epsilon} \eta K_{,i} K_{,i} ds\, dx. \qquad (2.3.33)
\end{aligned}
$$

To handle the last term in (2.3.33), we write

$$
\int_\Omega \int_{\epsilon f}^{\epsilon} \eta K_{,i} K_{,i} ds\, dx = \int_\Sigma \eta K_{,i} g n_i dS - \int_\Omega \int_{\epsilon f}^{\epsilon} K(\eta K_{,i})_{,i} ds\, dx,
$$

which becomes, after substituting the differential equation (2.3.4),

$$
\begin{aligned}
\int_\Omega \int_{\epsilon f}^{\epsilon} \eta K_{,i} K_{,i} ds\, dx = & \int_\Sigma \eta K_{,i} g n_i dS - \int_\Omega \int_{\epsilon f}^{\epsilon} K K_{,s} ds\, dx \\
& - \int_\Omega \int_{\epsilon f}^{\epsilon} v_{i,i} K^2 ds\, dx - \int_\Omega \int_{\epsilon f}^{\epsilon} v_i K K_{,i} ds\, dx.
\end{aligned}
$$

Hence,

$$
\begin{aligned}
\int_\Omega \int_{\epsilon f}^{\epsilon} \eta K_{,i} K_{,i} ds\, dx \leq & \int_\Sigma \eta K_{,i} g n_i dS - \frac{1}{2} \int_\Sigma g^2 n_t dS \\
& - \frac{1}{2} \int_\Sigma v_i n_i g^2 dS \\
& + \frac{1}{2} M \int_\Omega \int_{\epsilon f}^{\epsilon} K^2 ds\, dx. \qquad (2.3.34)
\end{aligned}
$$

We next find a bound for the last term in (2.3.34). From the triangle inequality we have

$$\begin{aligned}
\Big(\int_\Omega\int_{\epsilon f}^{\epsilon} K^2 ds\,dx\Big)^{1/2} \leq & \Big(\int_\Omega\int_{\epsilon f}^{\epsilon}(K-g)^2 ds\,dx\Big)^{1/2} \\
& + \Big(\int_\Omega \epsilon(1-f)g^2 dx\Big)^{1/2}, \\
\leq & \Big[\frac{4\epsilon^2}{\pi^2}\int_\Omega (1-f)^2\int_{\epsilon f}^{\epsilon} K_{,s}^2 ds\,dx\Big]^{1/2} \\
& + \Big(\int_\Omega \epsilon(1-f)g^2 dx\Big)^{1/2}, \\
\leq & \frac{4\epsilon}{\pi}J_2^{1/2} + \sqrt{2\epsilon}\Big(\int_\Omega g^2 dx\Big)^{1/2}.
\end{aligned}$$

Therefore,

$$\int_\Omega\int_{\epsilon f}^{\epsilon} K^2 ds\,dx \leq \frac{16\epsilon^2}{\pi^2}(1+\epsilon)J_2 + 2(1+\epsilon)\int_\Omega g^2 dx. \qquad (2.3.35)$$

We now substitute (2.3.34) and (2.3.35) into (2.3.33) to obtain

$$\begin{aligned}
J_2 \leq & 2\int_\Sigma \eta K_{,i}(n_i K_{,t} - n_t K_{,i})dS - \delta\int_\Sigma \eta K_{,i}K_{,i}dS \\
& + \frac{2Q^2}{\eta_0}\Big[\int_\Sigma \eta K_{,i} g n_i dS - \frac{1}{2}\int_\Sigma g^2 n_t dS - \frac{1}{2}\int_\Sigma v_i n_i g^2 dS\Big] \\
& + \Big[2M^2 + \frac{MQ^2}{\eta_0}\Big]\Big[\frac{16\epsilon^2}{\pi^2}(1+\epsilon)J_2 + 2(1+\epsilon)\int_\Omega g^2 dx\Big]. \qquad (2.3.36)
\end{aligned}$$

Further use of the arithmetic - geometric mean inequality in (2.3.36) yields

$$\begin{aligned}
\Big[1 - \Big(2M^2 + \frac{MQ^2}{\eta_0}\Big)\frac{16\epsilon^2}{\pi^2}(1+\epsilon)\Big]J_2 \leq & \frac{1}{\alpha_1}\int_\Sigma \eta(n_i g_{,t} - n_t g_{,i})(n_i g_{,t} - n_t g_{,i})dS \\
+ \frac{Q^2}{\eta_0}\int_\Sigma g^2\Big[\frac{\eta n_i n_i}{\alpha_2} - n_t - v_i n_i\Big]dS & + 2(1+\epsilon)\Big(2M^2 + \frac{MQ^2}{\eta_0}\Big)\int_\Omega g^2 dx \\
+ \Big(\alpha_1 + \frac{\alpha_2 Q^2}{\eta_0} - \delta\Big)\int_\Sigma \eta K_{,i}K_{,i}dS, & \qquad (2.3.37)
\end{aligned}$$

where α_1 and α_2 are arbitrary, positive constants. We choose α_1, α_2 so that

$$\alpha_1 + \frac{\alpha_2 Q^2}{\eta_0} - \delta = 0, \qquad (2.3.38)$$

and assume ϵ is so small that

$$1 - \left(2M^2 + \frac{MQ^2}{\eta_0}\right)\frac{16\epsilon^2}{\pi^2}(1+\epsilon) > \alpha_3, \tag{2.3.39}$$

for some constant $\alpha_3 > 0$. Thus,

$$\begin{aligned}\alpha_3 J_2 \leq & \frac{1}{\alpha_1}\int_\Sigma \eta(n_i g_{,t} - n_t g_{,i})(n_i g_{,t} - n_t g_{,i})dS \\ & + \frac{Q^2}{\eta_0}\int_\Sigma g^2\left[\frac{\eta n_i n_i}{\alpha_2} - n_t - v_i n_i\right]dS \\ & + 2(1+\epsilon)\left(2M^2 + \frac{MQ^2}{\eta_0}\right)\int_\Omega g^2 dx.\end{aligned} \tag{2.3.40}$$

Inequality (2.3.40) is the desired bound for J_2 in terms of data.

If we now designate the data terms on the right hand sides of (2.3.27) and (2.3.40) (i.e. the bounds for J_1 and $\alpha_3 J_2$) by Q_1 and Q_2, respectively, then (2.3.19) becomes

$$\|h(\epsilon)\|^2 \leq 3\epsilon[Q_1 + \alpha_3^{-1}Q_2], \tag{2.3.41}$$

provided (2.3.39) is satisfied. We then obtain from (2.3.15) the continuous dependence inequality

$$\|h(t)\|^2 \leq 3\epsilon[Q_1 + \alpha_3^{-1}Q_2]e^{c(t-\epsilon)}. \tag{2.3.42}$$

The Backward in Time Problem

As usual, instead of investigating equation (2.3.1) backward in time, we study the backward equation forward in time. In this section, we shall compare the solutions (assumed to exist) of the following two ill posed problems:

$$\frac{\partial H}{\partial t} = -\big(\eta(\mathbf{x})H_{,i}\big)_{,i} + \big(v_i(\mathbf{x})H\big)_{,i}, \quad \text{in } \Omega \times (0,T), \tag{2.3.43}$$

$$H = 0, \quad \text{on } \Gamma \times [0,T], \tag{2.3.44}$$

$$H(\mathbf{x},0) = g(\mathbf{x}), \quad \mathbf{x} \in \Omega; \tag{2.3.45}$$

and

$$\frac{\partial K}{\partial t} = -\big(\eta(\mathbf{x})K_{,i}\big)_{,i} + \big(v_i(\mathbf{x})K\big)_{,i}, \quad \epsilon f(\mathbf{x}) < t < T,\ \mathbf{x} \in \Omega, \tag{2.3.46}$$

$$K = 0, \quad \epsilon f(\mathbf{x}) \leq t \leq T,\ \mathbf{x} \in \Gamma, \tag{2.3.47}$$

$$K(\mathbf{x}, \epsilon f(\mathbf{x})) = g(\mathbf{x}), \quad \mathbf{x} \in \Omega, \tag{2.3.48}$$

where $f(\mathbf{x})$ and $\eta(\mathbf{x})$ satisfy (2.3.7) and (2.3.8) as before. We again set $h = H - K$ so that h satisfies the equation

$$\frac{\partial h}{\partial t} = -\big(\eta(\mathbf{x})h_{,i}\big)_{,i} + \big(v_i(\mathbf{x})h\big)_{,i}, \tag{2.3.49}$$

for $\epsilon < t < T$. We shall now show that

$$\|h(t)\|^2 \le \epsilon^{\delta(t)} F(t), \tag{2.3.50}$$

for explicit functions $\delta(t)$ and $F(t)$ with $0 < \delta(t) < 1$. In order to establish an inequality of the form (2.3.50), we first derive for $\epsilon < t < T$ the inequality

$$\phi(t) \stackrel{\text{def}}{=} \|h(t)\|^2 \le \big[\phi(\epsilon)\big]^{\delta(t)} F(t), \tag{2.3.51}$$

with $0 < \delta(t) < 1$. (Here and throughout the notation $\stackrel{\text{def}}{=}$ means the quantity is defined as indicated.) Since the problems for H and K are both ill posed we anticipate that an inequality of the type (2.3.51) will be satisfied provided H and K belong to the appropriate constraint sets.

We note that $h = 0$ on Γ, $\epsilon \le t \le T$ and suppose that

$$v_i n_i = 0 \tag{2.3.52}$$

on the boundary of Ω. We should point out that this condition is compatible with a compressible fluid. We shall also assume that the functions $\eta(\mathbf{x})$ and $\mathbf{v}(\mathbf{x})$ satisfy the conditions that

$$\eta, |\eta_{,i}|, |v_i|, |v_{i,i}|, |v_{i,k}|, |v_{i,ik}| \le M_1, \tag{2.3.53}$$

for a positive constant M_1. The function η is the magnetic resistivity and physically η and $|\nabla\eta|$ are bounded, therefore, (2.3.53) is not restrictive. Moreover, in (2.3.43) - (2.3.45) or (2.3.46) - (2.3.48) the velocity field is *prescribed*, as is usual in the kinematic dynamo problem and thus, we may impose bounds on v_i and its derivatives. However, $v_{i,i}$ being small is consistent with a slightly compressible fluid.

We use a logarithmic convexity argument to establish (2.3.51). Consequently, we differentiate ϕ to obtain

$$\begin{aligned} \phi'(t) &= 2\int_\Omega h h_{,t}\,dx, \\ &= 2\int_\Omega \eta h_{,i}h_{,i} + \int_\Omega v_{i,i}h^2\,dx, \end{aligned} \tag{2.3.54}$$

after substituting the differential equation (2.3.49) and integrating by parts. From (2.3.54) we easily find that

$$\begin{aligned}-2\eta_0\|\nabla h\|^2 \geq& -2\int_\Omega \eta|\nabla h|^2 dx,\\ =& -\phi'(t)+\int_\Omega v_{i,i}h^2 dx,\\ \geq& -\phi' - M_1\phi.\end{aligned}\tag{2.3.55}$$

Returning to (2.3.54), we differentiate again to get

$$\phi''(t) = 4\int_\Omega \eta h_{,it}h_{,i}dx + 2\int_\Omega v_{i,i}hh_{,t}dx.\tag{2.3.56}$$

Integrating by parts and reintroducing the differential equation, we have

$$\begin{aligned}\phi''(t) =& 4\|h_{,t}\|^2 - 2\int_\Omega v_{i,i}h\big[-(\eta h_{,k})_{,k} + (v_k h)_{,k}\big]dx\\ &-4\int_\Omega v_i h_{,i}\big[-(\eta h_{,k})_{,k} + (v_k h)_{,k}\big]dx.\end{aligned}\tag{2.3.57}$$

After integration by parts, use of the boundary condition $h = 0$ and (2.3.52) as well as some simplification, the last two terms in (2.3.57) can be written as

$$\begin{aligned}I \stackrel{\text{def}}{=} & -2\int_\Omega v_{i,i}h\big[-(\eta h_{,k})_{,k} + (v_k h)_{,k}\big]dx\\ &-4\int_\Omega v_i h_{,i}\big[-(\eta h_{,k})_{,k} + (v_k h)_{,k}\big]dx,\\ =& -2\int_\Omega v_{i,ik}\eta hh_{,k}dx + 2\int_\Omega v_i\eta_{,i}|\nabla h|^2 dx\\ &-2\int_\Omega (v_{i,i})^2h^2 dx\\ &-6\int_\Omega v_k v_{i,i}hh_{,k}dx - 4\int_\Omega v_i v_k h_{,i}h_{,k}dx\\ &-4\int_\Omega v_{i,k}\eta h_{,i}h_{,k}dx.\end{aligned}\tag{2.3.58}$$

The arithmetic - geometric mean inequality and the bounds (2.3.53) then imply

$$I \geq -6M_1^2\|h\|^2 - 18M_1^2\|\nabla h\|^2,\tag{2.3.59}$$

and hence,

$$\phi'' \geq 4\|h_{,t}\|^2 - 6M_1^2\|h\|^2 - 18M_1^2\|\nabla h\|^2.\tag{2.3.60}$$

In view of (2.3.55) we then find that

$$\phi'' \geq 4\|h_{,t}\|^2 - \frac{9}{\eta_0} M_1^2 \phi' - \left(6M_1^2 + \frac{9M_1^3}{\eta_0} \right) \phi. \tag{2.3.61}$$

We form $\phi\phi'' - (\phi')^2$ using (2.3.54) and (2.3.61), obtaining the inequality

$$\phi\phi'' - (\phi')^2 \geq 4S^2 - \frac{9}{\eta_0} M_1^2 \phi\phi' - \left(6M_1^2 + \frac{9M_1^3}{\eta_0} \right) \phi^2, \tag{2.3.62}$$

where

$$S^2 = \|h\|^2 \|h_{,t}\|^2 - \left(\int_\Omega h h_{,t} dx \right)^2,$$

which is nonnegative by Schwarz's inequality. It follows that there exist positive constants k_1 and k_2 such that

$$\phi\phi'' - (\phi')^2 \geq -k_1 \phi\phi' - k_2 \phi^2. \tag{2.3.63}$$

This inequality integrates as in section 1.5 to yield

$$\phi(t) \leq e^{-k_2 t/k_1} \left[\phi(T) e^{k_2 T/k_1} \right]^\delta \left[\phi(\epsilon) e^{k_2 \epsilon/k_1} \right]^{1-\delta}, \tag{2.3.64}$$

where

$$\delta = \frac{\tau_2 - \tau}{\tau_2 - \tau_1}, \qquad \tau = e^{-k_1 t}, \tag{2.3.65}$$

and

$$\tau_2 = e^{-k_1 \epsilon}, \quad \tau_1 = e^{-k_1 T}.$$

If we now assume that

$$\|H(T)\|^2 + \|K(T)\|^2 \leq M_2^2, \tag{2.3.66}$$

then for $t < T$, (2.3.64) leads to an inequality of the form (2.3.51). Note that $\delta(t)$ as defined in (2.3.65) lies in the interval (0,1).

We proceed to establish that $\phi(\epsilon)$ is $O(\epsilon)$. As we showed in the previous subsection, we have

$$\|h(\epsilon)\|^2 \leq 3\epsilon [J_1 + J_2], \tag{2.3.67}$$

where

$$J_1 = \int_0^\epsilon \int_\Omega H_{,s}^2 dx\, ds \quad \text{and} \quad J_2 = \int_\Omega \int_{\epsilon f}^\epsilon K_{,s}^2 ds\, dx.$$

To bound J_1 we define

$$\rho(T,t) = \begin{cases} 1, & 0 < t < \epsilon, \\ (T-t)/(T-\epsilon), & \epsilon \leq t < T, \end{cases} \tag{2.3.68}$$

and observe that

$$J_1 \leq \int_0^T \int_\Omega \rho(T,s) H_{,s}^2 dx\, ds.$$

Substituting the differential equation satisfied by H and integrating by parts, we find

$$\begin{aligned}\int_0^T \int_\Omega \rho(T,s) H_{,s}^2 dx\, ds = &\int_0^T \int_\Omega (\rho H)_{,s} H_{,s} dx\, ds \\ &+ \frac{1}{2(T-\epsilon)} \int_0^T \int_\Omega \frac{\partial}{\partial s} H^2 dx\, ds, \\ = &\int_0^T \int_\Omega (\rho H)_{,si} \eta H_{,i} dx\, ds \\ &+ \int_0^T \int_\Omega (\rho H)_{,s} (v_i H)_{,i} dx\, ds \\ &+ \frac{1}{2(T-\epsilon)} \|H(T)\|^2. \end{aligned} \tag{2.3.69}$$

To bound the terms in (2.3.69) we first observe that we can obtain from (2.3.43) and (2.3.44)

$$\begin{aligned}\eta_0 \int_0^T \int_\Omega |\nabla H|^2 dx\, ds \leq &\int_0^T \int_\Omega \eta H_{,i} H_{,i} dx\, ds, \\ = &\frac{1}{2} \int_\Omega H^2 \Big|_0^T dx - \frac{1}{2} \int_0^T \int_\Omega v_{i,i} H^2 dx\, ds.\end{aligned}$$

Thus,

$$\int_0^T \int_\Omega |\nabla H|^2 dx\, ds \leq \frac{1}{2\eta_0} \|H(T)\|^2 + \frac{M_1}{2\eta_0} \int_0^T \int_\Omega H^2 dx\, ds. \tag{2.3.70}$$

Returning to (2.3.69), we note that the first term on the right hand side is equal to

$$-\frac{1}{2(T-\epsilon)} \int_0^T \int_\Omega \eta H_{,i} H_{,i} dx\, ds - \frac{1}{2} \int_{\Omega(0)} \eta H_{,i} H_{,i} dx,$$

and so is nonpositive, and that

$$\begin{aligned}\int_0^T \int_\Omega (\rho H)_{,s} (v_i H)_{,i} dx\, ds \leq &\frac{1}{2} \int_0^T \int_\Omega \rho H_{,s}^2 dx\, ds \\ &+ \frac{1}{2} \int_0^T \int_\Omega \rho (v_i H)_{,i} (v_k H)_{,k} dx\, ds \\ &- \frac{1}{2(T-\epsilon)} \int_0^T \int_\Omega v_{i,i} H^2 dx\, ds,\end{aligned}$$

after an application of the arithmetic - geometric mean inequality.

We thus find that

$$\int_0^T \int_\Omega \rho H_{,s}^2 dx\, ds \leq \int_0^T \int_\Omega \rho (v_i H)_{,i} (v_k H)_{,k} dx\, ds - \frac{1}{(T-\epsilon)} \int_0^T \int_\Omega v_{i,i} H^2 dx\, ds + \frac{1}{(T-\epsilon)} \|H(T)\|^2. \tag{2.3.71}$$

Upon expanding the first term on the right side, using the bounds (2.3.53), (2.3.70) and applying the arithmetic - geometric mean inequality again, we conclude

$$\begin{aligned} J_1 \leq & \int_0^T \int_\Omega \rho H_{,s}^2 dx\, ds, \\ \leq & \left(\frac{M_1}{T-\epsilon} + 2M_1^2 + \frac{M_1^3}{\eta_0} \right) \int_0^T \int_\Omega H^2 dx\, ds \\ & + \left\{ \frac{1}{T-\epsilon} + \frac{M_1^2}{\eta_0} \right\} \|H(T)\|^2. \end{aligned} \tag{2.3.72}$$

Recalling the constraint (2.3.66) and assuming that

$$\int_0^T \int_\Omega H^2 dx\, ds \leq M_3^2, \tag{2.3.73}$$

for some prescribed constant M_3, we see that (2.3.72) is the desired bound for J_1.

To bound J_2 we define

$$\rho(T,t) = \begin{cases} 1, & \epsilon f < t < \epsilon, \\ (T-t)/(T-\epsilon), & \epsilon \leq t < T, \end{cases}$$

and then write

$$\begin{aligned} J_2 = & \int_\Omega \int_{\epsilon f}^{\epsilon} K_{,s}^2 ds\, dx, \\ \leq & \int_\Omega \int_{\epsilon f}^{T} \rho(T,s) K_{,s}^2 ds\, dx. \end{aligned}$$

Substitution of the equation (2.3.46) followed by integration by parts and application of the arithmetic - geometric mean inequality leads to

$$\begin{aligned}\int_\Omega\int_{\epsilon f}^T \rho(T,s)K_{,s}^2 ds\,dx \leq &\frac{1}{(T-\epsilon)}\int_\Omega\int_{\epsilon f}^T \eta|\nabla K|^2 ds\,dx\\ &+\int_\Omega\int_{\epsilon f}^T \rho(v_iK)_{,i}(v_kK)_{,k}ds\,dx\\ &-2\int_\Sigma \eta K_{,s}K_{,i}n_i dS\\ &+\int_\Sigma \eta K_{,i}K_{,i}n_t dS. \end{aligned} \tag{2.3.74}$$

The first term on the right side of (2.3.74) can be bounded using the differential equation for K. We easily obtain an inequality analogous to (2.3.70), namely

$$\begin{aligned}\eta_0\int_\Omega\int_{\epsilon f}^T |\nabla K|^2 ds\,dx \leq &\int_\Omega\int_{\epsilon f}^T \eta|\nabla K|^2 ds\,dx,\\ \leq &\frac{1}{2}\|K(T)\|^2+\frac{1}{2}M_1\int_\Omega\int_{\epsilon f}^T K^2 ds\,dx\\ &+\int_\Sigma \eta K K_{,i}n_i dS-\frac{1}{2}\int_\Sigma v_in_iK^2 dS. \end{aligned} \tag{2.3.75}$$

Expanding the second term on the right in (2.3.74) and using (2.3.75), we find, after some simplification, that

$$\begin{aligned}\int_\Omega\int_{\epsilon f}^T \rho(T,s)K_{,s}^2 ds\,dx \leq &\left(\frac{M_1}{2(T-\epsilon)}+2M_1^2+\frac{M_1^3}{\eta_0}\right)\int_\Omega\int_{\epsilon f}^T K^2 ds\,dx\\ &+\left(\frac{1}{2(T-\epsilon)}+\frac{M_1^2}{\eta_0}\right)\|K(T)\|^2\\ &+\left(\frac{1}{(T-\epsilon)}+\frac{2M_1^2}{\eta_0}\right)\left(\int_\Sigma gK_{,i}n_i dS-\frac{1}{2}\int_\Sigma v_in_ig^2 dS\right)\\ &-2\int_\Sigma \eta K_{,s}K_{,i}n_i dS+\int_\Sigma \eta|\nabla K|^2 n_t dS. \end{aligned} \tag{2.3.76}$$

The last two surface integrals in (2.3.76) can be rewritten as

$$\begin{aligned}I \stackrel{\text{def}}{=} &-2\int_\Sigma \eta K_{,i}(K_{,t}n_i-K_{,i}n_t)dS-\int_\Sigma \eta K_{,i}K_{,i}n_t dS,\\ \leq &\,\alpha_1\int_\Sigma \eta(K_{,i}n_t-n_iK_{,t})(K_{,i}n_t-n_iK_{,t})dS\\ &-\alpha_2\int_\Sigma \eta K_{,i}K_{,i}n_t dS, \end{aligned} \tag{2.3.77}$$

for computable constants α_1, α_2. Thus, if we assume, in addition to (2.3.66),

$$\int_\Omega \int_{\epsilon f}^T K^2 ds\, dx \le M_4^2, \tag{2.3.78}$$

and

$$-\int_\Sigma \eta K_{,i} K_{,i} n_t dS \le M_5^2, \tag{2.3.79}$$

for constants M_4 and M_5, it follows from (2.3.78) and (2.3.79) that

$$J_2 \le \alpha_1 \int_\Sigma \eta (K_{,i} n_t - n_i K_{,t})(K_{,i} n_t - n_i K_{,t}) dS + \alpha_3 N^2, \tag{2.3.80}$$

where N^2 depends on the constraints necessary due to the fact that we are dealing with an improperly posed problem. Combining this result with (2.3.72) and substituting it back into (2.3.67), we conclude that $\|h(\epsilon)\|^2$ is $O(\epsilon)$ and have established an inequality of the type (2.3.50).

To summarize the results for the backward in time problem, let us define a class $\mathcal{M}$ of solutions: H and K belong to $\mathcal{M}$ if

$$\int_0^T \int_\Omega H^2 dx\, ds + \int_\Omega \int_{\epsilon f}^T K^2 dx\, ds + \|H(T)\|^2 + \|K(T)\|^2 \le M_6^2, \tag{2.3.81}$$

for some constant M_6, and if K also satisfies (2.3.79). We have thus established the following theorem which implies continuous dependence of solutions on the initial - time geometry.

Theorem 2.3.1. (Ames & Straughan (1995).)
If solutions H of (2.3.43) - (2.3.45) and K of (2.3.46) - (2.3.48) both belong to $\mathcal{M}$, then there exist explicit functions $F(t)$ and $\delta(t)$ with $0 < \delta < 1$ such that

$$\|H(t) - K(t)\|^2 \le \epsilon^{\delta(t)} F(t),$$

for $\epsilon \le t \le T$.

2.4 Continuous Dependence on the Initial - Time Geometry for the Navier - Stokes Equations

The first three sections of this chapter have concentrated on investigations of the continuous dependence of the solution on changes in the initial - time geometry for linear partial differential equations. However, many partial differential equations which describe real life situations are nonlinear and it is important to be able to extend the analysis to such nonlinear equations. In this and the next section we do tackle such questions for nonlinear systems of partial differential equations which arise in fluid dynamics.

Song (1988) examined the continuous dependence on the initial - time geometry problem for the Navier - Stokes equations forward in time. Let Ω be a bounded domain in $\mathbf{R}^2$ or $\mathbf{R}^3$ with boundary Γ. Song (1988) compared the solution to the boundary - initial value problem

$$\begin{aligned} &\frac{\partial u_i}{\partial t} + u_j \frac{\partial u_i}{\partial x_j} = \nu \Delta u_i - \frac{\partial p}{\partial x_i}, \qquad \text{in} \quad \Omega \times (0,T), \\ &\frac{\partial u_i}{\partial x_i} = 0, \qquad \text{in} \quad \Omega \times (0,T), \\ &u_i = 0, \qquad \text{on} \quad \Gamma \times [0,T), \\ &u_i(\mathbf{x},0) = g_i(\mathbf{x}), \qquad \mathbf{x} \in \Omega, \end{aligned} \tag{2.4.1}$$

with the solution v_i to the following boundary - initial value problem:

$$\begin{aligned} &\frac{\partial v_i}{\partial t} + v_j \frac{\partial v_i}{\partial x_j} = \nu \Delta v_i - \frac{\partial p^*}{\partial x_i}, \qquad \mathbf{x} \in \Omega, \quad \epsilon f(\mathbf{x}) < t < T, \\ &\frac{\partial v_i}{\partial x_i} = 0, \qquad \mathbf{x} \in \Omega, \quad \epsilon f(\mathbf{x}) < t < T, \\ &v_i = 0, \qquad \mathbf{x} \in \Gamma, \quad \epsilon f(\mathbf{x}) \le t < T, \\ &v_i\big(\mathbf{x}, \epsilon f(\mathbf{x})\big) = g_i(\mathbf{x}), \qquad \mathbf{x} \in \Omega. \end{aligned} \tag{2.4.2}$$

Song (1988) requires

$$\frac{\partial g_i}{\partial x_i} = 0, \qquad \text{in} \quad \Omega,$$

and

$$0 \le f(\mathbf{x}) \le 1.$$

Basically, Song (1988) establishes continuous dependence on the initial - time geometry by deriving an inequality of form

$$\int_\epsilon^t \int_\Omega \big[u_i(\mathbf{x},s) - v_i(\mathbf{x},s)\big]\big[u_i(\mathbf{x},s) - v_i(\mathbf{x},s)\big]\,dx\,ds \le K(t)\epsilon^{1/2}, \tag{2.4.3}$$

where $K(t)$ depends on the inital data g_i, the viscosity ν, the geometry of the domain Ω, and perhaps the surface Σ (which is the surface $t = \epsilon f(\mathbf{x})$).

When $\Omega \subset \mathbf{R}^2$ the number $K(t)$ has form

$$K(t) = \big[1 - e^{\gamma(t-\epsilon)}\big](k_1\|g\|^2 + k_2\|\nabla g\|^2)\exp\big[\epsilon^2(k_3\|g\|^2 + k_4\|\nabla g\|^2)\big];$$

the constants $k_1, \ldots, k_4$ and γ depend here on Ω and ν. In the case $\Omega \subset \mathbf{R}^3$ Song (1988) can derive (2.4.3) only provided a Reynold's number criterion holds, which essentially means for ν not too large.

The first result of continuous dependence on the initial - time geometry for the Navier - Stokes equations backward in time is by Payne (1992). He compares the solution to the boundary - initial value problem

$$\begin{aligned}
&\frac{\partial u_i}{\partial t} = u_j \frac{\partial u_i}{\partial x_j} - \nu \Delta u_i + \frac{\partial p}{\partial x_i}, \qquad \text{in} \quad \Omega \times (0,T), \\
&\frac{\partial u_i}{\partial x_i} = 0, \qquad \text{in} \quad \Omega \times (0,T), \\
&u_i = 0, \qquad \text{on} \quad \Gamma \times [0,T), \\
&u_i(\mathbf{x},0) = g_i(\mathbf{x}), \qquad \mathbf{x} \in \Omega,
\end{aligned} \tag{2.4.4}$$

with the solution v_i to the following boundary - initial value problem:

$$\begin{aligned}
&\frac{\partial v_i}{\partial t} = v_j \frac{\partial v_i}{\partial x_j} - \nu \Delta v_i + \frac{\partial p^*}{\partial x_i}, \qquad \mathbf{x} \in \Omega, \quad \epsilon f(\mathbf{x}) < t < T, \\
&\frac{\partial v_i}{\partial x_i} = 0, \qquad \mathbf{x} \in \Omega, \quad \epsilon f(\mathbf{x}) < t < T, \\
&v_i = 0, \qquad \mathbf{x} \in \Gamma, \quad \epsilon f(\mathbf{x}) \le t < T, \\
&v_i\big(\mathbf{x}, \epsilon f(\mathbf{x})\big) = g_i(\mathbf{x}), \qquad \mathbf{x} \in \Omega,
\end{aligned} \tag{2.4.5}$$

and requires of f that

$$\sup_{\mathbf{x}\in\Omega} |f(\mathbf{x})| \le 1,$$

and that the initial data are Dirichlet integrable, i.e.

$$\int_\Omega \frac{\partial g_i}{\partial x_j} \frac{\partial g_i}{\partial x_j}\, dx \le M,$$

for some constant M. The constraint sets of Payne (1992) are $\mathcal{M}_1$ and $\mathcal{M}_2$ where these sets are defined as follows. A function $r_i(\mathbf{x},t) \in \mathcal{M}_1$ if there is a prescribed constant M_1 such that

$$\sup_{\Omega\times[\epsilon f(\mathbf{x}),T]} r_i r_i \le M_1^2, \tag{2.4.6}$$

while $r_i(\mathbf{x},t) \in \mathcal{M}_2$ for some prescribed constant M_2 if

$$\sup_{\Omega\times[\epsilon f(\mathbf{x}),T]} \left(r_i r_i + \frac{\partial r_i}{\partial x_j} \frac{\partial r_i}{\partial x_j} \right) \le M_2^2. \tag{2.4.7}$$

Payne (1992) establishes the following theorem.

Theorem 2.4.1. (Payne (1992).)
If $u_i(\mathbf{x},t) \in \mathcal{M}_1$ and $v_i(\mathbf{x},t) \in \mathcal{M}_2$ then we may compute an explicit constant K and a function $\delta(t)$, with $0 \leq \delta \leq 1$, with K, δ being independent of u_i and v_i, such that for $\epsilon \leq t < T$,

$$\int_\epsilon^t \int_\Omega \big[u_i(\mathbf{x},s) - v_i(\mathbf{x},s)\big]\big[u_i(\mathbf{x},s) - v_i(\mathbf{x},s)\big]\, dx\, ds \leq K\epsilon^{\delta(t)}. \tag{2.4.8}$$

The proof of this theorem first uses a logarithmic convexity argument on the functional

$$F(t) = \int_\epsilon^t \int_\Omega w_i w_i \, dx\, ds + (T - t + k_1)\sigma^2,$$

where $w_i = u_i - v_i$, for appropriate constants k_1 and σ to derive an inequality of form

$$\int_\epsilon^t \int_\Omega w_i w_i \, dx\, ds \leq C\|w(\epsilon)\|^{2\delta(t)}. \tag{2.4.9}$$

The bound for $\|w(\epsilon)\|$ extends the functions u_i and v_i backward from $t = 0$ and $t = \epsilon f(\mathbf{x})$, respectively, by g_i, and denoting these extensions by $\tilde{u}_i, \tilde{v}_i$ we observe that with

$$\tilde{w}_i = \tilde{u}_i - \tilde{v}_i,$$

$$\|w(\epsilon)\| = \|\tilde{w}(\epsilon)\|. \tag{2.4.10}$$

Due to the fact that $\tilde{w}_i(\mathbf{x}, -\epsilon) = 0$ application of a Stekloff inequality gives

$$\|\tilde{w}(\epsilon)\|^2 \leq 2\epsilon \int_{-\epsilon}^\epsilon \int_\Omega \frac{\partial \tilde{w}_i}{\partial s}\frac{\partial \tilde{w}_i}{\partial s}\, dx\, ds$$

and then from this one may find

$$\|w(\epsilon)\|^2 \leq 4\epsilon(J_1 + J_2), \tag{2.4.11}$$

where

$$J_1 = \int_0^\epsilon \int_\Omega \frac{\partial u_i}{\partial s}\frac{\partial u_i}{\partial s}\, dx\, ds,$$

and

$$J_2 = \int_\Omega \int_{\epsilon f(\mathbf{x})}^\epsilon \frac{\partial v_i}{\partial s}\frac{\partial v_i}{\partial s}\, ds\, dx.$$

The J_α terms are bounded by introducing

$$\omega = \frac{T - t}{T - \epsilon},$$

and then

$$J_1 \leq \hat{J}_1, \tag{2.4.12}$$

where

$$\hat{J}_1 = \int_0^T \int_\Omega \omega^2 \frac{\partial u_i}{\partial s} \frac{\partial u_i}{\partial s} \, dx \, ds, \tag{2.4.13}$$

with a similar representation involving J_2 and $\hat{J}_2$.

To bound J_1 involves the differential equation, boundary conditions, several integrations by parts, and use of the arithmetic - geometric mean inequality as follows,

$$\begin{aligned}
\hat{J}_1 \leq & \int_0^T \int_\Omega \omega^2 \frac{\partial u_i}{\partial s} \Big(u_j \frac{\partial u_i}{\partial x_j} - \nu \Delta u_i \Big) \, dx \, ds, \\
\leq & \int_0^T \int_\Omega \omega^2 \frac{\partial u_i}{\partial s} u_j \frac{\partial u_i}{\partial x_j} \, dx \, ds + \frac{\nu}{(T-\epsilon)} \int_0^T \int_\Omega \omega \frac{\partial u_i}{\partial x_j} \frac{\partial u_i}{\partial x_j} \, dx \, ds, \\
\leq & \frac{1}{2} \alpha \hat{J}_1 + \frac{M_1^2}{2\alpha} \int_0^T \int_\Omega \omega^2 \frac{\partial u_i}{\partial x_j} \frac{\partial u_i}{\partial x_j} \, dx \, ds, \\
& + \frac{\nu}{(T-\epsilon)} \int_0^T \int_\Omega \omega \frac{\partial u_i}{\partial x_j} \frac{\partial u_i}{\partial x_j} \, dx \, ds,
\end{aligned}$$

where α is a positive constant such that $\alpha < 2$. From this inequality one may bound $\hat{J}_1$ as

$$\hat{J}_1 \leq \frac{2}{(2-\alpha)} \left[\frac{\nu}{(T-\epsilon)} + \frac{M_1^2}{2\alpha} \right] \int_0^T \int_\Omega \omega \frac{\partial u_i}{\partial x_j} \frac{\partial u_i}{\partial x_j} \, dx \, ds, \tag{2.4.14}$$

and the right hand side of this inequality is then bounded using integration by parts and the differential equation to find

$$\int_0^T \int_\Omega \omega \frac{\partial u_i}{\partial x_j} \frac{\partial u_i}{\partial x_j} \, dx \, ds \leq \frac{T M_1^2 m(\Omega)}{2\nu (T-\epsilon)} . \tag{2.4.15}$$

Upon insertion of (2.4.15) in (2.4.14) a bound is established for $\hat{J}_1$ which with the help of (2.4.12) leads to the required bound for J_1.

The analysis for J_2 is not dissimilar, but more involved. Nevertheless, Payne (1992) shows J_2 is also bounded. These bounds used in (2.4.11) together with (2.4.9) prove the theorem.

2.5 Continuous Dependence on the Initial - Time Geometry for Solutions to the Darcy and Brinkman Equations for Flow in Porous Media

In this section we again describe results for an initial - time geometry problem where the data are not all measured at $t = 0$ but over a series of curves $t = \epsilon f(x)$, and since we typically compute the solution assuming data given at $t = 0$ we need to know precisely how the solution varies with changes in the initial data curve.

We describe results of Payne & Straughan (1994) who investigated the initial - time geometry problem for the Darcy equations for flow in porous media and the extended version of Darcy's equations known as Brinkman's equations. A lucid exposition of both sets of equations is given in Nield & Bejan (1992). For the forward in time problem Payne & Straughan (1994) derive an *a priori* condition on the boundary and initial data such that a solution to the Darcy equations with data given on an initial - time curve $t = \epsilon f(x)$ satisfies a precise continuous dependence estimate in L^2 norm when compared with a solution to the same problem with initial data given at $t = 0$. A similar continuous dependence estimate is derived for the Brinkman system. Payne & Straughan (1994) also establish continuous dependence on the initial - time geometry for the solutions to the Darcy and Brinkman equations backward in time, these being improperly posed problems.

The governing equations describing the flow of fluid through anisotropic and isotropic porous media and the variety of boundary conditions which may apply are carefully detailed by Nield & Bejan (1992), and applications of such equations in penetrative convection contexts in geophysics are reviewed by Straughan (1993). Payne & Straughan (1994) study the Darcy and Brinkman equations for flow through an isotropic solid.

The Darcy equations comprise an equation of momentum, and an equation of energy balance, and are

$$v_i = -\frac{\partial p}{\partial x_i} + g_i \Theta, \tag{2.5.1}$$

$$\frac{\partial \Theta}{\partial t} + v_i \frac{\partial \Theta}{\partial x_i} = \Delta \Theta, \tag{2.5.2}$$

where v_i, p, g_i, Θ are velocity, pressure, gravity, and temperature, and standard indicial notation together with Einstein's summation convention is employed. The fluid is incompressible, and so the velocity field satisfies the equation

$$\frac{\partial v_i}{\partial x_i} = 0. \tag{2.5.3}$$

Equations (2.5.1) - (2.5.3) are defined on a bounded spatial domain $\Omega \subset \mathbf{R}^3$, the boundary of Ω being Γ.

If the porosity, defined as the ratio of the volume of fluid to the total volume of material, is close to one, or if the flow is near a solid boundary Nield & Bejan (1992) argue from a physical viewpoint that the above equations should be replaced by

$$v_i - \lambda\Delta v_i = -\frac{\partial p}{\partial x_i} + g_i\Theta, \tag{2.5.4}$$

$$\frac{\partial \Theta}{\partial t} + v_i\frac{\partial \Theta}{\partial x_i} = \Delta\Theta, \tag{2.5.5}$$

$$\frac{\partial v_i}{\partial x_i} = 0, \tag{2.5.6}$$

where λ is a positive viscosity coefficient. Equations (2.5.4) - (2.5.6) are called the Brinkman equations. Richardson & Straughan (1993) actually needed the extra derivative term in the Brinkman equations to establish nonlinear energy stability for the Bénard convection problem in a porous medium when the viscosity of the fluid is allowed to vary with the temperature as it does in reality. The gravity vector is taken to be constant and such that

$$|\mathbf{g}| \leq 1. \tag{2.5.7}$$

The Initial - Time Geometry Problem for Darcy's Equations, Forward in Time

To compare the solution to Darcy's equations (2.5.1) - (2.5.3) with data given at $t = 0$ with the corresponding solution when the data are prescribed at $t = \epsilon f(x)$ we let (v_i^*, p^*, Θ^*) solve the boundary - initial value problem

$$\left.\begin{aligned} v_i^* &= -\frac{\partial p^*}{\partial x_i} + g_i\Theta^*, \\ \frac{\partial v_i^*}{\partial x_i} &= 0, \\ \frac{\partial \Theta^*}{\partial t} &+ v_i^*\frac{\partial \Theta^*}{\partial x_i} = \Delta\Theta^*, \end{aligned}\right\} \tag{2.5.8}$$

on $\Omega \times (0,T)$, with the boundary conditions

$$v_i^* n_i = 0, \qquad \Theta^* = h(x,t), \qquad \text{on} \quad \Gamma \times (0,T), \tag{2.5.9}$$

and the initial condition

$$\Theta^*(x,0) = \Theta_0(x), \qquad x \in \Omega, \tag{2.5.10}$$

where h, Θ_0 are prescribed functions, n_i is the unit outward normal to Γ, and we let (v_i, p, Θ) be the solution to the boundary - initial value problem

$$\left.\begin{aligned} v_i &= -\frac{\partial p}{\partial x_i} + g_i\Theta, \\ \frac{\partial v_i}{\partial x_i} &= 0, \\ \frac{\partial \Theta}{\partial t} &+ v_i\frac{\partial \Theta}{\partial x_i} = \Delta\Theta, \end{aligned}\right\} \tag{2.5.11}$$

for $x \in \Omega$, $t > \epsilon f(x)$,

$$v_i n_i = 0, \qquad \Theta = h(x,t), \qquad x \in \Gamma,\ t > \epsilon f(x), \tag{2.5.12}$$

$$\Theta(x, \epsilon f(x)) = \Theta_0(x), \qquad x \in \Omega. \tag{2.5.13}$$

We shall assume the data curve Σ is such that

$$\left.\begin{aligned} |f(x)| \le 1, \quad x \in \Omega, \\ |\nabla f(x)| \le K, \quad x \in \Omega, \end{aligned}\right\} \tag{2.5.14}$$

where K is a constant.

Payne & Straughan (1994) assume the boundary data function h to be constant for $t \le \epsilon$. They argue that one may add a constant to Θ without changing (2.5.1) or (2.5.3). Then one must modify the pressure gradient to the form $-\partial(p + gz\tilde{\Theta})/\partial x_i$. It is then sufficient to take

$$h = 0, \qquad \text{for } t \le \epsilon. \tag{2.5.15}$$

The procedure is to define the difference solution (u_i, θ, π) by

$$u_i = v_i^* - v_i, \qquad \theta = \Theta^* - \Theta, \qquad \pi = p^* - p,$$

and then this solution satisfies the partial differential equations

$$\left.\begin{aligned} u_i &= -\frac{\partial \pi}{\partial x_i} + g_i\theta, \\ \frac{\partial u_i}{\partial x_i} &= 0, \\ \frac{\partial \theta}{\partial t} &+ v_i\frac{\partial \theta}{\partial x_i} + u_i\frac{\partial \Theta^*}{\partial x_i} = \Delta\theta, \end{aligned}\right\} \tag{2.5.16}$$

on $\Omega \times (\max\{0, \epsilon f(\mathbf{x})\}, T)$, and it satisfies the boundary conditions

$$u_i n_i = 0, \qquad \theta = 0, \qquad \text{on} \quad \Gamma \times [\epsilon, T). \tag{2.5.17}$$

We again let $\|\cdot\|$ denote the norm on $L^2(\Omega)$, and let Σ be the surface $t=\epsilon f(\mathbf{x})$, with dS being the element of surface integration on Σ.

The first step is to show that $\|\theta(t)\|$ may be bounded in terms of $\|\theta(\epsilon)\|$. This is done in the following theorem which requires the velocity field v_i^* to be bounded in part a) but not in part b).

Theorem 2.5.1. (Payne & Straughan (1994).)
a) If the velocity field v_i^* in (2.5.1) is bounded in $\bar{\Omega}\times[0,T]$ then $\exists$ a constant c which depends on T,ϵ and the maximum of the boundary and initial data for Θ^*, such that

$$\|\theta(t)\|^2\le c\|\theta(\epsilon)\|^2,\qquad t\in[\epsilon,T]. \tag{2.5.18}$$

b) If the *data* term $\mathcal{D}$ given by

$$\begin{aligned}
\mathcal{D}=&16\Big(\frac{k^2}{c^2}+\frac{\hat{c}}{c}\Big)^{1/2}\Big(\int_0^t\oint_\Gamma h^2dSd\eta\Big)^{1/2}\Big(\int_0^t\oint_\Gamma h^4|\nabla_s h|^2dSd\eta\Big)^{1/2}\\
&+\frac{3m^{1/3}}{4\alpha}\Big[\Big(\max_\Gamma\Big|\frac{\partial\psi}{\partial n}\Big|\Big)\oint_\Gamma h^6dS\Big]^{2/3}+3q\int_0^t\oint_\Gamma h^4dSd\eta\\
&+\|\Theta_0^*\|\Big[\Big(\max_\Gamma\Big|\frac{\partial\psi}{\partial n}\Big|\Big)\oint_\Gamma h^6dS\Big]^{1/2}+2\|\Theta_0^*\|_4^4\\
&+\frac{3^{7/3}m^{1/3}T^{1/3}}{4\beta^{4/3}}\Big[\Big(\max_\Gamma\Big|\frac{\partial\psi}{\partial n}\Big|\Big)\int_0^t\oint_\Gamma h^4h_{,\eta}^2dSd\eta\Big]^{2/3}\\
&+\frac{m^{1/2}T^{1/2}}{2\gamma\Omega_1^2}\max_{[0,t]}\Big\{\gamma_1\oint_\Gamma h^4|\nabla_s h|^2dS\\
&\qquad+9\oint_\Gamma\Big[b_1^1(g^{22})^2h^4\Big(\frac{\partial h}{\partial s^1}\Big)^2\\
&\qquad+b_2^2(g^{33})^2h^4\Big(\frac{\partial h}{\partial s^2}\Big)^2\Big]dS\\
&\qquad+\frac{1}{2}\oint_\Gamma\Big[g^{22}\Big\{18h\Big(\frac{\partial h}{\partial s^1}\Big)^2+6h^2\frac{\partial^2h}{\partial(s^1)^2}\Big\}\\
&\qquad+g^{33}\Big\{18h\Big(\frac{\partial h}{\partial s^2}\Big)^2+6h^2\frac{\partial^2h}{\partial(s^2)^2}\Big\}\\
&\qquad+\frac{3}{2}g^{22}\Big(\frac{\partial g_{33}}{\partial s^1}-\frac{\partial g_{22}}{\partial s^1}\Big)h^2\frac{\partial h}{\partial s^1}\\
&\qquad+\frac{3}{2}g^{33}\Big(\frac{\partial g_{22}}{\partial s^2}-\frac{\partial g_{33}}{\partial s^2}\Big)h^2\frac{\partial h}{\partial s^2}\Big]^2dS\Big\},
\end{aligned} \tag{2.5.19}$$

satisfies the bound

$$\mathcal{D}\le\frac{1}{k_1^4}, \tag{2.5.20}$$

where k_1 is a constant dependent only on Ω and is defined by

$$k_1 = \frac{1}{\Omega_1}\left[4 + \frac{1}{\lambda_1}\left(1 + \frac{6B}{\rho_0} + \frac{8B^2Q^2}{\rho_0^2}\right)\right],$$

if Ω is non-convex and

$$k_1 = \frac{4}{\Omega_1},$$

when Ω is convex, then

$$\|\theta(t)\|^2 \leq \|\theta(\epsilon)\|^2, \qquad t \in [\epsilon, T]. \tag{2.5.21}$$

In the statement of this theorem $\|\cdot\|_4$ is the norm on $L^4(\Omega)$, k, c, $\hat{c}$, q, β, α, γ, γ_1 are computable constants, Ω_1 is the constant in the Sobolev inequality

$$\left[\int_\Omega (u_i u_i)^2 dx\right]^{1/4} \leq \frac{1}{\Omega_1}\left[\int_\Omega u_i u_i dx + \int_\Omega u_{i,j} u_{i,j} dx\right]^{1/2},$$

m is the volume of Ω, b^α_β are related to the components of the second fundamental form of the surface Γ, ∇_s is the tangential derivative to Γ, s^1 and s^2 are mutually orthogonal surface coordinates on Γ, and ψ is the function defined by

$$\left.\begin{aligned} \Delta\psi &= -1, \qquad \text{in} \quad \Omega, \\ \psi &= 0, \qquad \text{on} \quad \Gamma. \end{aligned}\right\}$$

The constants B and Q are defined before inequality (2.5.30).

The proofs of parts a) and b) of the theorem are here merely sketched.

We commence by multiplying $(2.5.16)_1$ by u_i and integrating over Ω. Then with the aid of the Cauchy - Schwarz inequality and the bound (2.5.7),

$$\begin{aligned} \|\mathbf{u}\|^2 &= \int_\Omega \theta g_i u_i \, dx, \\ &\leq \|\theta\| \|\mathbf{u}\|, \end{aligned}$$

and so,

$$\|\mathbf{u}\| \leq \|\theta\|, \qquad t \in [\epsilon, T]. \tag{2.5.22}$$

Multiplication of $(2.5.16)_3$ by θ, and integration over Ω leads to

$$\frac{1}{2}\frac{d}{dt}\|\theta\|^2 = -\|\nabla\theta\|^2 + \int_\Omega u_i \Theta^* \theta_{,i} \, dx, \quad t \in [\epsilon, T]. \tag{2.5.23}$$

To establish part a) note that if v_i^* is bounded in $\bar{\Omega} \times [0,T]$ then by the parabolic maximum principle, $\Theta^* \leq \Theta_m$, where Θ_m is the maximum value of Θ^* on $\{\Gamma \times [0,T]\} \cup \{\Omega \times \{0\}\}$. Thus

$$\begin{aligned}\int_\Omega u_i \Theta^* \theta_{,i}\, dx &\leq \Theta_m \|\mathbf{u}\| \|\nabla\theta\|,\\ &\leq \Theta_m \|\theta\| \|\nabla\theta\|,\end{aligned}$$

where (2.5.22) has been used.Upon use of the arithmetic-geometric mean inequality there follows

$$\int_\Omega u_i \Theta^* \theta_{,i}\, dx \leq \frac{1}{4}\Theta_m^2 \|\theta\|^2 + \|\nabla\theta\|^2. \tag{2.5.24}$$

Inequality (2.5.24) used in (2.5.23) yields

$$\frac{d}{dt}\|\theta\|^2 \leq \frac{1}{2}\Theta_m^2 \|\theta\|^2.$$

This integrates to give

$$\|\theta(t)\|^2 \leq \exp\left\{\frac{1}{2}\Theta_m^2 (T-\epsilon)\right\} \|\theta(\epsilon)\|^2, \qquad \forall t \in [\epsilon, T], \tag{2.5.25}$$

which establishes part a).

The proof of part b) is involved and we refer to Payne & Straughan for full details.

However, the proof first derives an estimate for

$$\int_\Omega u_i \Theta^* \theta_{,i}\, dx$$

in the form

$$\int_\Omega u_i \Theta^* \theta_{,i}\, dx \leq \frac{1}{\Omega_1} \|\nabla\theta\| \|\Theta^*\|_4 \big(\|\nabla\mathbf{u}\|^2 + \|\theta\|^2\big)^{1/2}. \tag{2.5.26}$$

A bound for $\|\nabla\mathbf{u}\|^2$ is then established and this is

$$\|\nabla\mathbf{u}\|^2 \leq 4\|\nabla\theta\|^2 + \frac{2B}{\rho_0}\Big(3 + \frac{2Q^2B}{\rho_0}\Big)\|\theta\|^2, \tag{2.5.27}$$

if Ω is non - convex, whereas if Ω is convex then

$$\|\nabla\mathbf{u}\|^2 \leq 2\|\nabla\theta\|^2. \tag{2.5.28}$$

For Ω non - convex (2.5.27) is used in (2.5.26) to see that

$$\int_\Omega u_i\Theta^*\theta_{,i}\,dx \le \frac{1}{\Omega_1}\|\nabla\theta\|\|\Theta^*\|_4\Big[\Big(1+\frac{6B}{\rho_0}+\frac{4B^2Q^2}{\rho_0^2}\Big)\|\theta\|^2+4\|\nabla\theta\|^2\Big]^{1/2},$$

and then Poincaré's inequality

$$\|\theta\|^2 \le \frac{1}{\lambda_1}\|\nabla\theta\|^2,$$

allows us to show that

$$\int_\Omega u_i\Theta^*\theta_{,i}\,dx \le k_1\|\Theta^*\|_4\|\nabla\theta\|^2. \tag{2.5.29}$$

The constant k_1 is given by

$$k_1 = \frac{1}{\Omega_1}\Big[4+\frac{1}{\lambda_1}\Big(1+\frac{6B}{\rho_0}+\frac{4B^2Q^2}{\rho_0^2}\Big)\Big],$$

if Ω is non-convex and

$$k_1 = \frac{2}{\Omega_1},$$

when Ω is convex, where ρ_0 and λ_1 are constants dependent on Γ, while

$$Q = \max_\Omega |\mathbf{x}|,$$

and

$$B = \sup_\Gamma \sqrt{b_{\mu\omega}b^{\mu\omega}}.$$

Inequality (2.5.29) may then be employed in (2.5.23) to find

$$\frac{1}{2}\frac{d}{dt}\|\theta\|^2 \le -\|\nabla\theta\|^2\big(1-k_1\|\Theta^*\|_4\big). \tag{2.5.30}$$

Thus, if the condition

$$\|\Theta^*\|_4 \le \frac{1}{k_1}, \tag{2.5.31}$$

holds, we may derive from (2.5.30)

$$\|\theta(t)\|^2 \le \|\theta(\epsilon)\|^2, \qquad t\in[\epsilon,T]. \tag{2.5.32}$$

The remainder of the proof involves deriving an *a priori* estimate for $\|\Theta^*\|_4$ and this in turn is used in conjunction with (2.5.31) to produce a truly *a priori* bound. This is involved but basically shows

$$\|\Theta^*\|_4^4 \le \mathcal{D},$$

where $\mathcal{D}$ is defined in (2.5.19). We refer to Payne & Straughan (1994) for further details.

To proceed from (2.5.32) requires an analysis of the initial - time geometry zone, i.e. the region between $t = \epsilon f(\mathbf{x})$ and $t = \epsilon$. Payne & Straughan (1994) present two results which impose a weak bound for either the temperature gradient or velocity, but only in the small region $[0, \epsilon]$. These are contained in the next theorem.

Theorem 2.5.2. (Payne & Straughan (1994).)
a) If $\exists$ a constant M such that

$$\begin{aligned} &|\nabla\Theta^*| \le M, \quad \forall t \in [0, \epsilon], \qquad |\nabla\Theta| \le M, \quad \forall t \in [\epsilon f, \epsilon], \\ &\text{and} \qquad \int_\Sigma v_i n_i p \, dS \le M, \end{aligned} \tag{2.5.33}$$

then

$$\|\theta(\epsilon)\|^2 \le K\epsilon, \qquad t \in [\epsilon, T], \tag{2.5.34}$$

for a computable constant K.
b) If $\exists$ a constant M such that

$$\begin{aligned} v_i^* v_i^* &\le \frac{M^2}{\epsilon^\omega}, \quad \forall t \in [0, \epsilon], \qquad 0 < \omega < \frac{3}{2}, \\ v_i v_i &\le \frac{M^2}{\epsilon^\omega}, \quad \forall t \in [\epsilon f, \epsilon], \qquad 0 < \omega < \frac{3}{2}, \end{aligned} \tag{2.5.35}$$

then

$$\|\theta(\epsilon)\|^2 \le K\epsilon^{3/2-\omega}, \qquad t \in [\epsilon, T]. \tag{2.5.36}$$

The proofs are not given in detail but commence by observing that

$$\Theta^*(x, \epsilon) = \Theta_0^*(x, 0) + \int_0^\epsilon \Theta^*_{,\eta} d\eta,$$

$$\Theta(x, \epsilon) = \Theta_0(x, \epsilon f(\mathbf{x})) + \int_{\epsilon f(\mathbf{x})}^\epsilon \Theta_{,\eta} d\eta,$$

and so,

$$\begin{aligned} \|\theta(\epsilon)\|^2 &= \int_\Omega \Big(\int_0^\epsilon \Theta^*_{,\eta} d\eta - \int_{\epsilon f(\mathbf{x})}^\epsilon \Theta_{,\eta} d\eta \Big)^2 dx, \\ &\le 3\epsilon(J_1 + J_2), \end{aligned} \tag{2.5.37}$$

where J_1 and J_2 are given by,

$$J_1 = \int_0^\epsilon \int_\Omega {\Theta^*}^2_{,\eta} \, dx \, d\eta, \tag{2.5.38}$$

and

$$J_2 = \int_\Omega \int_{\epsilon f(\mathbf{x})}^{\epsilon} \Theta_{,\eta}^2 \, d\eta \, dx. \tag{2.5.39}$$

The proof shows J_1 and J_2 may be bounded by data and then a combination of (2.5.32) and (2.5.37) shows that $\|\theta(\epsilon)\|^2$ may be bounded by ϵ which is the continuous dependence on the initial - time geometry estimate sought.

For further details of the proof of part a) we refer to Payne & Straughan (1994).

The proof of part b) given by Payne & Straughan (1994) begins with the triangle inequality to see that

$$\|\theta(\epsilon)\| \leq \|\Theta^*(\epsilon) - \Theta_0\| + \|\Theta(\epsilon) - \Theta_0\|. \tag{2.5.40}$$

The estimate for $\|\Theta(\epsilon) - \Theta_0\|$ is trickier and only this is treated. We write

$$\begin{aligned}
\|\Theta(\epsilon) - \Theta_0\|^2 &= \int_\Omega (\Theta - \Theta_0) \int_{\epsilon f(\mathbf{x})}^{\epsilon} \frac{\partial}{\partial t} (\Theta - \Theta_0) \, dt \, dx, \\
&= \int_\Omega (\Theta - \Theta_0) \Big[\int_{\epsilon f(\mathbf{x})}^{\epsilon} (\Delta\Theta - v_i \Theta_{,i}) d\eta \Big] dx,
\end{aligned}$$

where in the last line substitution has been made from the differential equation. Next an integration by parts observing $\Theta = 0$ on Γ for $t \leq \epsilon$, followed by use of the Cauchy - Schwarz inequality gives

$$\begin{aligned}
&\|\Theta(\epsilon) - \Theta_0\|^2 \\
&\quad = - \int_\Omega (\Theta - \Theta_0)_{,i} \Big[\int_{\epsilon f(\mathbf{x})}^{\epsilon} (\Theta_{,i} - v_i \Theta) d\eta \Big] dx, \\
&\quad \leq \Big(\epsilon \int_\Omega (1 - f) |\nabla(\Theta - \Theta_0)|^2 dx \Big)^{1/2} \Big\{ \Big[\int_\Omega \int_{\epsilon f(\mathbf{x})}^{\epsilon} |\nabla\Theta|^2 d\eta \, dx \Big]^{1/2} \\
&\qquad + \Big[\int_\Omega \int_{\epsilon f(\mathbf{x})}^{\epsilon} v_i v_i \Theta^2 d\eta \, dx \Big]^{1/2} \Big\}, \\
&\quad \leq (2\epsilon)^{1/2} \|\nabla(\Theta - \Theta_0)\| \Big\{ \Big[\int_\Omega \int_{\epsilon f(\mathbf{x})}^{\epsilon} |\nabla\Theta|^2 d\eta \, dx \Big]^{1/2} \\
&\qquad + \Big[\int_\Omega \int_{\epsilon f(\mathbf{x})}^{\epsilon} v_i v_i \Theta^2 d\eta \, dx \Big]^{1/2} \Big\},
\end{aligned} \tag{2.5.41}$$

where $|f| \leq 1$, has been used.

To continue from (2.5.41) note that for $\epsilon f \leq t \leq \epsilon$,

$$\begin{aligned}\|\nabla(\Theta(t)-\Theta_0)\|^2 &= \int_\Omega \int_{\epsilon f(\mathbf{x})}^t \frac{\partial}{\partial\eta} |\nabla(\Theta-\Theta_0)|^2 d\eta\, dx \\ &\quad - \int_\Sigma (\Theta-\Theta_0)_{,i}(\Theta-\Theta_0)_{,i} n_t\, dS, \\ &= 2\int_\Omega \int_{\epsilon f(\mathbf{x})}^t (\Theta-\Theta_0)_{,i}(\Theta-\Theta_0)_{,i\eta}\, d\eta\, dx \\ &\quad - \int_\Sigma (\Theta-\Theta_0)_{,i}(\Theta-\Theta_0)_{,i} n_t\, dS,\end{aligned}$$

and so substitution from the differential equation and integration by parts leads to

$$\begin{aligned}\|\nabla(\Theta(t)-\Theta_0)\|^2 &= 2\int_\Sigma (\Theta-\Theta_0)_{,i}(\Theta-\Theta_0)_{,t} n_i\, dS \\ &\quad - \int_\Sigma (\Theta-\Theta_0)_{,i}(\Theta-\Theta_0)_{,i} n_t\, dS \\ &\quad - 2\int_\Omega \int_{\epsilon f(\mathbf{x})}^t \Theta_{,\eta}^2 d\eta\, dx - 2\int_\Omega \int_{\epsilon f(\mathbf{x})}^t \Theta_{,\eta} v_i \Theta_{,i} d\eta\, dx \\ &\quad + 2\int_\Omega \int_{\epsilon f(\mathbf{x})}^t \Theta_{,\eta} \Delta\Theta_0 d\eta\, dx.\end{aligned}$$

By rearrangment and use of the Cauchy - Schwarz inequality and the arithmetic - geometric mean inequality together with the bounds (2.5.35) one may then establish

$$\begin{aligned}\|\nabla(\Theta(t)-\Theta_0)\|^2 &\le \int_\Sigma \left[\frac{\partial}{\partial n}(\Theta-\Theta_0)\right]^2 n_i n_i n_t\, dS \\ &\quad + \frac{2M^2}{\epsilon^\omega} \int_\Omega \int_{\epsilon f(\mathbf{x})}^t |\nabla(\Theta-\Theta_0)|^2 d\eta\, dx \\ &\quad + 4M^2 \epsilon^{1-\omega} \int_\Omega |\nabla\Theta_0|^2 dx \\ &\quad + 2\epsilon \int_\Omega (\Delta\Theta_0)^2 dx.\end{aligned} \tag{2.5.42}$$

Put now

$$\phi(t) = \int_\Omega \int_{\epsilon f(\mathbf{x})}^t |\nabla(\Theta-\Theta_0)|^2 d\eta\, dx,$$

and

$$Q = 4M^2 \int_\Omega |\nabla\Theta_0|^2 dx + 2\epsilon^\omega \int_\Omega (\Delta\Theta_0)^2 dx.$$

From (2.5.42) we note that since the first term on the right of inequality (2.5.42) is non-positive we may discard it to find

$$\phi' \leq \frac{2M^2}{\epsilon^\omega}\,\phi + \epsilon^{1-\omega}Q, \tag{2.5.43}$$

and an integration of this yields

$$\int_\Omega \int_{\epsilon f(\mathbf{x})}^{t} |\nabla(\Theta - \Theta_0)|^2 d\eta\, dx \leq \frac{Qe^{4M^2}}{2M^2}\,\epsilon, \tag{2.5.44}$$

and so from (2.5.43)

$$\|\nabla\big(\Theta(\epsilon) - \Theta_0\big)\|^2 \leq \hat{Q}\epsilon^{1-\omega}, \tag{2.5.45}$$

where $\hat{Q}$ is a data term which may be chosen independently of ϵ.

Note also that from use of the triangle inequality and estimate (2.5.44) we see that

$$\begin{aligned}\Big(\int_\Omega \int_{\epsilon f(\mathbf{x})}^{\epsilon} |\nabla\Theta|^2 d\eta\, dx\Big)^{1/2} \leq & \Big(\int_\Omega \int_{\epsilon f(\mathbf{x})}^{\epsilon} |\nabla(\Theta - \Theta_0)|^2 d\eta\, dx\Big)^{1/2} \\ & + \Big(\int_\Omega \int_{\epsilon f(\mathbf{x})}^{\epsilon} |\nabla\Theta_0|^2 d\eta\, dx\Big)^{1/2}, \\ \leq & \,\epsilon^{1/2}\bigg[\sqrt{\frac{Qe^{4M^2}}{2M^2}} + 2\Big(\int_\Omega |\nabla\Theta_0|^2 dx\Big)^{1/2}\bigg]. \end{aligned} \tag{2.5.46}$$

To handle the last term on the right of (2.5.41) the restrictions of the theorem, i.e. (2.5.35), are used to find

$$\begin{aligned}\Big(\int_\Omega \int_{\epsilon f(\mathbf{x})}^{\epsilon} v_i v_i \Theta^2 d\eta\, dx\Big)^{1/2} \leq & \frac{M}{\epsilon^{\omega/2}}\bigg[\Big(\int_\Omega \int_{\epsilon f(\mathbf{x})}^{\epsilon} |\Theta - \Theta_0|^2 d\eta\, dx\Big)^{1/2} \\ & + \Big(2\epsilon \int_\Omega \Theta_0^2 dx\Big)^{1/2}\bigg]. \end{aligned} \tag{2.5.47}$$

Also

$$\begin{aligned}\int_\Omega \int_{\epsilon f(\mathbf{x})}^{\epsilon} (\Theta - \Theta_0)^2 d\eta\, dx = & -\int_\Omega \int_{\epsilon f(\mathbf{x})}^{\epsilon} (\Theta - \Theta_0)^2 (\epsilon - \eta)_{,\eta} d\eta\, dx, \\ = & \,2\int_\Omega \int_{\epsilon f(\mathbf{x})}^{\epsilon} (\Theta - \Theta_0)\frac{\partial \Theta}{\partial \eta}(\epsilon - \eta) d\eta\, dx, \\ = & \,2\int_\Omega \int_{\epsilon f(\mathbf{x})}^{\epsilon} (\Theta - \Theta_0)\Big\{\Delta(\Theta - \Theta_0) \end{aligned}$$

$$
\begin{aligned}
&\qquad - v_i(\Theta - \Theta_0)_{,i} + \Delta\Theta_0 - v_i\Theta_{0,i}\Big\}(\epsilon - \eta)d\eta\, dx \\
&= -2\int_\Omega \int_{\epsilon f(\mathbf{x})}^{\epsilon} (\epsilon - \eta)(\Theta - \Theta_0)_{,i}(\Theta - \Theta_0)_{,i} d\eta\, dx, \\
&\qquad + 2\int_\Omega \int_{\epsilon f}^{\epsilon} (\epsilon - \eta)(\Theta - \Theta_0)(\Delta\Theta_0 - v_i\Theta_{0,i}) d\eta\, dx.
\end{aligned}
$$

Hence, dropping a non-positive term on the right and use of the Cauchy - Schwarz inequality leads to

$$
\begin{aligned}
\int_\Omega \int_{\epsilon f(\mathbf{x})}^{\epsilon} (\Theta - \Theta_0)^2 d\eta\, dx \leq\, & 4\sqrt{\frac{2}{3}}\epsilon^{3/2}\Big(\int_\Omega (\Delta\Theta_0)^2 dx\Big)^{1/2} \\
& \times \Big(\int_\Omega \int_{\epsilon f(\mathbf{x})}^{\epsilon} (\Theta - \Theta_0)^2 d\eta\, dx\Big)^{1/2}, \\
& + 4\sqrt{\frac{2}{3}} M \epsilon^{3/2-\omega/2}\Big(\int_\Omega (\nabla\Theta_0)^2 dx\Big)^{1/2} \\
& \times \Big(\int_\Omega \int_{\epsilon f(\mathbf{x})}^{\epsilon} (\Theta - \Theta_0)^2 d\eta\, dx\Big)^{1/2}.
\end{aligned}
$$

Thus

$$
\begin{aligned}
\Big(\int_\Omega \int_{\epsilon f(\mathbf{x})}^{\epsilon} (\Theta - \Theta_0)^2 d\eta\, dx\Big)^{1/2} \leq\, & 4\sqrt{\frac{2}{3}}\epsilon^{3/2}\Big(\int_\Omega (\Delta\Theta_0)^2 dx\Big)^{1/2} \\
& + 4\sqrt{\frac{2}{3}} M \epsilon^{3/2-\omega/2}\Big(\int_\Omega (\nabla\Theta_0)^2 dx\Big)^{1/2},
\end{aligned}
$$

and then from (2.5.47) we find

$$
\Big(\int_\Omega \int_{\epsilon f(\mathbf{x})}^{\epsilon} v_i v_i \Theta^2 d\eta\, dx\Big)^{1/2} \leq \frac{M}{\epsilon^{\omega/2}}\Big(L_1\epsilon^{1/2} + L_2\epsilon^{3/2-\omega/2}\Big), \tag{2.5.48}
$$

where

$$
L_1 = \sqrt{2}\Big(\int_\Omega \Theta_0^2 dx\Big)^{1/2}, \qquad L_2 = 4M\sqrt{\frac{2}{3}}\Big(\int_\Omega |\nabla\Theta_0|^2 dx\Big)^{1/2}.
$$

The bounds (2.5.45), (2.5.46) and (2.5.48) are next used in (2.5.41) to obtain

$$
\begin{aligned}
\int_\Omega \big(\Theta(\epsilon) - \Theta_0\big)^2 dx \leq\, & (2\epsilon)^{1/2}\hat{Q}^{1/2}\epsilon^{1/2-\omega/2}(\hat{Q}\epsilon^{1/2} \\
& + L_1\epsilon^{1/2-\omega/2} + L_2\epsilon^{3/2-\omega}), \\
\leq\, & k\epsilon^{3/2-\omega},
\end{aligned} \tag{2.5.49}
$$

for a computable constant k.

A similar estimate may be produced for $\|\Theta^*(\epsilon) - \Theta_0\|$. Then use of the two bounds in (2.5.40) establishes the theorem.

The Initial Time Geometry Problem for Brinkman's Equations, Forward in Time

To compare the solution to Brinkman's equations (2.5.4) - (2.5.6) where the data are given at $t = 0$ with the corresponding solution for data given at $t = \epsilon f(\mathbf{x})$ we compare the solution to the following boundary - initial value problem:

$$\left.\begin{aligned} v_i^* - \lambda\Delta v_i^* &= -\frac{\partial p^*}{\partial x_i} + g_i\Theta^*,\\ \frac{\partial v_i^*}{\partial x_i} &= 0,\\ \frac{\partial \Theta^*}{\partial t} + v_i^*\frac{\partial \Theta^*}{\partial x_i} &= \Delta\Theta^*, \end{aligned}\right\} \tag{2.5.50}$$

on $\Omega \times (0,T)$, for some positive time T,

$$v_i^* = 0, \qquad \Theta^* = h(x,t), \qquad \text{on} \quad \Gamma \times (0,T), \tag{2.5.51}$$

$$\Theta^*(x,0) = \Theta_0(x), \qquad x \in \Omega, \tag{2.5.52}$$

for h, Θ_0 prescribed, with the corresponding solution to the boundary - initial value problem

$$\left.\begin{aligned} v_i - \lambda\Delta v_i &= -\frac{\partial p}{\partial x_i} + g_i\Theta,\\ \frac{\partial v_i}{\partial x_i} &= 0,\\ \frac{\partial \Theta}{\partial t} + v_i\frac{\partial \Theta}{\partial x_i} &= \Delta\Theta, \end{aligned}\right\} \tag{2.5.53}$$

for $x \in \Omega$, $t > \epsilon f(\mathbf{x})$,

$$v_i = 0, \qquad \Theta = h(x,t), \qquad x \in \Gamma,\ t > \epsilon f(\mathbf{x}), \tag{2.5.54}$$

$$\Theta(x, \epsilon f(\mathbf{x})) = \Theta_0(x), \qquad x \in \Omega. \tag{2.5.55}$$

Payne & Straughan (1994) again assume

$$\left.\begin{aligned} &|f(x)| \le 1, \quad x \in \Omega,\\ &|\nabla f(x)| \le K, \quad x \in \Omega, \end{aligned}\right\}$$

for K a constant, and take the boundary data function h to be $h = 0$ for $t \le \epsilon$.

The difference solution (u_i, θ, π) for the Brinkman equations satisfies the partial differential equations

$$\left.\begin{aligned} u_i - \lambda\Delta u_i &= -\frac{\partial \pi}{\partial x_i} + g_i\theta, \\ \frac{\partial u_i}{\partial x_i} &= 0, \\ \frac{\partial \theta}{\partial t} + v_i\frac{\partial \theta}{\partial x_i} + u_i\frac{\partial \Theta^*}{\partial x_i} &= \Delta\theta, \end{aligned}\right\} \tag{2.5.56}$$

on $\Omega \times [\epsilon, T)$, with

$$u_i = 0, \qquad \theta = 0, \qquad \text{on} \quad \Gamma \times [\epsilon, T). \tag{2.5.57}$$

The following theorem of Payne & Straughan (1994) shows that $\|\theta(t)\|$ may be bounded in terms of $\|\theta(\epsilon)\|$. They firstly require v_i^* to be bounded but then give a result which does not. It is important to realize that unlike the situation for Darcy's equations the following theorem is valid for all data.

Theorem 2.5.3. (Payne & Straughan (1994).)
a) If the velocity field v_i^* in (2.5.50) is bounded in $\bar{\Omega} \times [0, T]$ then $\exists$ a constant c which depends on T, ϵ and the maximum of the boundary and initial data for Θ^*, such that

$$\|\theta(t)\|^2 \leq c\|\theta(\epsilon)\|^2, \qquad t \in [\epsilon, T]. \tag{2.5.58}$$

b) There exists a constant c_1 depending only on ϵ, T, Ω and the data such that

$$\|\theta(t)\|^2 \leq c_1\|\theta(\epsilon)\|^2, \qquad t \in [\epsilon, T]. \tag{2.5.59}$$

We observe that the proof of part a) is identical to that of part a) for the Darcy model.

Payne & Straughan (1994) give two analyses for part b). The first commences with the equation

$$\frac{1}{2}\frac{d}{dt}\|\theta\|^2 = -\|\nabla\theta\|^2 + \int_\Omega \Theta^* u_i \theta_{,i}\, dx,$$

which extends as

$$\frac{1}{2}\frac{d}{dt}\|\theta\|^2 \leq -\|\nabla\theta\|^2 + \|\Theta^*\|_4\|\mathbf{u}\|_4\|\nabla\theta\|.$$

We refer the reader to Payne & Straughan (1994) for details of the remainder of this proof.

The second proof of part b) given by Payne & Straughan (1994) starts with

$$\frac{1}{2}\frac{d}{dt}\|\theta\|^2 = -\|\nabla\theta\|^2 - \int_\Omega \Theta^*_{,i} u_i \theta \, dx.$$

Use of the Cauchy - Schwarz inequality gives

$$\frac{1}{2}\frac{d}{dt}\|\theta\|^2 \le -\|\nabla\theta\|^2 + \|\nabla\Theta^*\| \|\mathbf{u}\|_4 \|\theta\|_4.$$

For the Brinkman system one can bound $\|\nabla\mathbf{u}\|$ in terms of $\|\theta\|$ and this is of enormous benefit, i.e.

$$\|\nabla\mathbf{u}\|^2 \le \frac{1}{4\lambda}\|\theta\|^2.$$

The Sobolev inequality together with Poincaré's inequality and the above bound for $\|\nabla\mathbf{u}\|^2$ show that

$$\frac{1}{2}\frac{d}{dt}\|\theta\|^2 \le -\|\nabla\theta\|^2 + \frac{1}{4\lambda\Omega_1^2}\Big(1 + \frac{1}{\lambda_1}\Big)\|\nabla\Theta^*\| \|\theta\| \|\nabla\theta\|. \qquad (2.5.60)$$

The Dirichlet integral term may be removed from the last term by using the arithmetic - geometric mean inequality to find

$$\frac{d}{dt}\|\theta\|^2 \le A\|\nabla\Theta^*\|^2 \|\theta\|^2, \qquad (2.5.61)$$

where

$$A = \frac{1}{32\lambda^2\Omega_1^4}\Big(\frac{\lambda_1 + 1}{\lambda_1}\Big)^2.$$

After integration this inequality yields

$$\|\theta(t)\|^2 \le \exp\left(A\int_\epsilon^t \|\nabla\Theta^*\|^2 d\eta\right) \|\theta(\epsilon)\|^2. \qquad (2.5.62)$$

Completion of the a *priori* bound needs an estimate of

$$\int_\epsilon^t \|\nabla\Theta^*\|^2 d\eta$$

in terms of data. To achieve this we introduce the function R which satisfies

$$\begin{aligned} \Delta R &= 0, \qquad \text{in } \Omega, \\ R &= h, \qquad \text{on } \Gamma, \end{aligned}$$

and then commence with the identity,

$$\int_\Omega (\Theta^* - R)\Big(\frac{\partial \Theta^*}{\partial t} + v_i^* \frac{\partial \Theta^*}{\partial x_i} - \Delta \Theta^*\Big)dx = 0. \tag{2.5.63}$$

From this expression Payne & Straughan (1994) are able to show that

$$\begin{aligned} \frac{1}{2}\|\Theta^*\|^2 + \int_0^t \|\nabla\Theta^*\|^2 d\eta \le{}& \frac{1}{2}\|\Theta_0^*\|^2 + \int_\Omega R\Theta^*\, dx - \int_\Omega R_0\Theta_0^*\, dx \\ & - \int_0^t \int_\Omega R_{,\eta}\Theta^*\, dx\, d\eta + \int_0^t \oint_\Gamma h \frac{\partial R}{\partial n}\, dS\, d\eta \\ & + k_2 \int_0^t \|\Theta^*\|^2 d\eta, \end{aligned} \tag{2.5.64}$$

and then upon use of the arithmetic - geometric mean inequality they show

$$\int_0^t \|\nabla\Theta^*\|^2 d\eta \le k_2 \int_0^t \|\Theta^*\|^2 d\eta + \text{data}. \tag{2.5.65}$$

Part b) of the theorem follows by estimating $\int_0^t \|\Theta^*\|^2 ds$ in terms of data.

Payne & Straughan (1994) then utilize the results of the last theorem to establish continuous dependence of the solution to Brinkman's equations on the initial time geometry. Their results are given in the next theorem whose proof we do not include.

Theorem 2.5.4. (Payne & Straughan (1994).)
a) If $\exists$ a constant M such that

$$\begin{aligned} v_i^* v_i^* &\le \frac{M^2}{\epsilon^\omega}, \quad \forall t \in [0, \epsilon], \qquad 0 < \omega < \frac{3}{2}, \\ v_i v_i &\le \frac{M^2}{\epsilon^\omega}, \quad \forall t \in [\epsilon f(\mathbf{x}), \epsilon], \qquad 0 < \omega < \frac{3}{2}, \end{aligned} \tag{2.5.66}$$

then

$$\|\theta(\epsilon)\|^2 \le K\epsilon^{3/2-\omega}, \qquad t \in [\epsilon, T]. \tag{2.5.67}$$

b) If $\exists$ a constant M such that

$$\begin{aligned} &\int_0^\epsilon \|\nabla\Theta^*\|_4^4 dt \le M, \qquad |\nabla\Theta| \le M, \quad \forall t \in [\epsilon f(\mathbf{x}), \epsilon], \\ &\text{and} \quad \int_\Sigma v_i n_i p\, dS \le M, \qquad \int_\Sigma v_i v_{i,j} n_j\, dS \le M, \end{aligned} \tag{2.5.68}$$

then

$$\|\theta(\epsilon)\|^2 \le K\epsilon, \qquad t \in [\epsilon, T], \tag{2.5.69}$$

for a computable constant K.

The Initial Time Geometry Problem for Darcy and Brinkman's Equations, Backward in Time

Payne & Straughan (1994) also carry out analyses analogous to those for the forward in time problem but when time is taken negative. They change t to $-t$ and thus for Darcy's equations consider the problems

$$\left.\begin{aligned} v_i^* &= -\frac{\partial p^*}{\partial x_i} + g_i\Theta^*, \\ \frac{\partial v_i^*}{\partial x_i} &= 0, \\ \frac{\partial \Theta^*}{\partial t} &= v_i^*\frac{\partial \Theta^*}{\partial x_i} - \Delta\Theta^*, \end{aligned}\right\} \tag{2.5.70}$$

on $\Omega \times (0,T)$,

$$v_i^* n_i = 0, \qquad \Theta^* = h(x,t), \qquad \text{on} \quad \Gamma\times(0,T), \tag{2.5.71}$$

$$\Theta^*(\mathbf{x},0) = \Theta_0(\mathbf{x}); \qquad \mathbf{x}\in\Omega, \tag{2.5.72}$$

where h, Θ_0 are prescribed functions, and

$$\left.\begin{aligned} v_i &= -\frac{\partial p}{\partial x_i} + g_i\Theta, \\ \frac{\partial v_i}{\partial x_i} &= 0, \\ \frac{\partial \Theta}{\partial t} &= v_i\frac{\partial \Theta}{\partial x_i} - \Delta\Theta, \end{aligned}\right\} \tag{2.5.73}$$

for $x\in\Omega$, $t>\epsilon f(\mathbf{x})$,

$$v_i n_i = 0, \qquad \Theta = h(x,t), \qquad x\in\Gamma,\ t>\epsilon f(\mathbf{x}), \tag{2.5.74}$$

$$\Theta(x,\epsilon f(\mathbf{x})) = \Theta_0(\mathbf{x}), \qquad \mathbf{x}\in\Omega. \tag{2.5.75}$$

The analogous problems for the Brinkman system of equations are

$$\left.\begin{aligned} v_i^* - \lambda\Delta v_i^* &= -\frac{\partial p^*}{\partial x_i} + g_i\Theta^*, \\ \frac{\partial v_i^*}{\partial x_i} &= 0, \\ \frac{\partial \Theta^*}{\partial t} &= v_i^*\frac{\partial \Theta^*}{\partial x_i} - \Delta\Theta^*, \end{aligned}\right\} \tag{2.5.76}$$

on $\Omega \times (0,T)$,

$$v_i^* = 0, \qquad \Theta^* = h(x,t), \qquad \text{on} \quad \Gamma \times (0,T), \tag{2.5.77}$$

$$\Theta^*(\mathbf{x},0) = \Theta_0(\mathbf{x}); \qquad \mathbf{x} \in \Omega, \tag{2.5.78}$$

for h, Θ_0 prescribed functions, and for the problem with data given on $t = \epsilon f(\mathbf{x})$,

$$\left.\begin{aligned} v_i - \lambda\Delta v_i &= -\frac{\partial p}{\partial x_i} + g_i\Theta, \\ \frac{\partial v_i}{\partial x_i} &= 0, \\ \frac{\partial \Theta}{\partial t} &= v_i\frac{\partial \Theta}{\partial x_i} - \Delta\Theta, \end{aligned}\right\} \tag{2.5.79}$$

for $x \in \Omega$, $t > \epsilon f(\mathbf{x})$,

$$v_i = 0, \qquad \Theta = h(x,t), \qquad x \in \Gamma,\ t > \epsilon f(\mathbf{x}), \tag{2.5.80}$$

$$\Theta(\mathbf{x}, \epsilon f(\mathbf{x})) = \Theta_0(\mathbf{x}), \qquad \mathbf{x} \in \Omega. \tag{2.5.81}$$

The difference solution (u_i, θ, π) for equations (2.5.70) - (2.5.75) solves

$$\left.\begin{aligned} u_i &= -\frac{\partial \pi}{\partial x_i} + g_i\theta, \\ \frac{\partial u_i}{\partial x_i} &= 0, \\ \frac{\partial \theta}{\partial t} &= v_i\frac{\partial \theta}{\partial x_i} + u_i\frac{\partial \Theta^*}{\partial x_i} - \Delta\theta, \end{aligned}\right\} \tag{2.5.82}$$

on $\Omega \times [\epsilon, T)$, and

$$u_i n_i = 0, \qquad \theta = 0, \qquad \text{on} \quad \Gamma \times [\epsilon, T). \tag{2.5.83}$$

Payne & Straughan (1994) assume the solutions belong to the constraint set which satisfies

$$|\nabla\Theta^*| \le M, \qquad |\mathbf{v}| \le M, \tag{2.5.84}$$

where M is a known bound. They use a logarithmic convexity method as a first part in establishing continuous dependence on the initial - time geometry and the result of this is given in the theorem below.

Theorem 2.5.5. (Payne & Straughan (1994).)
Let (v_i, p, Θ) and (v_i^*, p^*, Θ^*) be solutions to (2.5.73) - (2.5.75) and (2.5.70) - (2.5.72), respectively, which lie in the class governed by the constraint (2.5.84). Provided

$$\int_\epsilon^T \|\theta\|^2 d\eta \le C,$$

for C known, then the difference solution θ satisfies

$$\int_\epsilon^t \|\theta\|^2 d\eta \le K(T-\epsilon+k_1)^{1-Q}\|\theta(\epsilon)\|^{2(1-Q)}, \qquad t\in[\epsilon,T), \tag{2.5.85}$$

where Q, K, k_1 are computable constants with $0 \le Q < 1$.

The above theorem is employed to derive a continuous dependence on the initial - time geometry result. This is contained in the following theorem.

Theorem 2.5.6. (Payne & Straughan (1994).)
If in addition to the constraints (2.5.84) we have

$$|\nabla\Theta| \le M, \qquad \text{in} \quad t\in[\epsilon f,\epsilon], \tag{2.5.86}$$

then

$$\int_\epsilon^t \|\theta\|^2 d\eta \le C\epsilon^{2(1-Q)}, \qquad t\in[\epsilon,T), \tag{2.5.87}$$

where C is a computable constant.

Payne & Straughan (1994) also extend the above results to a solution to the Brinkman system of equations, backward in time.

2.6 Continuous Dependence on the Spatial Geometry

Thus far in this chapter we have considered problems where errors have been made in the geometry where the initial data are measured. However, there is another very important class of problem involving geometric errors, namely that of estimating changes in the solution when the boundary itself changes. A study of continuous dependence on changes in the spatial geometry is of much practical importance in numerical analysis. For instance in a finite element analysis of a problem for a partial differential equation on an irregular shaped domain one typically approximates the boundary by a series of intersecting planes, or in a two - dimensional problem the region is triangulated and the boundary is approximated by straight lines. One can ask if the solution on the triangulated region depends continuously on changes of the domain from the original shape.

The first study of continuous dependence on the spatial geometry in improperly posed problems was by Crooke & Payne (1984) who investigated the behavior of the solution to the backward heat equation under changes of the spatial boundary itself.

We begin our account with an exposition of work of Payne (1987b) on continuous dependence on the spatial geometry for a solution to the heat equation forward in time.

Continuous Dependence on the Spatial Geometry for a Solution to the Heat Equation, Forward in Time

Let now Ω_1 and Ω_2 be bounded regions in $\mathbf{R}^n$ with boundaries Γ_1 and Γ_2, respectively. Denote by Ω the domain

$$\Omega = \Omega_1 \cap \Omega_2, \tag{2.6.1}$$

and suppose the origin is selected in Ω. The domains Ω_1 and Ω_2 are throughout supposed star shaped with respect to an interior origin. The situation is depicted in figure 2.

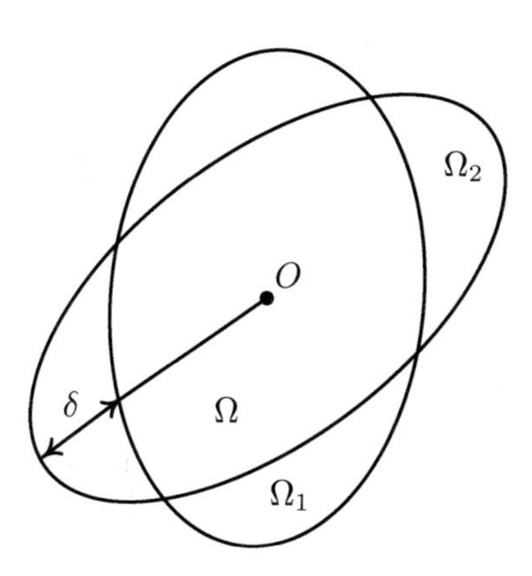

Figure 2. Geometry of the Domains Ω_1 and Ω_2.

In order to compare the solution to the heat equation on a given domain with the equivalent solution when the domain is slightly altered, Payne

(1987b) studies the case where u solves the boundary - initial value problem

$$\begin{aligned} &\frac{\partial u}{\partial t} = \Delta u, \qquad \text{in} \quad \Omega_1 \times (0,T), \\ &u = 0, \qquad \text{on} \quad \Gamma_1 \times [0,T), \\ &u(\mathbf{x},0) = g_1(\mathbf{x}), \qquad \mathbf{x} \in \Omega_1, \end{aligned} \tag{2.6.2}$$

and v solves the boundary - initial value problem

$$\begin{aligned} &\frac{\partial v}{\partial t} = \Delta v, \qquad \text{in} \quad \Omega_2 \times (0,T), \\ &v = 0, \qquad \text{on} \quad \Gamma_2 \times [0,T), \\ &v(\mathbf{x},0) = g_2(\mathbf{x}), \qquad \mathbf{x} \in \Omega_2. \end{aligned} \tag{2.6.3}$$

For simplicity he restricts attention to the case where $g_1 \equiv g_2$ on Ω.

Thus on $\Omega \times (0,T)$ the difference function $w = u - v$ satisfies

$$\frac{\partial w}{\partial t} = \Delta w, \tag{2.6.4}$$

with

$$w(\mathbf{x},0) = 0, \qquad \mathbf{x} \in \Omega.$$

The next step is to use an a *priori* inequality for which a solution to (2.6.4) which is zero at $t = 0$ satisfies

$$\|w(t)\|^2 \le k \int_0^t \oint_{\partial\Omega} w^2(x,s)\, dA\, ds, \tag{2.6.5}$$

where $\|\cdot\|$ here denotes the $L^2(\Omega)$ norm, $\partial\Omega$ is the boundary of Ω, dA denotes integration over $\partial\Omega$, and k depends only on Ω.

To complete a continuous dependence on geometry estimate requires a bound for the right hand side of (2.6.5). To this end Crooke & Payne (1984) have shown that one may compute a constant c such that

$$\begin{aligned} \int_0^t \oint_{\partial\Omega} w^2(x,s)\, dA\, ds \le c\delta\Big[&\int_0^t \int_{\Omega_1} |\nabla u|^2 dx\, ds \\ &+ \int_0^t \int_{\Omega_2} |\nabla v|^2 dx\, ds \Big]. \end{aligned} \tag{2.6.6}$$

The number δ in (2.6.6) is a measure of the departure of Ω_2 from Ω_1. If one allows a radius arm originating at the origin to rotate through 360° then δ is essentially the maximum distance from Γ_2 to Γ_1 on this radius, cf. figure 2.

Let $\|\cdot\|_1$ denote the $L^2(\Omega_1)$ norm then multiplication of $(2.6.2)_1$ by u_t and integration by parts shows

$$\frac{1}{2}\frac{d}{dt}\|\nabla u\|_1^2 = -\int_{\Omega_1}\Big(\frac{\partial u}{\partial t}\Big)^2 dx\,, \tag{2.6.7}$$

but by the Cauchy - Schwarz inequality

$$-\Big(\int_{\Omega_1} u\frac{\partial u}{\partial t}\,dx\Big)^2 \geq -\|u\|_1^2\int_{\Omega_1}\Big(\frac{\partial u}{\partial t}\Big)^2 dx\,,$$

and so from (2.6.7)

$$\begin{aligned}\frac{1}{2}\frac{d}{dt}\|\nabla u\|_1^2 &\leq -\Big(\int_{\Omega_1} u\frac{\partial u}{\partial t}\,dx\Big)^2\Big/\|u\|_1^2,\\ &= -\Big(\int_{\Omega_1} u\Delta u\,dx\Big)^2\Big/\|u\|_1^2,\end{aligned}$$

using the differential equation, and then after a further integration by parts

$$\frac{1}{2}\frac{d}{dt}\|\nabla u\|_1^2 \leq -\|\nabla u\|_1^4\Big/\|u\|_1^2. \tag{2.6.8}$$

But from Poincaré's inequality

$$-\|\nabla u\|_1^2 \leq -\lambda_1\|u\|_1^2,$$

where λ_1 is the least eigenvalue in the membrane problem for Ω_1. Hence we find from (2.6.8)

$$\frac{d}{dt}\|\nabla u\|_1^2 \leq -2\|\nabla u\|_1^2,$$

and this integrates to show

$$\|\nabla u\|_1^2 \leq \|\nabla g_1\|_1^2\, e^{-2\lambda_1 t}. \tag{2.6.9}$$

An analogous calculation for v establishes that

$$\|\nabla v\|_2^2 \leq \|\nabla g_2\|_2^2\, e^{-2\beta_1 t}, \tag{2.6.10}$$

where $\|\cdot\|_2$ denotes the norm on $L^2(\Omega_2)$ and β_1 is the least eigenvalue in the membrane problem for Ω_2.

Estimates (2.6.9) and (2.6.10) in (2.6.6), together with (2.6.5) lead to

$$\|w(t)\|^2 \leq Ke^{-\mu t}\,\delta, \tag{2.6.11}$$

for K, μ positive constants. This inequality establishes continuous dependence on the spatial geometry. Result (2.6.11) is due to Payne (1987b).

Continuous Dependence on the Spatial Geometry for a Solution to the Heat Equation, Backward in Time

The analysis of the subtitle is contained in the paper of Crooke & Payne (1984) which began the study of continuous dependence on the spatial geometry. This is a somewhat technically involved paper making heavy use of a variety of ingenious *a priori* estimates. A *very brief* account is given here.

Let u be a solution to the boundary - initial value problem

$$\begin{aligned} &\frac{\partial u}{\partial t} = -\Delta u, \qquad \text{in} \quad \Omega_1 \times (0,T), \\ &u = f_1, \qquad \text{on} \quad \Gamma_1 \times [0,T), \\ &u(\mathbf{x},0) = g_1(\mathbf{x}), \qquad \mathbf{x} \in \Omega_1, \end{aligned} \tag{2.6.12}$$

whereas v is a solution to the boundary - initial value problem

$$\begin{aligned} &\frac{\partial v}{\partial t} = -\Delta v, \qquad \text{in} \quad \Omega_2 \times (0,T), \\ &v = f_2, \qquad \text{on} \quad \Gamma_2 \times [0,T), \\ &v(\mathbf{x},0) = g_2(\mathbf{x}), \qquad \mathbf{x} \in \Omega_2. \end{aligned} \tag{2.6.13}$$

Define $w = u - v$, and then the paper of Crooke & Payne (1984) shows how one may derive a bound of form

$$\int_0^t \int_\Omega w^2 dx\, ds \le \int_0^{2t} \oint_{\partial\Omega} w^2\, dA\, ds + \text{data}. \tag{2.6.14}$$

They further show how to develop an inequality like

$$\begin{aligned} \Big(\int_0^t \oint_{\partial\Omega} w^2\, dA\, ds\Big)^{1/2} &\le \Big(\int_0^t \oint_{\partial\Omega} (\hat{f}_1 - \hat{f}_2)^2\, dA\, ds\Big)^{1/2} \\ &+ \delta^{1/2}\Big(\int_0^t \int_{\Omega_1} |\nabla u|^2\, dx\, ds + \int_0^t \int_{\Omega_2} |\nabla v|^2\, dx\, ds\Big)^{1/2}, \end{aligned} \tag{2.6.15}$$

where $\hat{f}_1, \hat{f}_2$ are extensions of f_1 and f_2 to the boundary $\partial\Omega$. For example, if we consider a piece of $\partial\Omega$ which coincides with Γ_2 then $\hat{f}_1$ is the value of f_1 on Γ_1 but projected along a radius arm to its equivalent position on Γ_2. Further bounds demonstrate that $\int_0^t \int_{\Omega_1} |\nabla u|^2\, dx\, ds$ may be estimated in terms of boundary integrals of f_1 and its derivatives and $\|u\|_1^2$ and $\int_0^{2t} \|u\|_1^2 ds$. The constraint set of Crooke & Payne (1984) is those functions which satisfy

$$\int_{\Omega_1} u^2 dx \le M_1^2, \qquad \int_{\Omega_2} v^2 dx \le M_2^2,$$

and thus they bound

$$\int_0^t \int_{\Omega_1} |\nabla u|^2 \, dx \, ds \quad \text{and} \quad \int_0^t \int_{\Omega_2} |\nabla v|^2 \, dx \, ds.$$

This estimate together with (2.6.15) in (2.6.14) leads to an inequality of form

$$\begin{aligned}\int_0^t \int_\Omega w^2 dx \, ds \leq & \, k_1 \sqrt{\int_\Omega (g_1 - g_2)^2 dx} \\ & + k_2 \sqrt{\int_0^t \oint_{\partial\Omega} (\hat{f}_1 - \hat{f}_2)^2 dA \, ds} \\ & + k_3 \delta^{1/2},\end{aligned} \tag{2.6.16}$$

where $k_1, \dots, k_3$ may be calculated according to the recipe in Crooke & Payne (1984), and may depend on the constraint set.

Estimate (2.6.16) is the inequality which establishes continuous dependence on the spatial geometry for a solution to the backward heat equation.

Continuous Dependence on the Spatial Geometry for a Solution to the Heat Equation Backward in Time with Prescribed Neumann Boundary Data

When the Dirichlet data of (2.6.12) and (2.6.13) are changed to Neumann data the equivalent problems are

$$\begin{aligned} &\frac{\partial u}{\partial t} = -\Delta u, && \text{in} \quad \Omega_1 \times (0, T), \\ &\frac{\partial u}{\partial n_1} = f_1, && \text{on} \quad \Gamma_1 \times [0, T), \\ &u(\mathbf{x}, 0) = g_1(\mathbf{x}), && \mathbf{x} \in \Omega_1, \end{aligned} \tag{2.6.17}$$

and

$$\begin{aligned} &\frac{\partial v}{\partial t} = -\Delta v, && \text{in} \quad \Omega_2 \times (0, T), \\ &\frac{\partial v}{\partial n_2} = f_2, && \text{on} \quad \Gamma_2 \times [0, T), \\ &v(\mathbf{x}, 0) = g_2(\mathbf{x}), && \mathbf{x} \in \Omega_2. \end{aligned} \tag{2.6.18}$$

where $\mathbf{n}_1$ and $\mathbf{n}_2$ denote the unit outward normals to Γ_1 and Γ_2, respectively. Persens (1986) established continuous dependence on spatial geometry for solutions to (2.6.17) and (2.6.18). He supposed the solutions lie in the class governed by the constraint set

$$\int_0^T \left(\int_{\Omega_1} u^2 dx + \int_{\Omega_2} v^2 dx \right) ds \leq M^2.$$

If $w = u - v$, then the type of result established by Persens (1986) has form

$$\int_0^T \int_\Omega w^2 dx\, ds \le k_1 \|g_1 - g_2\| + k_2 \sqrt{\int_0^{4t} \int_{\Gamma_1} (f_1 - \hat{f}_2)^2 dA\, ds} + k_3 \delta + k_4 \sigma, \tag{2.6.19}$$

where k_i again depend on the constraint, and the number σ is here a measure of the maximum deviation between the directions of $\mathbf{n}_1$ and $\mathbf{n}_2$ at suitably defined "equivalent" points.

Continuous Dependence on the Spatial Geometry for a Solution to the Heat Equation Backward in Time with Dirichlet Boundary Data, but when Ω_i are Time Dependent

Persens (1986) also considers the spatial geometry problem for (2.6.12) and (2.6.13), but he allows the actual spatial domains to vary in time, i.e. $\Omega_1 = \Omega_1(t)$, $\Omega_2 = \Omega_2(t)$.

The sort of result achieved by Persens (1986) is of form (2.6.16) as established by Crooke & Payne (1984) when Ω_i are not time dependent. Payne (1987a) remarks that Persens (1986) analysis encounters surprisingly little difficulty beyond that faced by Crooke & Payne (1984), a fact which is somewhat remarkable.

Continuous Dependence on the Spatial Geometry for a Solution to the Heat Equation Backward in Time on an Unbounded Exterior Region

Payne & Straughan (1994b) (see also Payne & Straughan (1990a)) tackle the equivalent problem to that of Crooke & Payne (1984), i.e. they investigate the behavior of solutions to (2.6.12) and (2.6.13), but Ω_1 and Ω_2 are now unbounded regions, each being exterior to a bounded region Ω_1^*, Ω_2^* in $\mathbf{R}^n$. They define

$$\Omega = \Omega_1 \cap \Omega_2,$$

and set $w = u - v$ but due to the nature of the unbounded region they work with a weighted L^2 integral over Ω. The weight is necessary in order to avoid imposition of solution decay at infinity.

To establish their result Payne & Straughan (1994b) first transform w to a function which vanishes at $t = 0$, namely

$$\phi(\mathbf{x}, t) = \int_0^t w(\mathbf{x}, s) ds,$$

and then ϕ satisfies the boundary - initial value problem

$$
\begin{aligned}
&\frac{\partial \phi}{\partial t} + \Delta \phi = g_1 - g_2, \qquad \text{in} \quad \Omega \times (0,T),\\
&\phi = \int_0^t w\, ds, \qquad \text{on} \quad \partial\Omega \times [0,T],\\
&\phi(\mathbf{x},0) = 0, \qquad \mathbf{x} \in \Omega.
\end{aligned}
\tag{2.6.20}
$$

The main steps in the analysis of Payne & Straughan (1994b) are now outlined.

Firstly an auxiliary function $H(x,t)$ is defined so that

$$
\begin{aligned}
&\frac{\partial H}{\partial t} - \Delta H = 0, \qquad \text{in} \quad \Omega \times (0,T),\\
&H = \phi, \qquad \text{on} \quad \partial\Omega \times [0,T],\\
&H(\mathbf{x},0) = 0, \qquad \mathbf{x} \in \Omega.
\end{aligned}
\tag{2.6.21}
$$

Then another transformed variable ψ is introduced which is zero on both $\partial\Omega$ and initially, so that

$$
\psi(\mathbf{x},t) = \phi(\mathbf{x},t) - H(\mathbf{x},t)
$$

satisfies

$$
\begin{aligned}
&\frac{\partial \psi}{\partial t} + \Delta \psi = -2\frac{\partial H}{\partial t} + g_1 - g_2, \qquad \text{in} \quad \Omega \times (0,T),\\
&\psi = 0, \qquad \text{on} \quad \partial\Omega \times [0,T],\\
&\psi(\mathbf{x},0) = 0, \qquad \mathbf{x} \in \Omega.
\end{aligned}
\tag{2.6.22}
$$

The basic continuous dependence result of Payne & Straughan (1994b) derives an inequality for the weighted norm

$$
\int_0^t \int_\Omega e^{-\alpha r}(u-v)^2 dx\, ds,
$$

for a constant α to be selected. The analysis begins with the triangle inequality to find

$$
\begin{aligned}
\left(\int_0^t \int_\Omega e^{-\alpha r}(u-v)^2 dx\, ds\right)^{1/2} \leq & \left(\int_0^t \int_\Omega e^{-\alpha r}\left(\frac{\partial \psi}{\partial s}\right)^2 dx\, ds\right)^{1/2}\\
& + \left(\int_0^t \int_\Omega e^{-\alpha r}\left(\frac{\partial H}{\partial s}\right)^2 dx\, ds\right)^{1/2}.
\end{aligned}
\tag{2.6.23}
$$

The constraint set of Payne & Straughan (1994b) which is natural for an unbounded domain problem (and is weaker than L^2 integrability) consists of those functions u, v such that

$$\sup_{\Omega_1\times[0,T]} |u| + \sup_{\Omega_2\times[0,T]} |v| \leq M. \tag{2.6.24}$$

After (2.6.23) the method proceeds via five stages which are now outlined.

Stage 1. Here an a *priori* bound is derived of form

$$\int_0^t \int_\Omega e^{-\alpha r}\Big(\frac{\partial H}{\partial s}\Big)^2 dx\, ds \leq k_1 \int_0^t \oint_{\partial\Omega} (u-v)^2 dA\, ds, \tag{2.6.25}$$

for an explicit constant k_1.

Stage 2. This stage derives an inequality of form

$$\begin{aligned}\int_0^t \int_\Omega e^{-\alpha r}\Big(\frac{\partial \psi}{\partial s}\Big)^2 dx\, ds \leq k_2 &\int_0^{2t} \int_\Omega e^{-\alpha r}\psi^2 dx\, ds\\ &+ k_3 \int_0^{2t} \int_\Omega e^{-\alpha r}\Big(\frac{\partial H}{\partial s}\Big)^2 dx\, ds\\ &+ k_4 \int_\Omega e^{-\alpha r}(g_1-g_2)^2 dx,\end{aligned} \tag{2.6.26}$$

and then uses (2.6.25) to show that

$$\begin{aligned}\int_0^t \int_\Omega e^{-\alpha r}\Big(\frac{\partial \psi}{\partial s}\Big)^2 dx\, ds \leq K_1 &\int_0^{2t} \int_\Omega e^{-\alpha r}\psi^2 dx\, ds\\ &+ K_2 \int_0^{2t} \oint_{\partial\Omega} (u-v)^2 dA\, ds\\ &+ K_3 \int_\Omega e^{-\alpha r}(g_1-g_2)^2 dx,\end{aligned} \tag{2.6.27}$$

where here and henceforth the constants K_i are all explicitly computable.

Stage 3. This stage is quite interesting and involves use of the Lagrange identity method to bound

$$\int_0^{2t} \int_\Omega e^{-\alpha r}\psi^2 dx\, ds$$

in terms of

$$\alpha \int_0^{8t} \int_\Omega e^{-\alpha r}\psi^2 dx\, ds, \qquad \int_0^{8t} \oint_{\partial\Omega} (u-v)^2 dA\, ds,$$

and

$$\int_\Omega e^{-\alpha r}(g_1-g_2)^2dx,$$

and then derives an inequality of form

$$\int_0^{2t}\int_\Omega e^{-\alpha r}\psi^2dx\,ds\le F\Big(\int_0^{8t}\oint_{\partial\Omega}(u-v)^2dA\,ds\,,\,\sup_\Omega|g_1-g_2|^2\Big),\quad(2.6.28)$$

where the functional form of F may have a variety of forms.

Stage 4. This is a modification of the equivalent stage in the proof of Crooke & Payne (1984) and derives a bound for

$$\int_0^{8t}\oint_{\partial\Omega}(u-v)^2dA\,ds$$

in terms of

$$\int_0^{8t}\oint_{\Gamma_1}(f_1-\hat{f}_2)^2dA\,ds$$

and the geometry variation measure δ. Here $\hat{f}_2$ is an extension of f_1 to Γ_1 and δ is essentially the measure of the variation of Ω_1 and Ω_2 described earlier.

Stage 5. The procedure of the final stage is to combine stages 1 to 4 and determine a continuous dependence estimate of form

$$\begin{aligned}\int_0^t\int_\Omega e^{-\alpha r}(u-v)^2dx\,ds\le C_1\int_0^T\oint_{\Gamma_1}(f_1-\hat{f}_2)^2dA\,dt\\ C_2\Big\{\alpha_1\int_0^T\oint_{\Gamma_1}(f_1-\hat{f}_2)^2dA\,dt+\alpha_2\delta\\ +\Big[\alpha_3\Big(\int_0^T\oint_{\Gamma_1}(f_1-\hat{f}_2)^2dA\,dt+\alpha_4\delta\Big)^2\\ \alpha_5\sup_{\mathbf{x}\in\Omega}|g_1-g_2|^2\Big]^{1/2}\Big\}^{1/3},\end{aligned}\qquad(2.6.29)$$

where $c_1,c_2,\ \alpha_1,\ldots,\alpha_5$ are computable constants which depend on the geometry, the constant α, and on the constraint set bound.

If the right hand side of (2.6.29) is denoted by $\mathcal{F}$, we may observe $\mathcal{F}$ is a data term involving δ. For $S(R)$ being the ball $B(0,R)$ center 0, radius R (supposing R is sufficiently large that $\Omega\subset B(0,R)$) an unweighted estimate may be produced which has form

$$\int_0^t\int_{S(R)}(u-v)^2dx\,ds\le e^{\alpha R}\mathcal{F}.\qquad(2.6.30)$$

The constant α is chosen by minimizing the right hand side of (2.6.30).

The estimates (2.6.29) or (2.6.30) were derived by Payne & Straughan (1994b) and establish continuous dependence of the solution to the backward heat equation on changes in the geometry when Ω_i are unbounded exterior regions.

Analysis of Continuous Dependence on the Spatial Geometry for Solutions to the Navier - Stokes Equations

Thus far in this chapter we have described work on continuous dependence on the spatial geometry for problems involving *linear* partial differential equations. As we pointed out in the sections on continuous dependence on the initial - time geometry many real life situations involve nonlinear partial differential equations and so it is natural to expect the development of analyses of continuous dependence on the spatial geometry for such nonlinear equations. To date we are aware of only one such investigation, that of Payne (1994) who establishes continuous dependence on the spatial geometry for a solution to the improperly posed problem of the Navier - Stokes equations backward in time. The area of continuous dependence on the spatial geometry for nonlinear partial differential equations is one which could profitably be developed in the future.

We now include an account of Payne's (1994) work. He again chooses two intersecting, but not equal, spatial domains $\Omega_1, \Omega_2 (\subset \mathbf{R}^3)$ with C^1 boundaries Γ_1 and Γ_2. Suppose u_i is a solution to the boundary - initial value problem

$$\begin{aligned} &\frac{\partial u_i}{\partial t} = u_j \frac{\partial u_i}{\partial x_j} - \nu \Delta u_i + \frac{\partial p}{\partial x_i}, \qquad \text{in} \quad \Omega_1 \times (0,T), \\ &\frac{\partial u_i}{\partial x_i} = 0, \qquad \text{in} \quad \Omega_1 \times (0,T), \\ &u_i = 0, \qquad \text{on} \quad \Gamma_1 \times [0,T], \\ &u_i(\mathbf{x},0) = f_i(\mathbf{x}), \qquad \mathbf{x} \in \Omega_1, \end{aligned} \tag{2.6.31}$$

while v_i solves an analogous problem on Ω_2, i.e.

$$\begin{aligned} &\frac{\partial v_i}{\partial t} = v_j \frac{\partial v_i}{\partial x_j} - \nu \Delta v_i + \frac{\partial q}{\partial x_i}, \qquad \text{in} \quad \Omega_2 \times (0,T), \\ &\frac{\partial v_i}{\partial x_i} = 0, \qquad \text{in} \quad \Omega_2 \times (0,T), \\ &v_i = 0, \qquad \text{on} \quad \Gamma_2 \times [0,T], \\ &v_i(\mathbf{x},0) = g_i(\mathbf{x}), \qquad \mathbf{x} \in \Omega_2. \end{aligned} \tag{2.6.32}$$

The object is again to study how a measure of

$$w_i = u_i - v_i,$$

depends on changes in Ω_2 from Ω_1, supposing the region of difference of domains

$$(\Omega_1 \cup \Omega_2) \cap \Omega^c$$

is small. To make the analysis not overly complicated Payne (1994) also supposes that

$$f_i(\mathbf{x}) \equiv g_i(\mathbf{x}), \qquad \text{in} \quad \Omega.$$

The constraint sets of Payne (1994) are $\mathcal{M}_1$ and $\mathcal{M}_2$ defined below. A function $r_i \in \mathcal{M}_1$ if $\exists$ a constant M_1 such that

$$\sup_{\Omega_1 \times (0,T)} \ r_i r_i \leq M_1^2, \tag{2.6.33}$$

whereas a function $r_i \in \mathcal{M}_2$ if $\exists$ a constant M_2 such that

$$\sup_{\Omega_2 \times (0,T)} \ \big[r_i r_i + r_{i,j} r_{i,j} + r_{i,t} r_{i,t} \big] \leq M_2^2. \tag{2.6.34}$$

Payne (1994) remarks that by a more careful analysis one could weaken these constraint sets. He develops a continuous dependence estimate for the function

$$\int_0^t (t-s)^2 \int_\Omega w_i w_i \, dx \, ds.$$

If we put $\pi = p - q$ the function $w_i(\mathbf{x}, t)$ satisfies the boundary - initial value problem

$$\begin{aligned}
&\frac{\partial w_i}{\partial t} = u_j \frac{\partial w_i}{\partial x_j} + w_j \frac{\partial v_i}{\partial x_j} - \nu \Delta w_i + \frac{\partial \pi}{\partial x_i}, \qquad && \text{in} \quad \Omega \times (0,T), \\
&\frac{\partial w_i}{\partial x_i} = 0, \qquad \text{in} \quad \Omega \times (0,T), && \\
&w_i = u_i - v_i, \qquad \text{on} \quad \partial\Omega \times [0,T], && \\
&w_i(\mathbf{x}, 0) = 0, \qquad \mathbf{x} \in \Omega. &&
\end{aligned} \tag{2.6.35}$$

Payne's (1994) analysis is very technical and involves a logarithmic convexity argument. We only provide a *very brief* outline plus a statement of the final result.

The inhomogeneous boundary data cause a problem and so the first step is to transform to a function which has zero boundary data. To do this Payne (1994) introduces an auxiliary function H_i which $\forall t \in (0,T)$ solves the linear problem

$$\begin{aligned}
&\nu \Delta H_i = \frac{\partial S}{\partial x_i}, \qquad \frac{\partial H_i}{\partial x_i} = 0, \qquad \text{in} \quad \Omega, \\
&H_i(\mathbf{x}, t) = w_i(\mathbf{x}, t), \qquad \text{on} \quad \partial\Omega.
\end{aligned} \tag{2.6.36}$$

Since (2.6.36) has a unique solution Payne (1994) then concludes $H_i \equiv w_i$ on Ω for $t = 0$, and so

$$H_i(\mathbf{x}, 0) = 0, \qquad \mathbf{x} \in \Omega. \tag{2.6.37}$$

He then sets

$$\psi_i = w_i - H_i,$$

and the function ψ_i satisfies the boundary - initial value problem

$$\begin{aligned}
&\frac{\partial \psi_i}{\partial t} = u_j \frac{\partial \psi_i}{\partial x_j} + \psi_j \frac{\partial v_i}{\partial x_j} - \nu \Delta \psi_i + \frac{\partial P}{\partial x_i} + L_i, \qquad \text{in} \quad \Omega \times (0, T),\\
&\frac{\partial \psi_i}{\partial x_i} = 0, \qquad \text{in} \quad \Omega \times (0, T),\\
&\psi_i = 0, \qquad \text{on} \quad \partial\Omega \times [0, T],\\
&\psi_i(\mathbf{x}, 0) = 0, \qquad \mathbf{x} \in \Omega,
\end{aligned} \tag{2.6.38}$$

where $P = \pi - S$ and L_i is defined by

$$L_i = -\frac{\partial H_i}{\partial t} + u_j \frac{\partial H_i}{\partial x_j} + H_j \frac{\partial v_i}{\partial x_j}.$$

The triangle inequality is used to see that

$$\begin{aligned}
\left[\int_0^t (t-s)^2 \int_\Omega w_i w_i \, dx \, ds\right]^{1/2} \leq & \left[\int_0^t (t-s)^2 \int_\Omega \psi_i \psi_i \, dx \, ds\right]^{1/2}\\
& + \left[\int_0^t (t-s)^2 \int_\Omega H_i H_i \, dx \, ds\right]^{1/2},
\end{aligned}$$

and the analysis consists of bounding the two expressions on the right hand side.

The continuous dependence on the spatial geometry result is given in the following theorem.

Theorem 2.6.1. (Payne (1994).)
If $u_i(\mathbf{x}, t) \in \mathcal{M}_1$ and $v_i(\mathbf{x}, t) \in \mathcal{M}_2$ then it is possible to compute an explicit constant K and a function $\alpha(t)$, with $0 < \alpha(t) \leq 1$, α and K being independent of u_i and v_i, such that $\forall t \in [0, T)$,

$$\int_0^t (t-s)^2 \int_\Omega w_i w_i \, dx \, ds \leq K \delta^\alpha, \tag{2.6.39}$$

where δ is the maximum distance along a ray between Γ_1 and Γ_2.

We close this chapter with a brief account of the only other studies of continuous dependence on the spatial geometry of which we are aware; these are again for linear partial differential equations.

Continuous Dependence on the Spatial Geometry for Solutions to Poisson's Equation

Persens (1986) studied solutions u, v which solve

$$\begin{aligned} \Delta u &= F_1, && \text{in} \quad \Omega_1, \\ u &= f_1, && \text{on} \quad \Sigma_1, \\ \frac{\partial u}{\partial n} &= g_1, && \text{on} \quad \Sigma_1, \end{aligned}$$

and

$$\begin{aligned} \Delta v &= F_2, && \text{in} \quad \Omega_2, \\ v &= f_2, && \text{on} \quad \Sigma_2, \\ \frac{\partial v}{\partial n} &= g_2, && \text{on} \quad \Sigma_2, \end{aligned}$$

where Σ_1 is a smooth piece of Γ_1 and Σ_2 is a smooth piece of Γ_2. The two sets of data are given on only a piece of the respective boundaries. The result of Persens (1986) involves a suitably defined subdomain Ω_k of the "common" domain Ω and essentially states

$$\int_{\Omega_k} (u-v)^2 dx \le K(a_1\epsilon^2 + a_2\delta^{1/2})^{1-\alpha},$$

where a_1, a_2, K are constants, α is a constant with $0 \le \alpha \le 1$, ϵ is an estimate of the error in an L^2 norm of an extension of the Cauchy data, and δ is as before a measure of the geometric error in variation between Ω_1 and Ω_2.

Continuous Dependence on the Spatial Geometry in Linear Elastodynamics

This work is again due to Persens (1986) who does not require any definiteness of the elastic coefficients. He compares the solutions to the boundary - initial value problems,

$$\begin{aligned} \rho(\mathbf{x})\frac{\partial^2 u_i}{\partial t^2} &= \frac{\partial}{\partial x_j}\Big(a_{ijkh}\frac{\partial u_k}{\partial x_h}\Big), && \text{in} \quad \Omega_1 \times (0,T), \\ u_i &= f_i^1, && \text{on} \quad \Gamma_1 \times [0,T], \\ u_i(\mathbf{x},0) &= g_i^1(\mathbf{x}), && \mathbf{x} \in \Omega_1, \\ \frac{\partial u_i}{\partial t}(\mathbf{x},0) &= h_i^1(\mathbf{x}), && \mathbf{x} \in \Omega_1, \end{aligned}$$

and

$$
\begin{aligned}
&\rho(\mathbf{x})\frac{\partial^2 v_i}{\partial t^2} = \frac{\partial}{\partial x_j}\Big(a_{ijkh}\frac{\partial v_k}{\partial x_h}\Big), \qquad \text{in} \quad \Omega_2 \times (0,T),\\
&v_i = f_i^2, \qquad \text{on} \quad \Gamma_2 \times [0,T],\\
&v_i(\mathbf{x},0) = g_i^2(\mathbf{x}), \qquad \mathbf{x} \in \Omega_2,\\
&\frac{\partial v_i}{\partial t}(\mathbf{x},0) = h_i^2(\mathbf{x}), \qquad \mathbf{x} \in \Omega_2.
\end{aligned}
$$

The elastic coefficients are not supposed sign definite but are required to satisfy the major symmetry

$$
a_{ijkh} = a_{khij}\,.
$$

Persens' (1986) technique involves use of the Lagrange identity method and the constraint set requires $u_i \in H^2(\Omega_1)$, $v_i \in H^2(\Omega_2)$. The basic result of Persens (1986) which establishes continuous dependence on the spatial geometry has form

$$
\int_0^t \int_\Omega \rho(\mathbf{x})(u_i - v_i)(u_i - v_i)\,dx\,ds \le k_1\epsilon_1 + k_2\epsilon_2 + k_3\delta^{1/2},
$$

where k_1, k_2, k_3 are constants which depend on the constraint, ϵ_1 is an L^2 measure of the error in the initial data, ϵ_2 is an L^2 measure of suitably extended boundary data, and δ is again the maximum distance along a ray between the boundaries Γ_1 and Γ_2.

3

Continuous Dependence on Modeling Backward in Time

3.1 Singular Perturbation in Improperly Posed Problems

The task of stabilizing ill posed problems against errors made in formulating the mathematical model of a physical process is a difficult one, primarily because it is not possible to characterize precisely errors such as those made in treating a physical quantity as a continuum, in assuming inexact physical laws, or in approximating the model equation. Recalling that it is usually necessary to constrain the solution to an ill posed problem in order to recover continuous dependence on various types of data, we also observe that such stabilizing constraints should not only be realizable and as weak as possible, but also should simultaneously stabilize the problem against all possible sources of error. If these errors are hard to characterize, the job of determining the appropriate constraints is compounded. In this chapter we shall illustrate the difficulties involved in attempting to regularize problems against these modeling errors, and we commence with a study of a class of singularly perturbed partial differential equations. It is perhaps worth noting at this juncture, that continuous dependence on modeling has also been referred to in the literature as structural stability, as we stressed in section 1.1 of this monograph.

Perhaps the first attempt at some kind of continuous dependence on modeling investigation for ill posed problems was made by Payne & Sather (1967) who compared the solution of the inhomogeneous backward heat conduction problem with that of a perturbed Cauchy problem for an inhomogeneous elliptic equation. In order to accomplish this they, therefore, considered the following pair of boundary - initial value problems:

$$
\begin{aligned}
&\frac{\partial^2 v}{\partial x^2}+\frac{\partial v}{\partial t}=F(x,t), \qquad \text{in} \quad \Omega,\\
&v(0,t)=v(1,t)=0, \qquad\qquad (3.1.1)\\
&v(x,0)=f(x);
\end{aligned}
$$

$$\begin{aligned}
&\epsilon\frac{\partial^2 u}{\partial t^2}+\frac{\partial^2 u}{\partial x^2}+\frac{\partial u}{\partial t}=F(x,t), \qquad \text{in} \quad \Omega,\\
&u(0,t)=u(1,t)=0, \\
&u(x,0)=f(x), \qquad \frac{\partial u}{\partial t}(x,0)=g(x);
\end{aligned} \tag{3.1.2}$$

where Ω is the rectangle $0 < x < 1$, $0 < t < T$ and $0 < \epsilon < \epsilon_0$. Noting that both of these problems are ill posed, Payne & Sather (1967) pose the question: if v and u are solutions to (3.1.1) and (3.1.2), respectively, what restrictions must be imposed on the class of admissible solutions to ensure that u converges to v as $\epsilon \to 0$ in some sense?

Using a logarithmic convexity argument, Payne & Sather (1967) show that if

$$\int_0^T \int_0^1 u^2 dx d\eta \leq M^2,$$

for some prescribed constant M and if

$$\int_0^T \int_0^1 v^2 dx d\eta \leq M_0^2,$$

for some constant M_0, then there exists a constant K, independent of ϵ, such that

$$\int_0^t \int_0^1 (u-v)^2 dx d\eta \leq K\epsilon^{1-\alpha}, \qquad 0 \leq t < T, \tag{3.1.3}$$

where $\alpha = t/T$. An immediate consequence of (3.1.3) is that as $\epsilon \to 0$, u converges to v in L^2 on compact subdomains of Ω.

In this section we first describe the analysis of Payne & Sather (1967) for the case $F(x,t) = 0$ and then present an alternative comparison of the solutions to multidimensional versions of problems (3.1.1) and (3.1.2).

Let us set $w = u - v$ and observe that w satisfies the problem

$$\begin{aligned}
&\epsilon\frac{\partial^2 w}{\partial t^2}+\frac{\partial^2 w}{\partial x^2}+\frac{\partial w}{\partial t}=-\epsilon\frac{\partial^2 v}{\partial t^2}, \qquad \text{in} \quad \Omega,\\
&w(0,t)=w(1,t)=0, \\
&w(x,0)=0, \qquad \frac{\partial w}{\partial t}(x,0)=g(x).
\end{aligned} \tag{3.1.4}$$

Consider the functional

$$\phi(t) = \int_0^t \int_0^1 w^2 dx d\eta + \epsilon Q^2, \tag{3.1.5}$$

where

$$Q^2 = k_1 \int_0^1 \Big(\frac{\partial w}{\partial t}(x,0)\Big)^2 dx + k_2 \int_0^T \int_0^1 v^2 dx dt. \tag{3.1.6}$$

The constants k_1 and k_2 are independent of ϵ so that Q, which involves only data terms, is bounded independently of ϵ. We proceed to show that F satisfies the inequality

$$\phi\phi'' - (\phi')^2 \geq -c\phi^2, \tag{3.1.7}$$

for a computable constant c.

Differentiating (3.1.5) with respect to t, we have

$$\phi'(t) = 2\int_0^t \int_0^1 w\frac{\partial w}{\partial \eta}\, dx d\eta, \tag{3.1.8}$$

and a second differentiation leads to

$$\phi''(t) = 2\int_0^t \int_0^1 \left(\frac{\partial w}{\partial \eta}\right)^2 dx d\eta + 2\int_0^t \int_0^1 w\frac{\partial^2 w}{\partial \eta^2} dx d\eta. \tag{3.1.9}$$

Substituting the differential equation in the second integral of (3.1.9) and integrating by parts, we obtain

$$\begin{aligned}
\phi''(t) =& 2\int_0^t \int_0^1 \left(\frac{\partial w}{\partial \eta}\right)^2 dx d\eta \\
&+ 2\int_0^t \int_0^1 w\left[-\frac{1}{\epsilon}\left(\frac{\partial^2 w}{\partial x^2} + \frac{\partial w}{\partial \eta}\right) - \frac{\partial^2 v}{\partial \eta^2}\right] dx d\eta, \\
=& 4\int_0^t \int_0^1 \left(\frac{\partial w}{\partial \eta}\right)^2 dx d\eta \\
&- 2\int_0^t \int_0^1 \left[\left(\frac{\partial w}{\partial \eta}\right)^2 - \frac{1}{\epsilon}\left(\frac{\partial w}{\partial x}\right)^2 + \frac{1}{\epsilon} w\frac{\partial w}{\partial \eta} + w\frac{\partial^2 v}{\partial \eta^2}\right] dx d\eta.
\end{aligned} \tag{3.1.10}$$

An expression for the second term in (3.1.10) can be found by considering the equation

$$0 = \int_0^t \int_0^1 (1 - e^{-\xi})(w_\eta + \frac{1}{2}\epsilon^{-1} w)(\epsilon w_{\eta\eta} + w_{xx} + w_\eta + \epsilon v_{\eta\eta}) dx d\eta, \tag{3.1.11}$$

where

$$\xi = \frac{t - \eta}{\epsilon},$$

and where the shorthand subscript notation for derivatives has been employed, e.g.

$$v_{\eta\eta} = \frac{\partial^2 v}{\partial \eta^2}.$$

Several integrations by parts lead to

$$
\begin{aligned}
-\frac{1}{2}\int_0^t\int_0^1 [w_\eta^2 - \epsilon^{-1}w_x^2 + \epsilon^{-1}ww_\eta + wv_{\eta\eta}]dxd\eta = \\
= -\frac{1}{2}\int_0^1 (1-e^{-t/\epsilon})\epsilon w_t^2(x,0)dx + \frac{1}{2}\int_0^t\int_0^1 e^{-\xi}wv_{\eta\eta}dxd\eta \\
- \int_0^t\int_0^1 (1-e^{-\xi})\epsilon wv_{\eta\eta\eta}dxd\eta.
\end{aligned}
$$

If we substitute this result in (3.1.10), we find

$$
\begin{aligned}
\phi'' =& 4\int_0^t\int_0^1 w_\eta^2 dxd\eta - 2\int_0^1 (1-e^{-t/\epsilon})\epsilon w_t^2(x,0)dx \\
&+ 2\int_0^t\int_0^1 e^{-\xi}wv_{\eta\eta}dxd\eta - 4\int_0^t\int_0^1 (1-e^{-\xi})\epsilon wv_{\eta\eta\eta}dxd\eta.
\end{aligned}
$$

Use of the arithmetic-geometric mean inequality results in the bound

$$
\begin{aligned}
\phi'' \geq& 4\int_0^t\int_0^1 w_\eta^2 dxd\eta - 2\epsilon\int_0^1 w_t^2(x,0)dx \\
&- 3\int_0^t\int_0^1 w^2 dxd\eta - \int_0^t\int_0^1 e^{-2\xi}v_{\eta\eta}^2 dxd\eta \\
&- 2\epsilon\int_0^t\int_0^1 (1-e^{-\xi})^2 v_{\eta\eta\eta}^2 dxd\eta.
\end{aligned}
\tag{3.1.12}
$$

Payne & Sather (1967) now observe that

$$
\int_0^t\int_0^1 e^{-2\xi}v_{\eta\eta}^2 dxd\eta \leq \frac{1}{2}\epsilon(1-e^{-2t/\epsilon})\left(\max_{0\leq\tau\leq t}\int_0^1 \big[v_{\eta\eta}(x,\tau)\big]^2 dx\right),
$$

and show that there exist constants α_1 and α_2 independent of ϵ such that

$$
\max_{0\leq\tau\leq t}\int_0^1 \big[v_{\eta\eta}(x,\tau)\big]^2 dx \leq \alpha_1 Q^2, \tag{3.1.13}
$$

and

$$
\int_0^t\int_0^1 v_{\eta\eta\eta}^2 dxd\eta \leq \alpha_2 Q^2, \tag{3.1.14}
$$

where Q^2 is given by (3.1.6). Hence, it follows that

$$
\phi'' \geq 4\int_0^t\int_0^1 w_\eta^2 dxd\eta - 3\int_0^t\int_0^1 w^2 dxd\eta - c_1\epsilon Q^2, \tag{3.1.15}
$$

for a computable constant c_1 independent of ϵ. We now form the expression $\phi\phi'' - (\phi')^2$ using (3.1.5), (3.1.8) and (3.1.15) to find

$$\begin{aligned}\phi\phi''-(\phi')^2 &\geq 4\Big(\int_0^t\int_0^1 w_\eta^2\,dx\,d\eta\Big)\Big(\int_0^t\int_0^1 w^2\,dx\,d\eta\Big) - 4\Big(\int_0^t\int_0^1 ww_\eta\,dx\,d\eta\Big)^2\\ &\quad + 4\epsilon Q^2\int_0^t\int_0^1 w_\eta^2\,dx\,d\eta - 3\phi\int_0^t\int_0^1 w^2\,dx\,d\eta - c_1\epsilon Q^2\phi. \qquad (3.1.16)\end{aligned}$$

Observing that

$$\Big(\int_0^t\int_0^1 w_\eta^2\,dx\,d\eta\Big)\Big(\int_0^t\int_0^1 w^2\,dx\,d\eta\Big) - \Big(\int_0^t\int_0^1 ww_\eta\,dx\,d\eta\Big)^2$$

is nonnegative by virtue of the Cauchy-Schwarz inequality, and recalling the definition of ϕ, we then conclude that

$$\phi\phi'' - (\phi')^2 \geq -c\phi^2,$$

for $0 \leq t < T$. Integration of this inequality as in chapter 1, inequality (1.5.9), yields

$$\phi(t) \leq e^{ct(T-t)/2}\phi(0)^{1-t/T}\phi(T)^{t/T}. \qquad (3.1.17)$$

Since $\phi(0) = \epsilon Q^2$, we note that if we assume

$$\int_0^T\int_0^1 w^2\,dx\,d\eta \leq M_1^2, \qquad (3.1.18)$$

it follows from (3.1.17) that there exists a constant M_2 independent of ϵ such that

$$\int_0^t\int_0^1 w^2\,dx\,d\eta \leq M_2\epsilon^{1-t/T}, \qquad (3.1.19)$$

for $t \in [0, T)$.

Let us observe that (3.1.18) is satisfied if we assume

$$\int_0^T\int_0^1 u^2\,dx\,d\eta \leq M^2 \quad \text{and} \quad \int_0^T\int_0^1 v^2\,dx\,d\eta \leq M_0^2,$$

for constants M and M_0 independent of ϵ. Consequently, if u and v satisfy these constraints, we have (3.1.3) which is the desired continuous dependence result of Payne & Sather (1967).

By employing the Green's function for the heat equation in two independent variables Payne & Sather (1967) are able to extend inequality (3.1.19) to establish that as $\epsilon \to 0$ then u converges to v pointwise and uniformly on compact subdomains of Ω.

An $n-$ Spatial Dimension Version of the Elliptic - Backward Heat Equation Problem

Continuous dependence on modeling results can be obtained for generalizations of problems (3.1.1) and (3.1.2) via the Lagrange identity method. We consider the following boundary-initial value problems:

Problem A

$$\begin{aligned}
&\frac{\partial v}{\partial t} + \Delta v = 0, \qquad \text{in} \quad \Omega \times [0,T),\\
&v = 0, \qquad \text{on} \qquad \partial\Omega \times [0,T),\\
&v(\mathbf{x},0) = f(\mathbf{x}), \qquad \mathbf{x} \in \Omega;
\end{aligned} \tag{3.1.20}$$

Problem B

$$\begin{aligned}
&\epsilon \frac{\partial^2 u}{\partial t^2} + \Delta u + \frac{\partial u}{\partial t} = 0, \qquad \text{in} \quad \Omega \times [0,T),\\
&u = 0, \qquad \text{on} \qquad \partial\Omega \times [0,T),\\
&u(\mathbf{x},0) = f(\mathbf{x}), \qquad \frac{\partial u}{\partial t}(\mathbf{x},0) = g(\mathbf{x}), \qquad \mathbf{x} \in \Omega.
\end{aligned} \tag{3.1.21}$$

In these problems Ω is a bounded open set in $\mathbf{R}^n$ with smooth boundary $\partial\Omega$. Defining

$$w = u - v,$$

we observe that w satisfies

$$\begin{aligned}
&\frac{\partial w}{\partial t} + \Delta w = -\epsilon \frac{\partial^2 u}{\partial t^2}, \qquad \text{in} \quad \Omega \times [0,T),\\
&w = 0, \qquad \text{on} \qquad \partial\Omega \times [0,T),\\
&w(\mathbf{x},0) = 0, \qquad \mathbf{x} \in \Omega.
\end{aligned} \tag{3.1.22}$$

Assume that $w^*(\mathbf{x},t)$ is any solution to the adjoint equation of (3.1.22). Then for such a $w^*(\mathbf{x},t)$ we have

$$0 = \int_0^t \int_\Omega \big\{ w^*(w_\eta + \Delta w + \epsilon u_{\eta\eta}) - w(-w^*_\eta + \Delta w^* + \epsilon u^*_{\eta\eta}) \big\} dx d\eta, \tag{3.1.23}$$

where u^* is a solution to the adjoint equation of (3.1.21). Integration of (3.1.23) leads to the equation

$$\begin{aligned}
0 = &\int_\Omega w(\mathbf{x},t) w^*(\mathbf{x},t) dx + \epsilon \int_\Omega \big\{ w^*(\mathbf{x},t) w_t(\mathbf{x},t) - w(\mathbf{x},t) w^*_t(\mathbf{x},t) \big\} dx\\
&+ \epsilon \int_0^t \int_\Omega (w^* v_{\eta\eta} - w v^*_{\eta\eta}) dx d\eta.
\end{aligned} \tag{3.1.24}$$

Here v^* is a solution of the forward heat equation. We choose

$$w^*(\mathbf{x}, \eta) = w(\mathbf{x}, 2t - \eta)$$

and

$$v^*(\mathbf{x}, \eta) = v(\mathbf{x}, 2t - \eta)$$

in (3.1.24) to obtain

$$\int_\Omega w^2(\mathbf{x}, t)dx + 2\epsilon \int_\Omega w(\mathbf{x}, t)w_t(\mathbf{x}, t)dx = \epsilon \int_0^t \int_\Omega (wv^*_{\eta\eta} - w^* v_{\eta\eta})dxd\eta.$$

Integrating with respect to t, we find that

$$\int_0^t \int_\Omega w^2 dxd\eta + \epsilon \int_\Omega w^2 dx = \epsilon \int_0^t \int_\Omega (t-\eta)(wv^*_{\eta\eta} - w^* v_{\eta\eta})dxd\eta. \quad (3.1.25)$$

Application of the Cauchy-Schwarz inequality in (3.1.25) yields the estimate

$$\begin{aligned} &\int_0^t \int_\Omega w^2 dxd\eta + \epsilon \int_\Omega w^2 dx \\ &\quad \leq 2\epsilon t \Big(\int_0^{2t} \int_\Omega w^2 dxd\eta \Big)^{1/2} \Big(\int_0^{2t} \int_\Omega v^2_{\eta\eta} dxd\eta \Big)^{1/2}. \end{aligned} \quad (3.1.26)$$

A bound for the last term in (3.1.26) must now be found. This can be accomplished by introducing the function $\gamma(t) \in C^4(t \geq 0)$ defined as

$$\begin{aligned} &\gamma(t) = 1, \ 0 \leq t \leq t_0; \\ &0 \leq \gamma(t) \leq 1, \ t_0 \leq t \leq t_1; \\ &\gamma(t) = 0, \ t \geq t_1, \end{aligned} \quad (3.1.27)$$

which also satisfies $\gamma^{(\alpha)}(t) = 0$ at $t = t_0, t_1$ for $\alpha = 1, \ldots, 4$. Then for $t_0 \geq T$ we have

$$\begin{aligned} \int_0^{2t} \int_\Omega v^2_{\eta\eta} dxd\eta \leq &\int_0^\infty \int_\Omega \gamma(\eta) v^2_{\eta\eta} dxd\eta \\ = &\frac{1}{2} \int_0^\infty \int_\Omega \gamma(\eta) \frac{\partial}{\partial \eta}(v_{,i\eta} v_{,i\eta}) dxd\eta \\ = &- \int_\Omega v_{,i\eta} v_{,i\eta} \Big|_{\eta=0} dx - \frac{1}{2} \int_0^\infty \int_\Omega \gamma'(\eta) v_{,i\eta} v_{,i\eta} dxd\eta \\ \leq &\frac{1}{2} \int_0^\infty \int_\Omega \gamma'(\eta) v_{,i\eta} \Delta v_{,i} dxd\eta \\ = &\frac{1}{4} \int_0^\infty \int_\Omega \gamma''(\eta) (\Delta v)^2 dxd\eta \\ = &- \frac{1}{8} \int_0^\infty \int_\Omega \gamma'''(\eta) v_{,i} v_{,i} dxd\eta \\ = &\frac{1}{16} \int_0^\infty \int_\Omega \gamma^{(4)}(\eta) v^2 dxd\eta \,, \end{aligned}$$

through repeated substitution of the differential equation (assumed to hold for $0 \leq t \leq t_1$) and integration by parts. Thus,

$$\int_0^{2t} \int_\Omega v_{\eta\eta}^2 dx d\eta \leq \frac{1}{16} \int_0^{t_1} \int_\Omega \gamma^{(4)}(\eta) v^2 dx d\eta. \tag{3.1.28}$$

Since $\gamma^{(4)}(t)$ is bounded for $t \geq 0$, it follows that if we require

$$\int_0^{t_1} \int_\Omega v^2 dx d\eta \leq M_0^2,$$

for a prescribed constant M_0, then a constant K can be computed so that

$$\int_0^{2t} \int_\Omega v_{\eta\eta}^2 dx d\eta \leq K M_0^2.$$

Thus, (3.1.26) becomes

$$\int_0^t \int_\Omega w^2 dx d\eta + \epsilon \int_\Omega w^2 dx \leq K_1 \epsilon t M_0 \left(\int_0^{2t} \int_\Omega w^2 dx d\eta \right)^{1/2}. \tag{3.1.29}$$

Assuming that we can prescribe a constant M independent of ϵ such that

$$\int_0^T \int_\Omega u^2 dx d\eta \leq M^2,$$

then there exists M_1 such that

$$\int_0^{2t} \int_\Omega w^2 dx d\eta \leq M_1^2,$$

when $2t \leq T$ and thus (3.1.29) implies

$$\int_0^t \int_\Omega w^2 dx d\eta + \epsilon \int_\Omega w^2 dx \leq K_2 \epsilon, \tag{3.1.30}$$

for $0 \leq t \leq \frac{1}{2}T$. This is the desired continuous dependence inequality. We are thus led to the conclusion that if the solutions of both problems A and B lie in the appropriate solution spaces, then the L^2 integral of the difference between the two solutions over compact subdomains is small if the parameter ϵ is small.

Elliptic Singular Perturbation Type Problems

Adelson (1973,1974) continued the investigation of singular perturbation in ill posed problems by considering a class of problems in which both the perturbed and unperturbed problems are Cauchy problems for elliptic equations. He was interested in the following system:

$$\begin{aligned} &\epsilon bLv + v = u, \qquad \text{in} \quad \Omega \subset \mathbf{R}^n, \\ &Lu \equiv \frac{\partial}{\partial x_j}\Big(a_{ij}\frac{\partial u}{\partial x_i}\Big) = F(x,\epsilon,v,u), \end{aligned} \tag{3.1.31}$$

$$u, v, \operatorname{grad} u, \operatorname{grad} v, \text{ prescribed on } \Gamma. \tag{3.1.32}$$

Here ϵ is a small positive parameter, b is a constant, L is a strongly elliptic operator, F is assumed to be Lipschitz in its last three arguments, and Γ is a portion of the boundary of Ω. Adelson (1973,1974) asked several questions about the system (3.1.31) - (3.1.32) among which are the following:

i) Can a reasonable approximate solution of this system be found in some neighborhood Ω_a of Γ if u and/or v are suitably constrained?

ii) If ϵ is very small, can a good enough approximation be obtained by setting $\epsilon = 0$ and approximating the solution w of the simpler Cauchy problem

$$\begin{aligned} &Lw = F(x,0,w,w), \\ &w, \operatorname{grad} w, \text{ prescribed on } \Gamma, \end{aligned} \tag{3.1.33}$$

assuming the data of w and v are close?

Using L^2 constraints, Adelson (1973,1974) was able to provide affirmative answers to these questions for the case $b < 0$. For $b > 0$ the necessary constraints on v were more severe and the resulting estimates less sharp. The details of his analysis are quite complicated but a little ingenuity led Adelson to a continuous dependence inequality of the form

$$\|v - w\|_{\Omega_\alpha}^2 \equiv \int_{\Omega_\alpha} (v-w)^2 dx \le C\epsilon^{\delta(\alpha)}, \tag{3.1.34}$$

where $\delta(\alpha)$ is a positive function of α for $0 \le \alpha < \alpha_1 < 1$ and $\delta(\alpha_1) = 0$, that depends on the imposed constraint, the sign of b, and the size of the subdomain $\Omega_\alpha \subset \Omega$. The inequality (3.1.34) indicates that for $0 \le \alpha < \alpha_1 < 1$, v will be arbitrarily close to w in L^2 over Ω_α if ϵ is sufficiently small. Thus, continuous dependence on the perturbation parameter ϵ is established.

To achieve the result (3.1.34) Adelson (1973,1974) employs the triangle inequality to write

$$\|v - w\|_{\Omega_\alpha} \le \|v - u\|_{\Omega_\alpha} + \|u - w\|_{\Omega_\alpha}, \tag{3.1.35}$$

and then obtains bounds for the two terms on the right separately. The analysis for the first term is more or less standard while logarithmic convexity arguments lead to bounds on the second term. More specifically, let $\mathcal{M}_1$ be the class of functions $\phi \in C^4(\Omega)$ such that

$$\int_\Omega \phi^2 dx \leq M_1^2, \tag{3.1.36}$$

for some prescribed constant M_1, and let $\mathcal{M}_2$ denote the class of functions $\psi \in C^4(\Omega)$ that satisfy

$$\int_\Omega [\psi^2 + |\nabla\psi|^2] dx \leq M_2^2, \tag{3.1.37}$$

for a prescribed constant M_2. Adelson (1973,1974) obtains the bounds

$$(i) \qquad \|u - v\|^2_{\Omega_\alpha} = O(\epsilon^2), \quad \text{if } b < 0 \text{ and } v \in \mathcal{M}_1, \tag{3.1.38}$$

$$(ii) \qquad \|u - v\|^2_{\Omega_\alpha} = O(\epsilon), \quad \text{if } b > 0 \text{ and } v \in \mathcal{M}_2, \tag{3.1.39}$$

for $0 \leq \alpha < 1$. He then proceeds to show that $\|u - w\|^2_{\Omega_\alpha}$ is of order ϵ to a positive power depending on the sign of b. Defining $\xi = u - w$ and

$$\phi(\alpha) = \int_0^\alpha (\alpha - \eta) \int_{\Omega_\eta} \left[a_{ij}\xi_{,i}\xi_{,j} + \xi L\xi\right] dx d\eta + Q,$$

where Q is data and $0 \leq \alpha \leq \alpha_1 < 1$, Adelson (1973,1974) shows that ϕ satisfies the differential inequality

$$\phi\phi'' - (\phi')^2 \geq -K_1\phi\phi' - K_2\phi^2,$$

for explicit constants K_1 and K_2. This leads to the bound

$$\|u - w\|^2_{\Omega_\alpha} \leq KQ^{1-\nu(\alpha)}\left[Q + \|u - w\|^2_{\Omega_{\alpha_1}}\right]^{\nu(\alpha)}, \tag{3.1.40}$$

where

$$0 \leq \nu(\alpha) = \frac{1 - e^{-K_1\alpha}}{1 - e^{-K_1\alpha_1}} < 1.$$

Under the appropriate data assumptions, it can then be shown that, either

$$\|u - w\|^2_{\Omega_\alpha} = O\Big(\epsilon^{2(1-\nu(\alpha))}\Big), \tag{3.1.41}$$

if $b < 0$, or

$$\|u - w\|^2_{\Omega_\alpha} = O\Big(\epsilon^{(1-\nu(\alpha))}\Big), \tag{3.1.42}$$

if $b > 0$, and in addition $v \in \mathcal{M}_2$ for $0 \leq \alpha < \alpha_1 < 1$.

Combining expressions (3.1.38) and (3.1.41) or (3.1.39) and (3.1.42) with the triangle inequality, Adelson (1973,1974) proved the result:

Theorem 3.1.1. (Adelson (1973,1974).)
If v is a solution to (3.1.31), w is a solution to (3.1.33) and the data satisfy the order conditions,

$$\begin{aligned} &\|u - v\|_\Gamma = O(\epsilon), \qquad &\|u - w\|_\Gamma = O(\epsilon), \\ &\|\nabla(u - v)\|_\Gamma = O(1), \qquad &\|\nabla(u - w)\|_\Gamma = O(1), \end{aligned} \tag{3.1.43}$$

then,
(i) for $v, w \in \mathcal{M}_1$ and $b < 0$,

$$\|v - w\|^2_{\Omega_\alpha} = O\Big(\epsilon^{2(1-\nu(\alpha))}\Big), \tag{3.1.44}$$

(ii) for $v \in \mathcal{M}_2, w \in \mathcal{M}_1$, and $b > 0$,

$$\|v - w\|^2_{\Omega_\alpha} = O\Big(\epsilon^{(1-\nu(\alpha))}\Big), \tag{3.1.45}$$

where $0 \leq \alpha < \alpha_1 < 1$, and $0 \leq \nu(\alpha) < 1$.

Adelson (1973) obtained some extensions of these results by allowing the function F to depend on derivatives of u and v as well as on the functions u and v themselves. In addition, he treated a class of higher order equations in Adelson (1974). His results are sensitive to signs of coefficients, the powers of ϵ appearing in the coefficients, and the geometry of Ω_α, which is described in section 3.4.

One reason for considering problems such as (3.1.31) and (3.1.33) is that they arise in our attempts to model physical phenomena. One such example appears in classical linear elastic plate theory. If $L \equiv \Delta$, $F \equiv 0$, and $b\epsilon = (KP)^{-1}$, then (3.1.31) models the deflection of a plate that is loaded with a very large uniform loading P in the plane of the plate. Here K is a combination of geometric and material constants. If P is negative, then the loading is tensile and then $b < 0$. The case $b > 0$ corresponds to a compressive loading. The question that arises is whether a good enough approximate solution can be obtained by setting $\epsilon = 0$ and solving the Cauchy problem for the Laplace equation, i.e. problem (3.1.33). Adelson's (1973) work addresses this question, establishing continuous dependence on the loading.

In the above analyses existence of solutions is assumed; this is in keeping with many other studies of improperly posed problems. If one is assured of the existence of the solution to the perturbed problem for a sequence of values ϵ_k tending to zero such that $0 < \epsilon_k \leq \epsilon_0$ and of the existence of the solution to the unperturbed problem, then the results of this section indicate that the first solution converges to the second in the appropriate norm through this sequence of values as $\epsilon_k \to 0$. Even though the solutions of these kinds of problems may fail to exist for some or perhaps all values of ϵ in the interval $0 \leq \epsilon \leq \epsilon_0$ under consideration, if small variations in the data over this range of ϵ values are allowed, then this difficulty may in part be overcome.

3.2 Modeling Errors for the Navier-Stokes Equations Backward in Time

The last section was placed first in this chapter because historically the work of Payne & Sather was the first on modeling in improperly posed problems and inspired much of what followed. However, a beautiful and practical illustration of the modeling problem may be obtained by studying how a solution to the Navier-Stokes equations will converge to a solution to the Stokes equations as the nonlinearities are "neglected" in a certain sense. The Navier-Stokes equations are studied backward in time by reversing time and investigating the appropriate system forward in time. Thus, the boundary-initial value problem for the Navier-Stokes equations *backward in time* may be written:

$$\left.\begin{aligned} &\frac{\partial u_i}{\partial t} - u_j \frac{\partial u_i}{\partial x_j} = -\nu \Delta u_i + \frac{\partial p}{\partial x_i}, \\ &\frac{\partial u_i}{\partial x_i} = 0, \end{aligned}\right\} \quad \text{in} \quad \Omega \times (0,T), \tag{3.2.1$_1$}$$

$$\begin{aligned} u_i &= 0, \qquad \text{on} \quad \partial\Omega \times [0,T], \\ u_i(x,0) &= \epsilon f_i(x), \qquad x \in \Omega, \end{aligned} \tag{3.2.1$_2$}$$

where Ω is a bounded region in $\mathbf{R}^3$ with a smooth boundary $\partial\Omega$. We shall say problem (3.2.1) is composed of the partial differential equations $(3.2.1)_1$ solved subject to the boundary and initial conditions $(3.2.1)_2$, and continue this usage of notation when the numbering of a boundary-initial value problem is split. Problem (3.2.1) is a Dirichlet boundary - initial value problem for the Navier-Stokes equations defined on a bounded domain Ω with p the unknown pressure and u_i the components of velocity. One question that arises is if ϵ is very small, can the solution to (3.2.1) be approximated adequately by ϵv_i, where v_i is a solution to the Stokes boundary -initial value problem *backward in time*:

$$\left.\begin{aligned}\frac{\partial v_i}{\partial t}&=-\nu\Delta v_i+\frac{\partial q}{\partial x_i},\\ \frac{\partial v_i}{\partial x_i}&=0,\end{aligned}\right\}\quad \text{in}\quad \Omega\times(0,T), \tag{3.2.2$_1$}$$

$$\begin{aligned} v_i&=0, &&\text{on}\quad \partial\Omega\times[0,T],\\ v_i(x,0)&=f_i(x), && x\in\Omega.\end{aligned} \tag{3.2.2$_2$}$$

The writers in Payne & Straughan (1989b) address the modeling question for the Navier-Stokes equations *backward in time*. Their analysis provides an L^2 bound for the difference $u_i - \epsilon v_i$ over a finite time interval assuming that solutions of the two improperly posed problems are suitably constrained. Logarithmic convexity arguments are used by Payne & Straughan (1989b) to derive bounds for the difference

$$w_i = u_i - \epsilon v_i,$$

which satisfies the boundary-initial value problem:

$$\left.\begin{aligned}\frac{\partial w_i}{\partial t}-u_j\frac{\partial w_i}{\partial x_j}-\epsilon u_j\frac{\partial v_i}{\partial x_j}&=-\Delta w_i+\frac{\partial P}{\partial x_i},\\ \frac{\partial w_i}{\partial x_i}&=0,\end{aligned}\right\}\quad \text{in}\quad \Omega\times(0,T), \tag{3.2.3$_1$}$$

$$\begin{aligned} w_i&=0, &&\text{on}\quad \partial\Omega\times[0,T],\\ w_i(x,0)&=\epsilon\big[f_i(x)-\hat f_i(x)\big], && x\in\Omega.\end{aligned} \tag{3.2.3$_2$}$$

Here the equations have been scaled so that the viscosity has the value one and it has been assumed that

$$u_i(x,0)=\epsilon f_i(x), \qquad x\in\Omega,$$

while

$$v_i(x,0)=\hat f_i(x), \qquad x\in\Omega.$$

The data f_i and $\hat f_i$ are assumed to be close in $L^2(\Omega)$. We point out that it is unreasonable to assume $f_i\equiv\hat f_i$ because u_i and v_i might not exist for the same data.

Since the backward in time problem is improperly posed we shall need two sets of functions to define constraint sets and hence obtain a bound on $\|w\|^2$. The constraint sets are introduced at this point. A function $\phi_i(x,t)$ belongs to the class $\mathcal{M}_1$ if

$$\sup_{(x,t)\in\Omega\times(0,T)} \phi_i\phi_i \le M_1^2, \tag{3.2.4}$$

while $\psi_i(x,t)$ belongs to $\mathcal{M}_2$ if

$$\sup_{(x,t)\in\Omega\times(0,T)}\Big(\|(\psi_{i,j}-\psi_{j,i})(\psi_{i,j}-\psi_{j,i})\|_{3/2}+\|\psi_{i,t}\psi_{i,t}\|_{3/2}\Big)\le M_2^2, \quad (3.2.5)$$

where $\|\cdot\|_p$ denotes the norm on $L^p(\Omega)$. In fact, in their analysis Payne & Straughan (1989b) require $u_i\in\mathcal{M}_1$ and $v_i\in\mathcal{M}_2$.

Payne & Straughan (1989b) apply a logarithmic convexity argument to the function $\phi(t)$ defined by

$$\phi(t)=\|w\|^2+\epsilon^4 Q, \tag{3.2.6}$$

where Q is a positive constant selected later. By differentiation and use of (3.2.3) one sees

$$\begin{aligned}\phi'(t)=&2\int_\Omega w_i w_{i,t}dx\\ =&2\int_\Omega w_{i,j}w_{i,j}dx+2\epsilon\int_\Omega w_i u_j v_{i,j}dx.\end{aligned} \tag{3.2.7}$$

A second differentiation yields

$$\begin{aligned}\phi''(t)=&4\int_\Omega w_{i,j}w_{i,jt}dx+2\epsilon\int_\Omega w_{i,t}u_j v_{i,j}dx\\ &+2\epsilon\int_\Omega w_i u_{j,t}v_{i,j}dx-2\epsilon\int_\Omega w_{i,j}u_j v_{i,t}dx,\end{aligned} \tag{3.2.8}$$

where an integration by parts has been carried out on the last term. Further integration by parts and use of (3.2.3) allows us to deduce

$$\begin{aligned}\phi''(t)=&4\int_\Omega w_{i,t}\big[w_{i,t}-u_j w_{i,j}-\frac{1}{2}\epsilon u_j(v_{i,j}-v_{j,i})\big]dx\\ &-2\epsilon\int_\Omega w_{i,j}w_j v_{i,t}dx-2\epsilon^2\int_\Omega w_{i,j}v_i v_{j,t}dx\\ &-2\epsilon^2\int_\Omega w_{i,j}v_j v_{i,t}dx.\end{aligned} \tag{3.2.9}$$

Payne & Straughan (1989b) next define

$$\zeta_i=w_{i,t}-\frac{1}{2}u_j w_{i,j}-\frac{\epsilon}{4}u_j(v_{i,j}-v_{j,i}), \tag{3.2.10}$$

so that ϕ'' can be rewritten as

$$\begin{aligned}\phi''(t)=&4\int_\Omega \zeta_i\zeta_i dx\\ &-\int_\Omega\big[u_j w_{i,j}+\frac{1}{2}\epsilon u_j(v_{i,j}-v_{j,i})\big]\big[u_k w_{i,k}+\frac{1}{2}\epsilon u_k(v_{i,k}-v_{k,i})\big]dx\\ &-2\epsilon\int_\Omega w_{i,j}w_j v_{i,t}dx-2\epsilon^2\int_\Omega w_{i,j}v_i v_{j,t}dx\\ &-2\epsilon^2\int_\Omega w_{i,j}v_j v_{i,t}dx.\end{aligned} \tag{3.2.11}$$

An application of the arithmetic - geometric mean inequality on the second term on the right of (3.2.11) allows us to obtain for some positive c,

$$\begin{aligned}\phi''(t) \geq & 4\int_\Omega \zeta_i\zeta_i dx - (1+c)\int_\Omega u_j w_{i,j} u_k w_{i,k} dx\\ & - \frac{\epsilon^2}{4c}(1+c)\int_\Omega u_j(v_{i,j}-v_{j,i})u_k(v_{i,k}-v_{k,i})dx\\ & - 2\epsilon\int_\Omega w_{i,j}w_j v_{i,t}dx - 2\epsilon^2\int_\Omega w_{i,j}v_i v_{j,t}dx\\ & - 2\epsilon^2\int_\Omega w_{i,j}v_j v_{i,t}dx. \end{aligned} \tag{3.2.12}$$

The u_i terms in the third integral on the right in (3.2.12) are written as $u_i = w_i + \epsilon v_i$ and then with use of the arithmetic - geometric mean inequality this can be bounded as follows

$$\begin{aligned}&\int_\Omega u_j(v_{i,j}-v_{j,i})u_k(v_{i,k}-v_{k,i})dx\\ &\quad \leq (1+c_1)\int_\Omega w_j(v_{i,j}-v_{j,i})w_k(v_{i,k}-v_{k,i})dx\\ &\qquad + \frac{\epsilon^2}{c_1}(1+c_1)\int_\Omega v_j(v_{i,j}-v_{j,i})v_k(v_{i,k}-v_{k,i})dx, \end{aligned} \tag{3.2.13}$$

for some positive constant c_1. This inequality is used in (3.2.12) and then we employ the restrictions that $u_i \in \mathcal{M}_1$ and $v_i \in \mathcal{M}_2$, to find

$$\begin{aligned}\phi''(t) \geq & 4\int_\Omega \zeta_i\zeta_i dx\\ & - \Big[(1+c)M_1^2 + \frac{\epsilon^2}{4c}(1+c)(1+c_1)M_2^2\Lambda + 2\epsilon\Lambda^{1/2}M_2\Big]\int_\Omega w_{i,j}w_{i,j}dx\\ & - 2^{3/2}|\Omega|^{1/6}\Lambda^{1/2}\epsilon^2 M_2^2\Big(\int_\Omega w_{i,j}w_{i,j}dx\Big)^{1/2}\\ & - \frac{\epsilon^4}{8cc_1}(1+c)(1+c_1)M_2^4|\Omega|^{1/3}\Lambda, \end{aligned} \tag{3.2.14}$$

where the Sobolev inequality

$$\|\psi\|_6^2 \leq \Lambda\int_\Omega \psi_{i,j}\psi_{i,j}dx, \tag{3.2.15}$$

has also been used. In (3.2.14), $|\Omega|$ is the volume of the domain Ω, the constant Λ is independent of the size of Ω, and the inequality

$$\|v\|_6^2 \leq \frac{1}{2}|\Omega|^{1/3} M_2^2, \tag{3.2.16}$$

has been utilized. Inequality (3.2.16) is established by observing that

$$\int_\Omega v_{i,j} v_{i,j}\, dx = \int_\Omega (v_{i,j} - v_{j,i})(v_{i,j} - v_{j,i})\, dx\,,$$

and then by applying Hölder's inequality to (3.2.15) with v_i in place of ψ_i and the above replacement, i.e.

$$\|v\|_6^2 \leq \frac{1}{2}\Lambda \int_\Omega (v_{i,j} - v_{j,i})(v_{i,j} - v_{j,i}) dx.$$

To progress beyond (3.2.14) we now need to find a bound for $\int_\Omega w_{i,j} w_{i,j} dx$ and thus return to equation (3.2.7) from which we obtain

$$\begin{aligned}
\int_\Omega w_{i,j} w_{i,j} dx &= \frac{1}{2}\phi' + \epsilon \int_\Omega w_{i,j} u_j v_i dx \\
&= \frac{1}{2}\phi' + \epsilon \int_\Omega w_{i,j} w_j v_i dx \\
&\quad - \epsilon^2 \int_\Omega w_i v_j (v_{i,j} - v_{j,i}) dx,
\end{aligned} \tag{3.2.17}$$

where it has been noted that $u_i = w_i + \epsilon v_i$. The arithmetic - geometric mean inequality is applied to the middle term on the right of (3.2.17) as

$$\epsilon \int_\Omega w_{i,j} w_j v_i\, dx \leq \frac{1}{2}\epsilon^2 \int_\Omega v_i v_i w_k w_k\, dx + \frac{1}{2}\int_\Omega w_{i,j} w_{i,j}\, dx\,,$$

and then when this is used in (3.2.17) there results

$$\int_\Omega w_{i,j} w_{i,j} dx \leq \phi' + \epsilon^2 \int_\Omega w_j w_j v_i v_i dx - 2\epsilon^2 \int_\Omega w_k v_j (v_{k,j} - v_{j,k}) dx\,,$$

and then Hölder's and Young's inequalities are used together with the following Sobolev inequality,

$$\|\psi\|_4^2 \leq \frac{1}{\sqrt{\gamma}} \|\psi\|^{1/2} \left(\int_\Omega \psi_{i,j} \psi_{i,j} dx \right)^{3/4},$$

and one estimates the Dirichlet integral of w_i as

$$\int_\Omega w_{i,j} w_{i,j} dx \leq 2\phi' + \frac{\epsilon^4}{\beta} M_2^4 |\Omega|^{1/3} \Lambda + (2\beta + K)\|w\|^2, \tag{3.2.18}$$

where the constant K is given by

$$K = \frac{27}{256\gamma^2} |\Omega|^2 \Lambda^4 M_2^8 \epsilon^8,$$

and where β is a positive constant. Inequality (3.2.18) is now substituted into (3.2.14) and further use is made of the arithmetic - geometric mean inequality to obtain the lower bound for ϕ''

$$\phi'' \geq 4 \int_\Omega \zeta_i \zeta_i dx - \beta_1 \int_\Omega w_i w_i dx - \beta_2 \epsilon^4 - \beta_3 \phi', \tag{3.2.19}$$

for computable constants β_i that are independent of ϵ.

Before forming the logarithmic convexity expression $\phi\phi'' - (\phi')^2$ the quantity ϕ' is rewritten using (3.2.10) as

$$\begin{aligned}\phi' =& 2 \int_\Omega w_i w_{i,t} dx \\ =& 2 \int_\Omega w_i \big[\zeta_i + \frac{\epsilon^2}{4} v_j (v_{i,j} - v_{j,i})\big] dx.\end{aligned}$$

Thus the term $(\phi')^2$ may be written in the form,

$$\begin{aligned}(\phi')^2 =& 4\Big(\int_\Omega w_i \zeta_i dx\Big)^2 + \epsilon^2 \phi' \int_\Omega w_i v_j (v_{i,j} - v_{j,i}) dx \\ & - \frac{\epsilon^4}{4} \Big(\int_\Omega w_i v_j (v_{i,j} - v_{j,i}) dx\Big)^2.\end{aligned} \tag{3.2.20}$$

We may now form the logarithmic convexity expression $\phi\phi'' - (\phi')^2$ using (3.2.6), (3.2.19), and (3.2.20) to obtain

$$\begin{aligned}\phi\phi'' - (\phi')^2 \geq & 4\Big\{\Big(\int_\Omega \zeta_i \zeta_i dx\Big)\Big(\int_\Omega w_i w_i dx\Big) - \Big(\int_\Omega w_i \zeta_i dx\Big)^2\Big\} \\ & - \phi(\beta_1 \|w\|^2 + \beta_2 \epsilon^4) - \beta_3 \phi\phi' \\ & - |\phi'| \Big\{\mu \|w\|^2 + \frac{\epsilon^4}{2\mu} M_2^4 |\Omega|^{1/3} \Lambda\Big\} \\ & + \frac{\epsilon^4}{4} \Big(\int_\Omega w_i v_j (v_{i,j} - v_{j,i}) dx\Big)^2,\end{aligned} \tag{3.2.21}$$

for a positive constant μ at our disposal. The last term we need to bound is $|\phi'|$. Manipulating (3.2.7), we can find that

$$|\phi'| \leq \phi' + k_1 \Big(\|w\|^2 + \frac{\beta_2}{\beta_1} \epsilon^4\Big), \tag{3.2.22}$$

where k_1 is a constant independent of ϵ for $\epsilon < 1$. If we now discard the first and last term on the right hand side of (3.2.21), both of which are positive, choose μ appropriately, and set

$$Q = \frac{\beta_2}{\beta_1}, \tag{3.2.23}$$

then using (3.2.22) in (3.2.21) we arrive at the inequality

$$\phi\phi'' - (\phi')^2 \geq -\alpha_1\phi\phi' - \alpha_2\phi^2, \tag{3.2.24}$$

where α_1 and α_2 are computable constants independent of ϵ. Expression (3.2.24) is the fundamental logarithmic convexity inequality (1.5.5) and by writing it as in (1.5.8) and integrating in the manner leading to (1.5.10) it follows that

$$\phi(t) \leq e^{-\alpha_2 t/\alpha_1}\big[\phi(0)\big]^{(\sigma-\sigma_1)/(1-\sigma_1)}\big[\phi(T)e^{\alpha_2 T/\alpha_1}\big]^{(1-\sigma)/(1-\sigma_1)}, \tag{3.2.25}$$

where $\sigma = e^{-\alpha_1 t}$ and $\sigma_1 = e^{-\alpha_1 T}$. Recalling the restrictions on u_i and v_i, i.e. $u_i \in \mathcal{M}_1$ and $v_i \in \mathcal{M}_2$ one may derive from (3.2.25) the estimate

$$\|w\|^2 \leq \big[M^2\big]^{(1-\sigma)/(1-\sigma_1)}\big[\epsilon^2\|f-\hat{f}\|^2 + Q\epsilon^4\big]^{(\sigma-\sigma_1)/(1-\sigma_1)}, \tag{3.2.26}$$

where M is a constant depending on M_1 and M_2, but not ϵ. Payne & Straughan (1989b) observed that this inequality may be interpreted as implying that if $\|f-\hat{f}\|$ is $O(\epsilon)$, then the convergence of the function u_i to the function v_i in $L^2(\Omega)$ is of order $\epsilon^{4(\sigma-\sigma_1)/(1-\sigma_1)}$ on compact subintervals of $(0, T)$.

Payne & Straughan (1989b) compare this result to the analogous one that can be obtained by comparing u_i to the zero solution, using a logarithmic convexity argument similar to the previous one. They conclude that their result is considerably sharper than this other inequality, at least for small t. The result (3.2.26) gives a justification for using the solution of the linearized Stokes problem as an approximation to the solution to the Navier-Stokes equations with small initial data, a practice frequently employed in solving applied fluid dynamics problems where the convective nonlinearity usually necessitates a computational approach.

We complete our discussion of continuous dependence on modeling for the Navier-Stokes equations by briefly reviewing the results of Ames & Payne (1993). Using logarithmic convexity arguments, they derive an inequality that implies solutions of the backward Navier-Stokes equations on an interior domain depend continuously on the body force. From this inequality, they are able to infer a variety of continuous dependence on modeling results, by formally identifying the body force with other quantities.

Further Modeling Results for Solutions to the Navier - Stokes Equations

Ames & Payne (1993) consider the Navier - Stokes system backward in time, which is

$$\left.\begin{aligned} &\frac{\partial u_i}{\partial t} - u_j\frac{\partial u_i}{\partial x_j} = \frac{\partial p}{\partial x_i} - \nu\Delta u_i + \mathcal{F}_i, \\ &\frac{\partial u_i}{\partial x_i} = 0, \end{aligned}\right\} \quad \text{in} \quad \Omega\times(0,T), \tag{3.2.27}$$

where $\mathcal{F}_i(x,t)$ is, in general, the prescribed body force. Defining (u_i,p) and (v_i,q) as two classical solutions of (3.2.27) corresponding to two different body forces $\mathcal{F}_i^1$ and $\mathcal{F}_i^2$, they show that if u_i and v_i are suitably constrained, then there exist a computable constant C and a function $\delta(t)$, $0<\delta(t)\leq 1$, such that for $0\leq t<T$,

$$\int_0^t\int_\Omega (t-\eta)w_iw_i\,dxd\eta \leq CQ^{\delta(t)}, \tag{3.2.28}$$

where $w_i=u_i-v_i$, and if $\mathcal{F}_i=\mathcal{F}_i^1-\mathcal{F}_i^2$,

$$Q=\int_0^T\int_\Omega \mathcal{F}_i\mathcal{F}_i\,dxdt. \tag{3.2.29}$$

The restrictions which are needed to derive (3.2.28) are that $u_i\in\mathcal{M}_1$ and $v_i\in\mathcal{M}_2$ where $\mathcal{M}_1$ is the set of vector valued functions $\psi(x,t)$ in which

$$\sup_{(x,t)\in\Omega\times(0,T)} \psi_i\psi_i \leq M_1^2,$$

for a prescribed constant M_1, while $\mathcal{M}_2$ is the set in which

$$\sup_{(x,t)\in\Omega\times(0,T)} \big[\psi_i\psi_i+\psi_{i,j}\psi_{i,j}\big] \leq M_2^2,$$

for a constant M_2.

In addition to proving continuous dependence on the body force Ames & Payne (1993) also derive results of continuous dependence on the viscosity and of continuous dependence on the boundary data. What is of concern in this chapter is continuous dependence on the modeling aspect and this is also examined by Ames & Payne (1993) who essentially derive another Navier - Stokes convergence to Stokes equations type of result.

To see how the inequality (3.2.28) can be used to obtain continuous dependence on modeling results, let us again suppose we wish to find the solution of the problem,

$$\left.\begin{aligned} &\frac{\partial u_i}{\partial t} - u_j\frac{\partial u_i}{\partial x_j} + \nu\Delta u_i = \frac{\partial p}{\partial x_i}, \\ &\frac{\partial u_i}{\partial x_i} = 0, \end{aligned}\right\} \quad \text{in} \quad \Omega\times(0,T), \tag{3.2.30$_1$}$$

$$\begin{aligned} u_i &= \epsilon f_i, && \text{on} \quad \partial\Omega \times [0,T], \\ u_i(x,0) &= 0, && x \in \Omega. \end{aligned} \tag{3.2.30$_2$}$$

The modeling aspect analysed by Ames & Payne (1993) may be summarized in the following way. One may ask the question if ϵ is very small, can the solution of (3.2.30) be approximated by ϵv_i where v_i is the solution of the linear problem,

$$\left.\begin{aligned} \frac{\partial v_i}{\partial t} + \nu\Delta v_i &= \frac{\partial q}{\partial x_i}, \\ \frac{\partial v_i}{\partial x_i} &= 0, \end{aligned}\right\} \quad \text{in} \quad \Omega \times (0,T), \tag{3.2.31$_1$}$$

$$\begin{aligned} v_i &= f_i, && \text{on} \quad \partial\Omega \times [0,T], \\ v_i(x,0) &= 0, && x \in \Omega? \end{aligned} \tag{3.2.31$_2$}$$

The approach taken in Ames & Payne (1993) to answer this question is to set

$$w_i = u_i - \epsilon v_i,$$

and observe that w_i satisfies the problem,

$$\left.\begin{aligned} \frac{\partial w_i}{\partial t} - u_j\frac{\partial w_i}{\partial x_j} - \epsilon w_j\frac{\partial v_i}{\partial x_j} + \nu\Delta w_i &= \frac{\partial P}{\partial x_i} + \epsilon^2 v_j\frac{\partial v_i}{\partial x_j}, \\ \frac{\partial w_i}{\partial x_i} &= 0, \end{aligned}\right\} \tag{3.2.32$_1$}$$

these equations holding in $\Omega \times (0,T)$, subject to the boundary and initial conditions,

$$\begin{aligned} w_i &= 0, && \text{on} \quad \partial\Omega \times [0,T], \\ w_i(x,0) &= 0, && x \in \Omega, \end{aligned} \tag{3.2.32$_2$}$$

with $P = p - \epsilon q$. In connection with continuous dependence on the body force Ames & Payne (1993) now identify $\mathcal{F}_i$ with the last term on the right of $(3.2.32)_1$, i.e. they take

$$\mathcal{F}_i = \epsilon^2 v_j v_{i,j}. \tag{3.2.33}$$

They are then able to interpret the continuous dependence estimate (3.2.28) as

$$\begin{aligned} \int_0^t \int_\Omega (t-\eta)(u_i - \epsilon v_i)(u_i - \epsilon v_i)\,dx\,d\eta &\leq C\epsilon^{4\delta}\Big(\int_0^T \int_\Omega v_j v_{i,j} v_k v_{i,k}\,dx\,dt\Big)^\delta, \\ &\leq C\epsilon^{4\delta}\big(M_2^4 T|\Omega|\big)^\delta. \end{aligned} \tag{3.2.34}$$

Consequently, if ϵ is very small, then (3.2.34) indicates that u_i and ϵv_i will be close in a certain measure and hence, it is reasonable to approximate the solution of the nonlinear problem by that of the linear one.

Observe that if, instead of approximating u_i by ϵv_i, we had used the zero vector as our approximation, then the relevant inequality obtained from (3.2.28) has form

$$\int_0^t \int_\Omega (t-\eta) u_i u_i \, dx d\eta \leq C_1 \epsilon^{2\delta}. \tag{3.2.35}$$

The estimate (3.2.34) is a sharper result than (3.2.35).

3.3 Modeling Errors for First and Second Order Operator Equations

The papers of Ames (1982a,b) can be interpreted as continuous dependence on modeling investigations. Ames compared solutions of Cauchy problems for certain classes of first and second order operator equations with solutions of associated perturbed problems. Neither the original problem nor the perturbed problem was required to be well posed. Hölder stability inequalities relating solutions of the perturbed and unperturbed problems were obtained by Ames using logarithmic convexity arguments. The class of problems and type of derivable results can be exemplified by a comparison of the solution of

$$\begin{aligned} \frac{du}{dt} + Mu &= F(t,u), \qquad t \in [0,T), \\ u(0) &= f_1, \end{aligned} \tag{3.3.1}$$

with the solution of the perturbed problem

$$\begin{aligned} \frac{dw}{dt} + Mw + \epsilon M^2 w &= F(t,w), \qquad t \in [0,T), \\ w(0) &= f_2. \end{aligned} \tag{3.3.2}$$

Here ϵ is a small positive parameter, M is a symmetric, time independent linear operator (bounded or unbounded) defined on a Hilbert space, and the nonlinear term F satisfies a uniform Lipschitz condition in its second argument. Ames showed that if u and w belong to the appropriate constraint sets and if the Cauchy data are close (in a suitably defined sense), then the difference $u - w$ in a certain measure is of order $\epsilon^{\mu(t)}$ for $0 \leq t < T$ where $0 < \mu(t) \leq 1$.

Ames studied first and second order operator equations in two separate papers. Since the methods of analysis are similar in the two investigations, we shall focus here on Cauchy problems for the second order operator equations, following Ames (1982a). The goal in this work is to establish continuous dependence results without referring to solution representation and without requiring the original problem or the related perturbed problem to be well posed.

To be more specific, we shall be interested in comparing the solution of an original problem of the form

$$\begin{aligned} &P\frac{d^2u}{dt^2} + L\frac{du}{dt} + Mu = F(t,u,\frac{du}{dt}), \qquad t \in [0,T), \\ &u(0) = f_1, \qquad \frac{du}{dt}(0) = g_1, \end{aligned} \tag{3.3.3}$$

with the solution of the perturbed problem

$$\begin{aligned} &P\frac{d^2w}{dt^2} + L\frac{dw}{dt} + Mw + \epsilon Nw = F(t,w,\frac{dw}{dt}), \qquad t \in [0,T), \\ &w(0) = f_2, \qquad \frac{dw}{dt}(0) = g_2, \end{aligned} \tag{3.3.4}$$

where ϵ denotes a small positive parameter lying in an interval $0 < \epsilon \leq \epsilon_0$. We propose to find a stabilizing constraint set such that if u and w both belong to this set and if the Cauchy data are "close" in an appropriately defined sense, then u and w will remain "close" over a finite time interval.

The logarithmic convexity method will be used to compare the solutions of (3.3.3) and (3.3.4). We will show that if u and w belong to the appropriate spaces of functions, then their difference in a suitably chosen measure is of order ϵ to some positive power which is a function of t for $0 \leq t < T$.

Let us define H to be a real Hilbert space with inner product $(\cdot,\cdot)$ and norm $\|\cdot\| = (\cdot,\cdot)^{1/2}$ and $D \subset H$ to be a dense linear subspace of H. Let P, L and M denote linear operators (bounded or unbounded) which map D into H and consider the problem

$$\begin{aligned} &P\frac{d^2u}{dt^2} + L\frac{du}{dt} + Mu = F(t,u,\frac{du}{dt}), \qquad t \in [0,T), \\ &u(0) = f_1, \qquad \frac{du}{dt}(0) = g_1. \end{aligned} \tag{3.3.5}$$

Here f_1 and g_1 belong to H and $T > 0$. We shall compare the solution of (3.3.5) with the solution of the problem

$$\begin{aligned} &P\frac{d^2w}{dt^2} + L\frac{dw}{dt} + Mw + \epsilon Nw = F(t,w,\frac{dw}{dt}), \qquad t \in [0,T), \\ &w(0) = f_2, \qquad \frac{dw}{dt}(0) = g_2. \end{aligned} \tag{3.3.6}$$

The initial data f_2 and g_2 belong to H, N is a linear operator mapping D into H, and ϵ is a small positive parameter. We adopt the following hypotheses:

(i) The operators P, L, M and N as well as the space H are independent of t;

(ii) P is symmetric and $\exists$ a constant $\lambda > 0$ such that

$$\lambda^2(P\phi, \phi) \geq \|\phi\|^2, \qquad \forall \phi \in D;$$

(iii) L is a symmetric, positive semi-definite operator;

(iv) M and N are symmetric;

(v) the solutions to (3.3.5) and (3.3.6) belong to $C^2([0,T); D)$;

(vi) the nonlinear term $F(t, z, z_t)$ satisfies, for $z_1, z_2 \in C^2([0,T); D)$, an inequality of the form

$$\begin{aligned}\left\|F\Big(t, z_1, \frac{dz_1}{dt}\Big) - F\Big(t, z_2, \frac{dz_2}{dt}\Big)\right\| \leq & K_1\big|(Py, y)^{1/2}\big| + K_2\Big|\Big(P\frac{dy}{dt}, \frac{dy}{dt}\Big)^{1/2}\Big| \\ & + K_3\big|(Ly, y)^{1/2}\big|,\end{aligned}$$

where $y = z_1 - z_2$ and the K_i are nonnegative constants.

In order to compare the solutions u and w, we define $v \equiv w - u$ so that $v \in C^2([0,T); D)$ satisfies the following problem:

Problem A

$$P\frac{d^2v}{dt^2} + L\frac{dv}{dt} + Mv = -\epsilon Nw + F\Big(t, w, \frac{dw}{dt}\Big) - F\Big(t, u, \frac{du}{dt}\Big), \quad t \in [0,T),$$

$$v(0) = f, \qquad \frac{dv}{dt}(0) = g.$$

Here the Cauchy data $f = f_2 - f_1$ and $g = g_2 - g_1$ are assumed to be small in the sense that there exist nonnegative constants k_i $(i = 1, \dots, 5)$ such that

$$\begin{aligned}(Pf, f) &\leq k_1\epsilon^2, \quad (Lf, f) \leq k_2\epsilon^2, \\ (Pg, g) &\leq k_3\epsilon^2, \quad |(Mf, f)| \leq k_4\epsilon^2,\end{aligned}$$

and

$$|(Nf, f)| \leq k_5\epsilon.$$

We proceed to show that the solution of Problem A depends Hölder continuously on the parameter ϵ in an appropriate measure for $0 \leq t < T$.

More exactly, we establish the following theorem:

Theorem 3.3.1. (Ames (1992a).)
Let u be a solution of (3.3.5) such that

$$\int_0^T \|Nu\|^2 d\eta \leq R_1^2$$

for a prescribed constant R_1. If v is a solution of Problem A which lies in the class of functions

$$\mathcal{M} = \{\phi \in C^2([0,T);D) : \int_0^T \big[(P\phi,\phi) + (T-\eta)(L\phi,\phi)\big] d\eta \leq R_2^2\}$$

where R_2 is an *a priori* constant independent of ϵ, then there exist computable constants C and R_3 independent of ϵ such that on any compact subset of $[0,T)$, v satisfies the inequality

$$\int_0^t \big[(Pv,v) + (t-\eta)(Lv,v)\big] d\eta \leq C\epsilon^{2[1-\delta(t)]} R_3^{2\delta(t)}, \tag{3.3.7}$$

where $0 \leq \delta(t) < 1$.

Proof.
Logarithmic convexity arguments applied to the functional

$$\begin{aligned}\phi(t) = &\int_0^t \big[(Pv,v) + (t-\eta)(Lv,v)\big] d\eta + (T-t)(Pf,f) \\ &+ \frac{1}{2}(T^2-t^2)(Lf,f) + Q^2,\end{aligned} \tag{3.3.8}$$

lead to the desired result. We shall show that if the constant term Q^2 is properly chosen, then as a function of t, ϕ satisfies a second order differential inequality of the form

$$\phi\phi'' - (\phi')^2 \geq -c_1\phi\phi' - c_2\phi^2, \tag{3.3.9}$$

for computable, nonnegative constants c_1 and c_2. As we shall see, Q^2 depends upon the initial data f and g as well as on an *a priori* bound on the norm of Nu.

Differentiating (3.3.8) we have

$$\begin{aligned}\frac{d\phi}{dt} =&(Pv,v) - (Pf,f) + \int_0^t (Lv,v)d\eta - t(Lf,f), \\ =&2\int_0^t (Pv_\eta,v)d\eta + 2\int_0^t (t-\eta)(Lv_\eta,v)d\eta,\end{aligned} \tag{3.3.10}$$

and

$$\begin{aligned}\frac{d^2\phi}{dt^2} =&2\int_0^t (Pv_{\eta\eta}, v)d\eta + 2\int_0^t (Pv_\eta, v_\eta)d\eta + 2\int_0^t (Lv_\eta, v)d\eta + 2(Pg, f),\\ =&2\int_0^t (Pv_\eta, v_\eta)d\eta - 2\int_0^t (Mv, v)d\eta - 2\epsilon\int_0^t (Nw, v)d\eta\\ &+ 2\int_0^t \Big(\big[F(\eta, w, w_\eta) - F(\eta, u, u_\eta)\big], v\Big)d\eta + 2(Pg, f). \end{aligned} \tag{3.3.11}$$

Expression (3.3.11) is obtained after substitution of the differential equation in the term $2\int_0^t (Pv_{\eta\eta}, v)d\eta$. Consider now the identity

$$\begin{aligned} 0 =& \int_0^t \Big(v_\eta, Pv_{\eta\eta} + Lv_\eta + Mv + \epsilon Nw - \big[F(\eta, w, w_\eta) - F(\eta, u, u_\eta)\big]\Big)d\eta,\\ =& \frac{1}{2}\big[(Pv_t, v_t) + (Mv, v) + \epsilon(Nv, v)\big] - \frac{1}{2}\big[(Pg, g) + (Mf, f) + \epsilon(Nf, f)\big]\\ &+ \int_0^t (Lv_\eta, v_\eta)d\eta + \epsilon\int_0^t (Nu, v_\eta)d\eta\\ &- \int_0^t \Big(\big[F(\eta, w, w_\eta) - F(\eta, u, u_\eta)\big], v_\eta\Big)d\eta. \end{aligned}$$

Thus,

$$\begin{aligned} -(Mv, v) =& (Pv_t, v_t) + \epsilon(Nv, v) + 2\int_0^t (Lv_\eta, v_\eta)d\eta + 2\epsilon\int_0^t (Nu, v_\eta)d\eta\\ &- 2\int_0^t \Big(\big[F(\eta, w, w_\eta) - F(\eta, u, u_\eta)\big], v_\eta\Big)d\eta - G_1, \end{aligned} \tag{3.3.12}$$

where

$$G_1 = (Pg, g) + (Mf, f) + \epsilon(Nf, f).$$

Using this result in (3.3.11), we see that

$$\begin{aligned} \frac{d^2\phi}{dt^2} =& 4\int_0^t \big[(Pv_\eta, v_\eta) + (t-\eta)(Lv_\eta, v_\eta)\big]d\eta\\ &+ 4\epsilon\int_0^t (t-\eta)(Nu, v_\eta)d\eta\\ &- 2\epsilon\int_0^t (Nu, v)d\eta\\ &+ 2\int_0^t \Big(\big[F(\eta, w, w_\eta) - F(\eta, u, u_\eta)\big], v\Big)d\eta\\ &- 4\int_0^t (t-\eta)\Big(\big[F(\eta, w, w_\eta) - F(\eta, u, u_\eta)\big], v_\eta\Big)d\eta + G_2, \end{aligned}$$

where

$$G_2 = 2(Pg, f) - 2tG_1.$$

Application of the Cauchy - Schwarz inequality to all but the first and last terms in the previous expression yields the inequality

$$\begin{aligned}
\frac{d^2\phi}{dt^2} \geq & 4\int_0^t \big[(Pv_\eta, v_\eta) + (t-\eta)(Lv_\eta, v_\eta)\big]\,d\eta \\
& - 4\epsilon T\Big(\int_0^t \|Nu\|^2 d\eta\Big)^{1/2}\Big(\int_0^t \|v_\eta\|^2 d\eta\Big)^{1/2} \\
& - 2\epsilon\Big(\int_0^t \|Nu\|^2 d\eta\Big)^{1/2}\Big(\int_0^t \|v\|^2 d\eta\Big)^{1/2} \\
& - 2\int_0^t \|F(\eta, w, w_\eta) - F(\eta, u, u_\eta)\|\,\|v\| d\eta \\
& - 4\int_0^t (t-\eta)\|F(\eta, w, w_\eta) - F(\eta, u, u_\eta)\|\,\|v_\eta\| d\eta + G_2. \quad (3.3.13)
\end{aligned}$$

Consider now the integral

$$J_1 = -4\int_0^t (t-\eta)\|F(\eta, w, w_\eta) - F(\eta, u, u_\eta)\|\,\|v_\eta\| d\eta.$$

To bound this integral, we make use of the arithmetic - geometric mean inequality and the hypotheses on F and the operator P. The following inequalities can be obtained:

$$\begin{aligned}
J_1 \geq & -4\lambda\int_0^t (t-\eta)\big|(Pv_\eta, v_\eta)^{1/2}\big|\Big[K_1\big|(Pv, v)^{1/2}\big| + K_2\big|(Pv_\eta, v_\eta)^{1/2}\big| \\
& + K_3\big|(Lv, v)^{1/2}\big|\Big]d\eta, \\
\geq & -\frac{2\lambda K_1}{a_1}\int_0^t (t-\eta)(Pv_\eta, v_\eta)d\eta - 2\lambda K_1 a_1 T\int_0^t (Pv, v)d\eta \\
& - 4\lambda K_2\int_0^t (t-\eta)(Pv_\eta, v_\eta)d\eta \\
& - \frac{2\lambda K_3}{a_2}\int_0^t (t-\eta)(Pv_\eta, v_\eta)d\eta - 2\lambda K_3 a_2\int_0^t (t-\eta)(Lv, v)d\eta, \\
\geq & - b_1\int_0^t (t-\eta)(Pv_\eta, v_\eta)d\eta - b_2\int_0^t (Pv, v)d\eta \\
& - b_3\int_0^t (t-\eta)(Lv, v)d\eta. \quad (3.3.14)
\end{aligned}$$

Here a_1 and a_2 are positive constants introduced by the application of the arithmetic - geometric mean inequality and the b_i $(i = 1, 2, 3)$ are positive, computable constants.

The integral

$$J_2 = -2\int_0^t \|F(\eta, w, w_\eta) - F(\eta, u, u_\eta)\| \, \|v\| d\eta$$

can be bounded in a similar manner. In fact, we have

$$\begin{aligned} J_2 \geq &- 2\lambda \int_0^t |(Pv, v)^{1/2}| \Big[K_1|(Pv, v)^{1/2}| + K_2|(Pv_\eta, v_\eta)^{1/2}| \\ &+ K_3|(Lv, v)^{1/2}|\Big] d\eta, \\ \geq &- 2\lambda K_1 \int_0^t (Pv, v) d\eta - 2\lambda K_2 \Big(\int_0^t (Pv, v) d\eta\Big)^{1/2} \Big(\int_0^t (Pv_\eta, v_\eta) d\eta\Big)^{1/2} \\ &- \lambda K_3 a_3 \int_0^t (Pv, v) d\eta - \frac{\lambda K_3}{a_3} \int_0^t (Lv, v) d\eta, \end{aligned} \tag{3.3.15}$$

where a_3 is a positive constant. In view of inequalities (3.3.14) and (3.3.15) expression (3.3.13) may be rewritten as

$$\begin{aligned} \frac{d^2\phi}{dt^2} \geq &\, 4 \int_0^t [(Pv_\eta, v_\eta) + (t-\eta)(Lv_\eta, v_\eta)] d\eta \\ &- 4\epsilon\lambda T \Big(\int_0^t \|Nu\|^2 d\eta\Big)^{1/2} \Big(\int_0^t (Pv_\eta, v_\eta) d\eta\Big)^{1/2} \\ &- 2\epsilon\lambda \Big(\int_0^t \|Nu\|^2 d\eta\Big)^{1/2} \Big(\int_0^t (Pv, v) d\eta\Big)^{1/2} - d_1 \int_0^t (Pv, v) d\eta \\ &- d_2 \Big(\int_0^t (Pv, v) d\eta\Big)^{1/2} \Big(\int_0^t (Pv_\eta, v_\eta) d\eta\Big)^{1/2} - d_3 \int_0^t (Lv, v) d\eta \\ &- b_1 \int_0^t (t-\eta)(Pv_\eta, v_\eta) d\eta - b_3 \int_0^t (t-\eta)(Lv, v) d\eta + G_2, \end{aligned} \tag{3.3.16}$$

for computable positive constants d_i $(i = 1, 2, 3)$.

We shall now proceed by showing that the term

$$J_3 = \int_0^t (t-\eta)(Pv_\eta, v_\eta) d\eta$$

in the previous inequality is bounded from above by an expression of the form $\alpha_1\phi' + \alpha_2\phi + \alpha_3 I$ where I involves initial data and the norm of Nu.

Equation (3.3.10) may be rewritten as

$$\begin{aligned}\phi' =& 2\int_0^t \int_0^\eta \frac{d}{d\sigma}(Pv_\sigma, v)d\sigma\, d\eta + 2\int_0^t (Pv_\sigma, v)\big|_{\sigma=0} d\eta \\ &+ 2\int_0^t (t-\eta)(Lv_\eta, v)d\eta, \\ =& 2\int_0^t (t-\eta)(Pv_\eta, v_\eta)d\eta + 2\int_0^t (t-\eta)(Pv_{\eta\eta}, v)d\eta \\ &+ 2\int_0^t (t-\eta)(Lv_\eta, v)d\eta + G_3,\end{aligned}$$

where

$$G_3 = 2t(Pg, f).$$

Then, after substitution of the differential equation, it follows that

$$\begin{aligned}2\int_0^t (t-\eta)(Pv_\eta, v_\eta)d\eta =& \phi' + 2\int_0^t (t-\eta)(Mv, v)d\eta \\ &- 2\int_0^t (t-\eta)\Big(\big[F(\eta, w, w_\eta) - F(\eta, u, u_\eta)\big], v\Big)d\eta \\ &+ 2\epsilon\int_0^t (t-\eta)(Nw, v)d\eta - G_3.\end{aligned}$$

The result of (3.3.12) leads to the equality

$$\begin{aligned}4\int_0^t (t-\eta)(Pv_\eta, v_\eta)d\eta =& \phi' - 2\int_0^t (t-\eta)^2(Lv_\eta, v_\eta)d\eta \\ &- 2\epsilon\int_0^t (t-\eta)^2(Nu, v_\eta)d\eta + 2\epsilon\int_0^t (t-\eta)(Nu, v)d\eta \\ &+ 2\int_0^t (t-\eta)^2\Big(\big[F(\eta, w, w_\eta) - F(\eta, u, u_\eta)\big], v_\eta\Big)d\eta \\ &- 2\int_0^t (t-\eta)\Big(\big[F(\eta, w, w_\eta) - F(\eta, u, u_\eta)\big], v\Big)d\eta + G_4.\end{aligned}$$

The term G_4 is given by

$$G_4 = G_1 t^2 - G_3.$$

Application of the Cauchy - Schwarz inequality and the arithmetic - geometric mean inequality as well as reference to the hypothesis on the operator

P yield

$$
\begin{aligned}
4\int_0^t &(t-\eta)(Pv_\eta, v_\eta)d\eta \\
&\le \phi' + a_3\lambda^2 \int_0^t (t-\eta)^2 (Pv_\eta, v_\eta)d\eta \\
&+ \frac{\epsilon^2}{a_3}\int_0^t (t-\eta)^2\|Nu\|^2 d\eta + \frac{\epsilon^2}{a_4}\int_0^t (t-\eta)\|Nu\|^2 d\eta \\
&+ a_4\lambda^2 \int_0^t (t-\eta)(Pv, v)d\eta \\
&+ 2\lambda \int_0^t (t-\eta)^2 \|F(\eta, w, w_\eta) - F(\eta, u, u_\eta)\| \left|(Pv_\eta, v_\eta)^{1/2}\right| d\eta \\
&+ 2\lambda \int_0^t (t-\eta) \|F(\eta, w, w_\eta) - F(\eta, u, u_\eta)\| \left|(Pv, v)^{1/2}\right| d\eta \\
&+ \bar{G}_4,
\end{aligned} \tag{3.3.17}
$$

where

$$\bar{G}_4 = G_1 T^2 - 2T(Pg, f)$$

and a_3, a_4 are positive constants. The last two integrals in (3.3.17) can be shown to satisfy the following inequalities:

$$
\begin{aligned}
2\lambda &\int_0^t (t-\eta)^2 \|F(\eta, w, w_\eta) - F(\eta, u, u_\eta)\| \left|(Pv_\eta, v_\eta)^{1/2}\right| d\eta \\
&\le \frac{\lambda K_1 T^2}{a_5}\int_0^t (Pv, v)d\eta + \lambda(K_1 a_5 + 2K_2 + K_3 a_6) \\
&\times \int_0^t (t-\eta)^2 (Pv_\eta, v_\eta)d\eta + \frac{\lambda K_3 T^2}{a_6}\int_0^t (Lv, v)d\eta,
\end{aligned} \tag{3.3.18}
$$

and

$$
\begin{aligned}
2\lambda &\int_0^t (t-\eta) \|F(\eta, w, w_\eta) - F(\eta, u, u_\eta)\| \left|(Pv, v)^{1/2}\right| d\eta \\
&\le \lambda\left(2K_1 T + \frac{K_2}{a_7} + \frac{K_3 T}{a_8}\right)\int_0^t (Pv, v)d\eta \\
&+ \lambda K_2 a_7 \int_0^t (t-\eta)^2 (Pv_\eta, v_\eta)d\eta + \lambda K_3 T a_8 \int_0^t (Lv, v)d\eta.
\end{aligned} \tag{3.3.19}
$$

Expressions (3.3.18) and (3.3.19) permit us to rewrite inequality (3.3.17) as

$$4\int_0^t (t-\eta)(Pv_\eta, v_\eta)d\eta \le \phi' + \gamma_0\epsilon^2 \int_0^t \|Nu\|^2 d\eta$$

$$+\gamma_1\int_0^t (Pv,v)d\eta + \gamma_2\int_0^t (Lv,v)d\eta$$
$$+\gamma_3\int_0^t (t-\eta)^2(Pv_\eta, v_\eta)d\eta + \bar{G}_4. \quad (3.3.20)$$

The γ_i $(i=0,1,2,3)$ are positive constants which depend on the quantities $\lambda, K_1, K_2, K_3, T$ and the constants a_j $(j=3,\dots,8)$ arising from application of the arithmetic - geometric mean inequality. If we let

$$H(t) = \int_0^t (t-\eta)^2(Pv_\eta, v_\eta)d\eta,$$

then

$$\frac{dH}{dt}(t) = 2\int_0^t (t-\eta)(Pv_\eta, v_\eta)d\eta,$$

and (3.3.20) becomes

$$2H'(t) \leq \phi' + \gamma_0\epsilon^2\int_0^t \|Nu\|^2 d\eta + \gamma_1\int_0^t (Pv,v)d\eta$$
$$+\gamma_2\int_0^t (Lv,v)d\eta + \bar{G}_4 + \gamma_3 H(t). \quad (3.3.21)$$

The above differential inequality can be written as

$$\frac{d}{dt}\left[2H(t)e^{-(\gamma_3 t/2)}\right] \leq e^{-(\gamma_3 t/2)}\bigg[\phi' + \gamma_0\epsilon^2\int_0^t \|Nu\|^2 d\eta + \gamma_1\int_0^t (Pv,v)d\eta$$
$$+\gamma_2\int_0^t (Lv,v)d\eta + \bar{G}_4\bigg]. \quad (3.3.22)$$

We now assume that u belongs to that class of functions satisfying

$$\int_0^T \|Nu\|^2 d\eta \leq R_1^2,$$

for a prescribed constant R_1. With this requirement, integration of (3.3.22) from 0 to t leads to an inequality of the form

$$H(t) \leq \frac{1}{2}e^{\gamma_3(t-\xi)/2}\phi(t) + \frac{1}{\gamma_3}(e^{\gamma_3 t/2}-1)(\gamma_0\epsilon^2 R_1^2 + \bar{G}_4)$$
$$+\frac{\gamma_1}{\gamma_3}\int_0^t (e^{\gamma_3(t-\eta)/2}-1)(Pv,v)d\eta$$
$$+\frac{\gamma_2}{\gamma_3}\int_0^t (e^{\gamma_3(t-\eta)/2}-1)(Lv,v)d\eta,$$

where $0 \le \xi \le t < T$. Thus, there exist positive constants δ_k $(k = 0, 1, 2, 3)$ such that

$$H(t) \le \delta_0 \phi + \delta_1 \big[\gamma_0 \epsilon^2 R_1^2 + \bar{G}_4\big] + \delta_2 \int_0^t (Pv, v) d\eta + \delta_3 \int_0^t (Lv, v) d\eta. \quad (3.3.23)$$

Recalling that

$$\int_0^t (Pv, v) d\eta \le \phi,$$

and that

$$\int_0^t (Lv, v) d\eta \le \phi' + (Pf, f) + t(Lf, f),$$

we obtain from (3.3.23) the result

$$H(t) \le \bar{\delta}_0 \phi + \delta_1 \big[\gamma_0 \epsilon^2 R_1^2 + \bar{G}_4\big] + \delta_3 \big[\phi' + (Pf, f) + T(Lf, f)\big].$$

Hence, it follows from (3.3.21) that

$$\begin{aligned} 2H'(t) &= 4 \int_0^t (t - \eta)(Pv_\eta, v_\eta) d\eta \\ &\le \alpha_1 \phi' + \alpha_2 \phi + \alpha_3 \big[\gamma_0 \epsilon^2 R_1^2 + \bar{G}_4 + (Pf, f) + T(Lf, f)\big], \end{aligned}$$

where α_1, α_2 and α_3 are positive constants. This inequality and the constraint on the solution u allow us to rewrite inequality (3.3.16) as

$$\begin{aligned} \frac{d^2\phi}{dt^2} \ge{}& 4 \int_0^t \big[(Pv_\eta, v_\eta) + (t - \eta)(Lv_\eta, v_\eta)\big] d\eta \\ & - 4\epsilon\lambda R_1 T \left(\int_0^t (Pv_\eta, v_\eta) d\eta \right)^{1/2} \\ & - 2\epsilon\lambda R_1 \left(\int_0^t (Pv, v) d\eta \right)^{1/2} \\ & - d_2 \left(\int_0^t (Pv, v) d\eta \right)^{1/2} \left(\int_0^t (Pv_\eta, v_\eta) d\eta \right)^{1/2} \\ & - d_4 \phi - d_5 \phi' - d_6 G, \end{aligned} \quad (3.3.24)$$

where

$$G = b_0 \epsilon^2 R_1^2 + b_1 (Pf, f) + b_2 (Lf, f) + b_3 (Pg, g) + b_4 \big|(Mf, f)\big| + b_5 \epsilon \big|(Nf, f)\big|.$$

Here the b_i $(i = 0, \dots, 5)$ are nonnegative constants.

Using expressions (3.3.8), (3.3.10) and (3.3.24) we form

$$\begin{aligned}\phi\phi'' - (\phi')^2 \geq & \, 4S^2 + 4Q_1^2 \int_0^t \big[(Pv_\eta, v_\eta) + (t-\eta)(Lv_\eta, v_\eta)\big]\, d\eta \\ & - 4\epsilon\lambda R_1 T\phi\Big(\int_0^t (Pv_\eta, v_\eta)d\eta\Big)^{1/2} \\ & - 2\epsilon\lambda R_1 \phi\Big(\int_0^t (Pv, v)d\eta\Big)^{1/2} \\ & - d_2\phi\Big(\int_0^t (Pv_\eta, v_\eta)d\eta\Big)^{1/2}\Big(\int_0^t (Pv, v)d\eta\Big)^{1/2} \\ & - d_4\phi^2 - d_5\phi\phi' - d_6\phi G,\end{aligned}$$

with

$$Q_1^2 = (T-t)(Pf, f) + \frac{1}{2}(T^2 - t^2)(Lf, f) + Q^2,$$

and

$$\begin{aligned}S^2 = & \Big(\int_0^t \big[(Pv_\eta, v_\eta) + (t-\eta)(Lv_\eta, v_\eta)\big]\, d\eta\Big) \\ & \times \Big(\int_0^t \big[(Pv, v) + (t-\eta)(Lv, v)\big]\, d\eta\Big) \\ & - \Big(\int_0^t \big[(Pv_\eta, v) + (t-\eta)(Lv_\eta, v)\big]\, d\eta\Big)^2.\end{aligned}$$

We note that both S^2 and S are nonnegative as a result of the Cauchy - Schwarz inequality. The term

$$D \equiv -d_2\phi\Big(\int_0^t (Pv_\eta, v_\eta)d\eta\Big)^{1/2}\Big(\int_0^t (Pv, v)d\eta\Big)^{1/2}$$

can now be bounded in the following way:

$$\begin{aligned}D \geq & - d_2\phi\big[S^2 + (\phi')^2\big]^{1/2} \\ \geq & - d_2\phi\big[S + |\phi'|\big] \\ \geq & - d_2\phi\big[S + \phi' + 2(Pf, f) + 2T(Lf, f)\big].\end{aligned}$$

If we complete the square on

$$4Q_1^2 \int_0^t (Pv_\eta, v_\eta)d\eta - 4\epsilon\lambda R_1 T\phi\Big(\int_0^t (Pv_\eta, v_\eta)d\eta\Big)^{1/2},$$

discard its nonnegative part as well as the nonnegative term

$$4Q_1^2 \int_0^t (t-\eta)(Lv_\eta, v_\eta)d\eta$$

and apply the arithmetic - geometric mean inequality to appropriate terms, we obtain the inequality

$$\begin{aligned}\phi\phi'' - (\phi')^2 \geq & 4S^2 - d_2\phi S - \frac{\epsilon^2 T^2 \lambda^2 R_1^2}{Q_1^2}\phi^2 \\ & - \frac{\lambda\phi}{\sqrt{\beta_1}}\left(\int_0^t (Pv, v)d\eta + \beta_1\epsilon^2 R_1^2\right) - d_4\phi^2 - (d_2 + d_5)\phi\phi' \\ & - \phi\big[d_6 G + 2d_2(Pf, f) + 2d_2 T(Lf, f)\big].\end{aligned}$$

The choice

$$\begin{aligned}Q^2 = & \beta_1\epsilon^2 R_1^2 + \beta_2(Pf, f) + \beta_3(Lf, f) \\ & + \beta_4(Pg, g) + \beta_5|(Mf, f)| + \beta_6\epsilon|(Nf, f)|,\end{aligned} \tag{3.3.25}$$

for some positive computable constants β_i $(i = 1, \ldots, 6)$ permits us to write

$$\phi\phi'' - (\phi')^2 \geq 4S^2 - d_2\phi S - c_0\phi^2 - c_1\phi\phi',$$

which upon a completion of squares leads to the fundamental inequality (3.3.9). As a consequence of (3.3.9) and the fact that $\phi(t) > 0$ for all $t \in [0, T)$, the functional $\phi(t)$ defined by (3.3.8) and (3.3.25) satisfies

$$\phi(t) \leq e^{-(c_2 t/c_1)}\big[\phi(0)\big]^{1-\delta(t)}\big[\phi(T)e^{(c_2 T/c_1)}\big]^{\delta(t)}, \tag{3.3.26}$$

where

$$\delta(t) = \frac{1 - e^{-c_1 t}}{1 - e^{-c_1 T}}.$$

In view of the bounds on the initial data, we see that

$$\phi(0) = T(Pf, f) + \frac{1}{2}T^2(Lf, f) + Q^2 = O(\epsilon^2).$$

If we assume that

$$v \in \mathcal{M} = \left\{\phi \in C^2([0, T); D) : \int_0^T \big[(P\phi, \phi) + (T-\eta)(L\phi, \phi)\big]d\eta \leq R_2^2\right\},$$

where R_2 is an *a priori* constant independent of ϵ, it follows that there exists a constant R_3 such that

$$\phi(T)e^{(c_2 T/c_1)} \leq R_3^2.$$

The assertion of the theorem then follows from (3.3.26).

A Special Case of the General Theory

Let us now consider the particular case of problems (3.3.5) and (3.3.6) in which $P = I$ (the identity operator), $L = 0$, $N = M^2$ and $F = 0$. We also assume that the operator M is negative semi - definite. We can then establish the theorem with a less restrictive constraint class for u, namely

$$\mathcal{N} = \Big\{\phi \in C^2([0,T);D) : \int_0^{t_1} \|\phi\|^2 d\eta \le R^2\Big\},$$

for some $t_1 > T$ and a prescribed constant R.

If M is negative semi-definite, the introduction of a suitable cutoff function permits us to bound the quantity

$$\int_0^t \|M^2 u\|^2 d\eta$$

in terms of the initial data and the integral

$$\int_0^{t_1} \|u\|^2 d\eta$$

for some $t_1 > T$. To see this, choose the function $\gamma(t) \in C^2(t \ge 0)$ defined as follows:

$$\begin{aligned}
&\gamma(t) = 1, \qquad 0 \le t \le t_0 \le T;\\
&0 \le \gamma(t) \le 1, \qquad t_0 \le t \le t_1;\\
&\gamma(t) = 0, \qquad t \ge t_1.
\end{aligned}$$

It follows that

$$\begin{aligned}
\int_0^t \|M^2 u\|^2 d\eta &\le \int_0^\infty \gamma(\eta)\|M^2 u\|^2 d\eta,\\
&= (M^2 u, Mu_t)\big|_{t=0} - \frac{1}{2}\int_0^\infty \gamma''(\eta)(M^2 u, Mu)d\eta\\
&\quad + \int_0^\infty \gamma(\eta)(M^2 u_\eta, Mu_\eta)d\eta. \qquad (3.3.27)
\end{aligned}$$

The equality in (3.3.27) is obtained by substituting the differential equation $u_{tt} + Mu = 0$ (now assumed to hold for $0 \le t < t_1$) and integrating by parts twice. The definiteness condition on M allows us to discard $\int_0^\infty \gamma(\eta)(M^2 u_\eta, Mu_\eta)d\eta$ from the bounding inequality since it is nonpositive. Then, setting

$$\hat{Q} = (M^2 f_1, M g_1)$$

and using the Cauchy - Schwarz inequality and the arithmetic - geometric mean inequality in (3.3.27) we find that

$$\begin{aligned}\int_0^\infty \gamma(\eta)\|M^2u\|^2 d\eta \leq & \hat{Q} + \frac{\alpha}{4}\int_0^\infty \gamma(\eta)\|M^2u\|^2 d\eta \\ & + \frac{1}{4\alpha}\int_0^{t_1} \frac{(\gamma'')^2}{\gamma}\|Mu\|^2 d\eta,\end{aligned} \tag{3.3.28}$$

for a positive constant α. Inequality (3.3.28) may be rewritten as

$$\left(1-\frac{\alpha}{4}\right)\int_0^\infty \gamma(\eta)\|M^2u\|^2 d\eta \leq \hat{Q} + \frac{1}{4\alpha}\int_0^{t_1} \frac{(\gamma'')^2}{\gamma}(M^2u, u) d\eta,$$

and then a second application of the Cauchy - Schwarz inequality and the arithmetic - geometric mean inequality yields

$$\left(1-\frac{\alpha}{4}-\frac{\beta}{8\alpha}\right)\int_0^\infty \gamma(\eta)\|M^2u\|^2 d\eta \leq \hat{Q} + \frac{1}{8\alpha\beta}\int_0^{t_1} \frac{(\gamma'')^4}{\gamma^3}\|u\|^2 d\eta, \tag{3.3.29}$$

where β is a positive constant. We observe that $\gamma(t)$ must be sufficiently continuous to ensure that the integral on the right hand side of (3.3.29) exists at $t = t_1$. If γ is assumed to behave like $(t - t_1)^p$ as t approaches t_1, then we need to choose p so that $3p \leq 4(p - 2)$ or $p \geq 8$. With such a choice, we see that $(\gamma'')^4/\gamma^3$ is bounded at $t = t_1$. Upon choosing α and β so that

$$1 - \frac{\alpha}{4} - \frac{\beta}{8\alpha} \geq k > 0,$$

(one possible choice is $\alpha = 1$, $\beta = 2$) and requiring $\hat{Q}$ to be bounded, it follows from (3.3.29) and (3.3.27) that if

$$\int_0^{t_1} \|u\|^2 d\eta \leq R^2,$$

for a prescribed constant R, then

$$\int_0^t \|M^2u\|^2 d\eta \leq R_1^2,$$

for a constant R_1 independent of ϵ.

A Physical Example of the Abstract Theory

To conclude this section we illustrate the abstract results with a specific example. We let $M = \Delta$, the Laplace operator and define Ω to be a bounded region in $\mathbf{R}^n$ (n typically is equal to 2) with a boundary $\partial\Omega$ smooth enough to ensure the existence of various integrals which arise in our computations. In addition, we assume that u, w and Δw vanish on $\partial\Omega$. The following two problems are then obtained:

Problem I

$$\begin{aligned}
&\frac{\partial^2 u}{\partial t^2} + \Delta u = 0, \qquad \text{in} \quad \Omega \times [0, T), \\
&u = 0, \qquad \text{on} \quad \partial\Omega \times [0, T), \\
&u(\mathbf{x}, 0) = f_1(\mathbf{x}), \qquad \frac{\partial u}{\partial t}(\mathbf{x}, 0) = g_1(\mathbf{x}), \qquad \mathbf{x} \in \Omega;
\end{aligned} \tag{3.3.30}$$

Problem II

$$\begin{aligned}
&\frac{\partial^2 w}{\partial t^2} + \Delta w + \epsilon \Delta^2 w = 0, \qquad \text{in} \quad \Omega \times [0, T), \\
&w = 0, \qquad \Delta w = 0, \qquad \text{on} \quad \partial\Omega \times [0, T), \\
&w(\mathbf{x}, 0) = f_2(\mathbf{x}), \qquad \frac{\partial w}{\partial t}(\mathbf{x}, 0) = g_2(\mathbf{x}), \qquad \mathbf{x} \in \Omega.
\end{aligned} \tag{3.3.31}$$

We observe that Problem II is well posed, while Problem I, the Cauchy problem for the Laplace equation, is improperly posed. In this case, u belongs to that class of functions which are continuous in $\bar{\Omega} \times [0, T]$, twice continuously differentiable in $[0, T)$ and $C^4(\Omega)$. Since Problem II is properly posed, we take w to be its classical solution. The f_i and g_i $(i = 1, 2)$ are assumed to be sufficiently regular prescribed functions.

The comparison of Problems I and II may be viewed in two different contexts. We might be interested in approximating the solution of the improperly posed problem by that of a well posed problem. In this case, (3.3.31) may be regarded as a comparison problem for (3.3.30) that is obtained via the quasireversibility method. Alternatively, our concern might be with the behaviour of the solution w as the parameter $\epsilon \to 0$. Thus, we might ask the question of whether in some sense the solution of Problem II converges to that of the simpler Problem I as $\epsilon \to 0$.

An important motivation for examining problems such as (3.3.30) and (3.3.31) is derived from the fact that they appear in mathematical models of physical processes. One such example occurs in linear plate theory. More specifically, let us consider the problem of the bending of a vibrating rectangular elastic plate under the action of a uniform compressive force in the middle plane of the plate. We denote by $w(x, y, t)$ the deflection of the plate away from its initial position in the region

$$\Omega = \{(x, y) : x \in (0, a), y \in (0, b)\}.$$

If we adopt the classical linear theory of vibrating elastic plates, then w will satisfy the equation

$$\rho_0 \frac{\partial^2 w}{\partial t^2} + k\Delta w + \hat{D}\Delta^2 w = 0, \tag{3.3.32}$$

where ρ_0, k and $\hat{D}$ are positive constants which denote the density of the plate (assumed to be uniform), the magnitude of the in planar forces, and the flexural rigidity of the plate, respectively. If we set

$$t^* = \sqrt{\frac{k}{\rho_0}}\, t\,,$$

equation (3.3.32) may be rewritten as

$$\frac{\partial^2 w}{\partial t^2} + \Delta w + \epsilon\Delta^2 w = 0, \tag{3.3.33}$$

where $\epsilon = \hat{D}/k$ and the *'s have been omitted. In order to complete the problem, we assume that the plate is simply supported and prescribe its initial displacement and initial velocity. Thus, we obtain a mathematical description of the form (3.3.31). Suppose we are interested in this problem for $\epsilon << 1$ (e.g. the case of a very thin plate). One question that arises is how the solution w compares to the solution of the problem obtained by setting $\epsilon = 0$. If we allow for small variations in the initial data, we are led to consideration of the ill-posed problem (3.3.30). We remark that in the present context, this problem describes the vibration of a membrane in compression.

If we assume that the solution of (3.3.30) exists, we can determine appropriate conditions on u and w so that $v = w - u$ depends continuously on the parameter ϵ. Assuming that the initial data are small in the sense indicated in our abstract theorem, we have the result

$$\int_0^t \int_\Omega (w-u)^2 d\sigma d\eta = O(\epsilon^{\delta(t)}), \tag{3.3.34}$$

where $\delta(t) = 2(1 - t/T)$ for $0 \leq t < T$. Thus, if ϵ is sufficiently small, (3.3.34) indicates that u will be arbitrarily close to w in the given measure on $[0, T)$.

3.4 Continuous Dependence on Modeling in the Cauchy Problem for Second Order Partial Differential Equations: Some Examples

Bennett (1986) considered three classes of ill-posed problems, obtaining continuous dependence on modeling results for a number of examples. These examples suggest that modeling errors may be the most critical type of error introduced in the formulation of a mathematical problem to analyze a physical process. In order to stabilize the problems he considered against modeling errors, Bennett found it necessary to impose much stronger constraints on solutions than are required to stabilize these problems against most other sources of error.

One class of problems Bennett studied consists of Cauchy problems for elliptic equations of the form

$$\frac{\partial}{\partial x_j}\left[a_{ij}(x,u,\nabla u)\frac{\partial u}{\partial x_i}\right] = F(x,u,\nabla u), \qquad \text{in} \quad \Omega \subset \mathbf{R}^n. \tag{3.4.1}$$

Bennett compared the solution of the Cauchy problem for this equation with small data to the solution of the related Cauchy problem for the equation

$$\frac{\partial}{\partial x_j}\left[a_{ij}(x,0,0)\frac{\partial v}{\partial x_i}\right] = 0, \tag{3.4.2}$$

assuming that the a_{ij} satisfy certain definiteness assumptions and that

$$F(x,u,p) = O(|u| + |p|), \tag{3.4.3}$$

for small $|u|$ and $|p|$.

A representative of this class is the Cauchy problem for the minimal surface equation. Bennett considered the problem

$$\frac{\partial}{\partial x_i}\left[(1+|\nabla v|^2)^{-1/2}\frac{\partial v}{\partial x_i}\right] = 0, \qquad \text{in} \quad \Omega \subset \mathbf{R}^n, \tag{3.4.4}$$

with

$$\int_\Gamma (v^2 + v_{,i}v_{,i})dS \leq \epsilon^2, \tag{3.4.5}$$

where Γ is a smooth portion of the boundary of Ω. The substitution $v = \epsilon u$ in (3.4.4) yields the equation

$$\Delta u = \epsilon^2(1+\epsilon^2|\nabla u|^2)^{-1}\frac{\partial u}{\partial x_i}\frac{\partial u}{\partial x_j}\frac{\partial^2 u}{\partial x_i \partial x_j} \qquad \text{in} \quad \Omega. \tag{3.4.6}$$

If ϵ is set equal to zero in this equation, the result is Laplace's equation

$$\Delta h = 0 \quad \text{in} \quad \Omega.$$

If $w = u - h$, then w will satisfy the equation

$$\Delta w = \epsilon^2(1 + \epsilon^2|\nabla u|^2)^{-1} \frac{\partial u}{\partial x_i} \frac{\partial u}{\partial x_j} \frac{\partial^2 u}{\partial x_i \partial x_j} \quad \text{in} \quad \Omega. \tag{3.4.7}$$

Bennett (1986) shows that if

$$\int_\Gamma (u_{,i}u_{,i})^2 \big|(u_{,k}u_{,jk} - u_{,j}\Delta u)n_j\big| dS = O(\epsilon^{-2}), \tag{3.4.8}$$

and if

$$\int_\Gamma (w^2 + w_{,i}w_{,i}) dS = O(\epsilon^2), \tag{3.4.9}$$

and if u and h are suitably constrained, then in a region Ω_β adjoining Γ, $v - \epsilon h$ satisfies a continuous dependence inequality of the form

$$\int_{\Omega_\beta} (v - \epsilon h)^2 dx \le K \epsilon^{4\nu(\beta)}, \qquad 0 < \beta < \beta_1 < 1, \tag{3.4.10}$$

where ν is a smooth function such that $\nu(0) = 1$, $\nu(\beta_1) = 0$, and $\nu'(\beta) < 0$. In order to obtain (3.4.10), Bennett imposed the restrictions that

$$\int_{\Omega_1} |\nabla u|^6 \big[1 + \epsilon^2|\nabla u|^2\big] dx = O(\epsilon^{-2}), \tag{3.4.11}$$

and

$$\int_{\Omega_1} u^2 dx = O(\epsilon^{-2}), \qquad \int_{\Omega_1} h^2 dx = O(\epsilon^{-2}). \tag{3.4.12}$$

The regions Ω_β are defined as follows: For each β in (0,1], the surface $h(x) = \beta$ intersects Ω and forms a closed region Ω_β whose boundary consists only of points of Γ and points of $h = \beta$. In addition, a unique surface passes through each point of Ω_1 and the surfaces $h = \beta$ satisfy the conditions

$$\begin{aligned}
&(i) \quad \alpha \le \mu \ \text{ implies } \ \Omega_\alpha \subset \Omega_\mu, \\
&(ii) \quad |\nabla h| \ge \gamma > 0 \ \text{ in } \ \Omega_1, \\
&(iii) \ \Delta h \le 0, \ \ |\Delta h| \le \delta\gamma^2 \ \text{ in } \ \Omega_1.
\end{aligned}$$

Payne (1970) gives examples of suitable functions h. For $n \ge 2$, a radial harmonic function can be chosen as $h(x)$.

We point out that if the powers of ϵ in (3.4.9) and (3.4.11) are reduced, a continuous dependence inequality similar to (3.4.10) can be obtained but with a lower power of ϵ. Even though the constraints (3.4.11) and (3.4.12) are sufficient conditions for continuous dependence, it is unlikely that continuous dependence could be established under significantly weaker restrictions such as the usual L^2 constraint that stabilizes other ill - posed problems against various sources of error.

Bennett's proof of (3.4.10) utilizes a logarithmic convexity argument along similar lines to that constructed by Payne (1970) in his treatment of the Cauchy problem for a class of elliptic equations; an exposition of the latter work is given in section 4.3. Bennett's (1986) argument consists of showing that for $\beta \in [0,1]$, the functional

$$F(\beta) = \int_0^\beta (\beta - \eta)\Big(\int_{\Omega_\eta} (w_{,i}w_{,i} + w\Delta w)dx\Big) d\eta + Q, \tag{3.4.13}$$

where the data term Q is defined by

$$Q = k_0 \int_\Gamma w^2 dS + k_1 \int_\Gamma w_{,i}w_{,i} dS + k_2\epsilon^2, \tag{3.4.14}$$

satisfies the fundamental inequality (1.5.5), namely

$$FF'' - (F')^2 \geq -c_1 FF' - c_2 F^2, \tag{3.4.15}$$

on the interval $(0, \beta_1)$ for explicit constants c_1 and c_2.

The path leading to (3.4.15) is now briefly sketched. The first and second derivatives of F are

$$F'(\beta) = \int_0^\beta \int_{\Omega_\eta} (w_{,i}w_{,i} + w\Delta w)dx\, d\eta, \tag{3.4.16}$$

and

$$F''(\beta) = \int_{\Omega_\beta} (w_{,i}w_{,i} + w\Delta w)dx. \tag{3.4.17}$$

Expression (3.4.16) is integrated by parts to find that

$$\begin{aligned} F'(\beta) &= \int_0^\beta \Big[\int_{\Gamma_\eta} ww_{,i}n_i dS + \int_{S_\eta} ww_{,i}h_{,i}|\nabla h|^{-1} dS\Big] d\eta, \\ &= \int_{\Omega_\beta} ww_{,i}h_{,i} dx + \int_0^\beta \int_{\Gamma_\eta} ww_{,i}n_i dS\, d\eta, \end{aligned} \tag{3.4.18}$$

where Γ_β and S_β have been defined by

$$\Gamma_\beta = \Gamma \cap \Omega_\beta, \qquad S_\beta = \{h(x) = \beta\} \cap \bar{\Omega}_\beta,$$

and where the observation has been employed that the component n_i of the unit normal is $h_{,i}|\nabla h|^{-1}$. Equation (3.4.18) may then be used to rewrite $F(\beta)$ as

$$\begin{aligned} F(\beta) =& Q + \int_0^\beta F'(\eta) d\eta \\ =& Q + \int_0^\beta \Big[\int_{\Omega_\eta} w w_{,i} h_{,i} dx + \int_0^\eta \int_{\Gamma_\xi} w w_{,i} n_i dS \, d\xi \Big] d\eta. \end{aligned}$$

An integration by parts and use of condition (iii) and the arithmetic - geometric mean inequality lead to

$$\begin{aligned} F(\beta) =& Q + \int_0^\beta \Big[\frac{1}{2} \int_{\Gamma_\eta} w^2 h_{,i} n_i dS + \frac{1}{2} \int_{S_\eta} w^2 |\nabla h| dS \\ & - \frac{1}{2} \int_{\Omega_\eta} w^2 \Delta h \, dx + \int_0^\eta \int_{\Gamma_\xi} w w_{,i} n_i dS \, d\xi \Big] d\eta, \\ \geq & \frac{1}{2} \int_{\Omega_\beta} |\nabla h|^2 w^2 dx - \mu_1 \int_\Gamma w^2 dS - \mu_2 \int_\Gamma w_{,i} w_{,i} dS + Q, \quad (3.4.19) \end{aligned}$$

for computable constants μ_1 and μ_2. The constants in (3.4.14) are now chosen so that

$$\begin{aligned} \int_{\Omega_\beta} |\nabla h|^2 w^2 dx + Q \leq & 2F(\beta) \\ \leq & (\delta + 1) \Big[\int_{\Omega_\beta} |\nabla h|^2 w^2 dx + Q \Big]. \qquad (3.4.20) \end{aligned}$$

Forming $FF'' - (F')^2$ and discarding a number of nonnegative terms from the right hand side it is found that

$$\begin{aligned} FF'' - (F')^2 \geq & \Big[\frac{1}{2} \int_{\Omega_\beta} |\nabla h|^2 w^2 dx \Big] \Big[\int_{\Omega_\beta} w_{,i} w_{,i} dx \Big] - \Big[\int_{\Omega_\beta} w w_{,i} h_{,i} dx \Big]^2 \\ & + F \int_{\Omega_\beta} w \Delta w dx - 2|F'| \Big| \int_0^\beta \int_{\Gamma_\eta} w w_{,i} n_i dS \, d\eta \Big|. \quad (3.4.21) \end{aligned}$$

In order to obtain inequality (3.4.15) from the inequality above, Bennett (1986) showed that the following inequalities hold for $\beta \in (0, \beta_1)$ provided (3.4.11) is imposed:

$$\Big| \int_{\Omega_\beta} w \Delta w dx \Big| \leq K_1 F, \qquad (3.4.22)$$

$$|F'| \leq F' + K_2 F, \qquad (3.4.23)$$

and

$$\int_{\Omega_\beta} \left[w_{,i}w_{,i} - 2|\nabla h|^{-2}(w_{,i}h_{,i})^2 \right] dx \geq -K_3F' - K_4F. \qquad (3.4.24)$$

Here the K_i $(i = 1, \ldots, 4)$ are all explicitly computable constants. To bound the last term in (3.4.21) Bennett also used the arithmetic - geometric mean inequality to demonstrate

$$\left| \int_0^\beta \int_{\Gamma_\eta} ww_{,i}n_i dS \, d\eta \right| \leq K_5F. \qquad (3.4.25)$$

Furthermore, use of the Cauchy - Schwarz inequality and the inequalities (3.4.24), (3.4.20), and (3.4.23) allows one to show that

$$\left[\int_{\Omega_\beta} |\nabla h|^2 w^2 dx \right] \left[\int_{\Omega_\beta} w_{,i}w_{,i} dx \right] - 2\left[\int_{\Omega_\beta} ww_{,i}h_{,i} dx \right]^2$$
$$\geq -2K_4FF' - 2(K_2K_4 + K_6)F^2. \qquad (3.4.26)$$

Use of (3.4.22), (3.4.23), (3.4.25) and (3.4.26) in (3.4.21) finally leads Bennett (1986) to the desired result, (3.4.15), for explicit constants c_1 and c_2. This inequality is integrated with Payne's transformation as in (1.5.10) to yield

$$F(\beta)\sigma^{-c_2/c_1^2} \leq \left[F(\beta_1)\sigma_1^{-c_2/c_1^2} \right]^{(1-\sigma)/(1-\sigma_1)} \left[F(0) \right]^{(\sigma-\sigma_1)/(1-\sigma_1)}, \qquad (3.4.27)$$

where

$$\sigma = e^{-c_1\beta}, \quad \sigma_1 = e^{-c_1\beta_1}.$$

Observing that $F(0) = Q$ which is $O(\epsilon^2)$ by assumption, one may now conclude that by imposition of the constraint (3.4.12), F satisfies the bound

$$F(\beta) \leq K\epsilon^{2[(\sigma-1)+(\sigma-\sigma_1)]/(1-\sigma_1)}. \qquad (3.4.28)$$

The constraint (3.4.12) is needed in order to ensure that $F(\beta_1)$ is small. Restricting the L^2 norms of u and h on Ω_1 allows the computation of a constant M independent of ϵ such that

$$F(\beta_1)\sigma_1^{-c_2/c_1^2} \leq M^2\epsilon^{-2}.$$

Thus, the result (3.4.10) follows from (3.4.20) and (3.4.28).

Bennett (1986) also treats the capillary surface equation, establishing continuous dependence on modeling results in a similar manner to that which he established for the minimal surface equation. He writes the capillary surface equation as

$$\frac{\partial}{\partial x_i}\Big[(1+|\nabla v|^2)^{-1/2}\frac{\partial v}{\partial x_i}\Big] = cv, \qquad \text{in} \quad \Omega, \tag{3.4.29}$$

where c is a positive constant, and he assumes that on the surface Γ the Cauchy data satisfy

$$\int_\Gamma (v^2 + |\nabla v|^2)dS \le \epsilon^2, \tag{3.4.30}$$

and then utilizes the substitution $v = \epsilon u$ to yield the perturbed equation

$$\Delta u - cu = \epsilon^2 \rho^2 \frac{\partial u}{\partial x_i}\frac{\partial u}{\partial x_j}\frac{\partial^2 u}{\partial x_i \partial x_j} + cu\Big(\frac{1}{\rho} - 1\Big), \tag{3.4.31}$$

where

$$\rho = \frac{1}{\sqrt{1+\epsilon^2|\nabla u|^2}}.$$

Observing that

$$\frac{1}{\rho} - 1 = O(\epsilon^2)|\nabla u|^2,$$

Bennett (1986) then compares the solution of (3.4.31) to the function h which satisfies

$$\Delta h = ch, \qquad \text{in} \quad \Omega. \tag{3.4.32}$$

He establishes that $v - \epsilon h = \epsilon(u - h) = \epsilon w$ satisfies the inequality

$$\int_{\Omega_\beta} (v - \epsilon h)^2 dx \le K\epsilon^{\nu(\beta)}, \qquad 0 < \beta < \beta_1, \tag{3.4.33}$$

with

$$\nu(\beta) = \frac{(6-p)(\sigma - \sigma_1)}{(1-\sigma_1)}, \quad \sigma = e^{-c_1\beta}, \quad \text{and} \quad \sigma_1 = e^{-c_1\beta_1},$$

provided u and h are solutions of (3.4.31) and (3.4.32) that are constrained in the following way:

$$(i) \qquad \int_\Gamma (w^2 + w_{,i}w_{,i})dS = O(\epsilon^{4-p}),$$

$$(ii) \qquad \int_\Gamma (u_{,i}u_{,i})^2\big|u_{,j}u_{,jk}n_k - u_{,j}\Delta u\, n_j\big|dS = O(\epsilon^{-p}),$$

(*iii*)

$$\int_{\Omega_1} \Big[|\nabla u|^6(1+\epsilon^2|\nabla u|^2)+u^2|\nabla u|^4(1+\epsilon^2|\nabla u|^2)^2+|u|\rho^{-1}|\nabla u|^2\Big]dx = O(\epsilon^{-p}),$$

and

$$(iv) \qquad \int_{\Omega_1} u^2 dx = O(\epsilon^{-2}), \qquad \int_{\Omega_1} h^2 dx = O(\epsilon^{-2}),$$

for $p < 6$.

The proof of (3.4.33) is similar to that of (3.4.10); the major difference is in obtaining the estimate for

$$\left| \int_{\Omega_\beta} w \Delta w \, dx \right|$$

which depends on finding a bound for the L^2 norm of Δw on Ω_β.

Bennett (1986) also studies continuous dependence on modeling for a class of second order nonlinear parabolic partial differential equations. A typical representative example is an end problem for a one dimensional nonlinear heat equation. In this example, the idea is to relate the solution of

$$\begin{aligned} \frac{\partial v}{\partial t} &= \frac{\partial}{\partial x}\left[\rho\left(\left|\frac{\partial v}{\partial x}\right|^2\right)\frac{\partial v}{\partial x}\right], \qquad \text{in } \Omega, \\ v, \frac{\partial v}{\partial x} &\quad \text{prescribed} \quad \text{on } \Gamma, \end{aligned} \tag{3.4.34}$$

to a solution of the problem

$$\begin{aligned} \frac{\partial h}{\partial t} &= \frac{\partial^2 h}{\partial x^2}, \qquad \text{in } \Omega, \\ h, \frac{\partial h}{\partial x} &\quad \text{prescribed} \quad \text{on } \Gamma. \end{aligned} \tag{3.4.35}$$

Here Ω is an open region in the first quadrant of the xt-plane whose boundary contains a segment Γ of the t axis given by $0 \le t \le T$. It is assumed that $\rho(q)$ with $q = v_x^2$ satisfies, for $q \ge 0$,

$$\frac{\rho + 2q\rho'(q) - 1}{\rho + 2q\rho'(q)} = O(q). \tag{3.4.36}$$

The Cauchy data on Γ are assumed to satisfy

$$\int_\Gamma \left(v^2 + \left(\frac{\partial v}{\partial x}\right)^2 + \left(\frac{\partial v}{\partial t}\right)^2\right) dS \le \epsilon^2,$$

for some small positive number ϵ. Using a weighted energy argument Bennett (1986) derives a continuous dependence inequality for classical solutions of (3.4.34) and (3.4.35) that are appropriately constrained in Ω_{β_1} where

$$\Omega_\beta = \{(x,t) : f(x,t) < \beta\} \cap \Omega.$$

It is assumed that

$$\Omega_\alpha \subseteq \Omega_\beta, \qquad \text{for} \qquad \alpha \leq \beta; \quad \alpha, \beta \in [0, \beta_1],$$

and

$$\frac{\partial f}{\partial x} \geq C_0 > 0 \qquad \text{in} \quad \Omega_{\beta_1}$$

as well as

$$\frac{\partial^2 f}{\partial x^2} \leq -C_1 < 0 \qquad \text{in} \quad \Omega_{\beta_1}.$$

In order to obtain this result, Bennett sets $v = \epsilon u$ and $w = u - h$. He then introduces the function $z = we^{-\lambda f}$ for some large positive constant λ, and defines the operator M by

$$M\phi \equiv \frac{\partial^2 \phi}{\partial x^2} - \frac{\partial \phi}{\partial t} + 2\lambda \frac{\partial f}{\partial x}\frac{\partial \phi}{\partial x} + \left[\lambda^2\left(\frac{\partial f}{\partial x}\right)^2 + \lambda\left(\frac{\partial^2 f}{\partial x^2} - \frac{\partial f}{\partial t}\right)\right]\phi.$$

The function z satisfies the equation

$$Mz = e^{-\lambda f} g, \tag{3.4.37}$$

where

$$g = \left[1 - \rho - 2q\rho'(q)\right]\frac{\partial^2 u}{\partial x^2}\,.$$

Integration and clever manipulation of the identity

$$\int_{\Omega_\beta} \left(2\lambda \frac{\partial f}{\partial x}\frac{\partial z}{\partial x} - \frac{\partial z}{\partial t}\right)(Mz - e^{-\lambda f} g)\,dx\,dt = 0 \tag{3.4.38}$$

leads to the desired inequality. More precisely, Bennett (1986) shows that if

$$\int_\Gamma \left[(u-h)^2 + \left(\frac{\partial u}{\partial x} - \frac{\partial h}{\partial x}\right)^2 + \left(\frac{\partial u}{\partial t} - \frac{\partial h}{\partial t}\right)^2\right] dS = O(\epsilon^{4-d}), \tag{3.4.39}$$

for $d < 6$, and if the constraints

$$\int_{\Omega_{\beta_1}} \left(\frac{\partial u}{\partial t}\right)^2 \left(\frac{\partial u}{\partial x}\right)^4 dx\,dt = O(\epsilon^{-d}), \tag{3.4.40}$$

and

$$\int_{\Omega_{\beta_1}} \left[u^2 + h^2 + \left(\frac{\partial u}{\partial x}\right)^2 + \left(\frac{\partial h}{\partial x}\right)^2 \right] dx\, dt = O(\epsilon^{-2}), \tag{3.4.41}$$

are imposed, then for $0 < \beta < \beta_1$,

$$\int_{\Omega_\beta} (v - \epsilon h)^2 dx\, dt = O\Big(\epsilon^{(6-d)(1-\beta/\beta_1)}\Big). \tag{3.4.42}$$

It is worth pointing out that Bennett's proof does not guarantee a continuous dependence result if $d = 6$.

Estimates similar to (3.4.42) can be obtained for an equation of the form

$$\frac{\partial v}{\partial t} = \frac{\partial}{\partial x}\Big[\rho(v, q)\frac{\partial v}{\partial x}\Big], \qquad \text{in } \Omega, \tag{3.4.43}$$

where

$$\frac{\partial \rho}{\partial v} = O(1); \qquad \frac{\rho + 2q(\partial\rho/\partial q) - 1}{\rho + 2q(\partial\rho/\partial q)} = O(q). \tag{3.4.44}$$

The proof of these estimates uses a weighted energy argument not unlike the argument employed for the equation (3.4.34). The details of these arguments can be found in Bennett (1986).

Bennett (1986) is also able to adapt a weighted energy method to handle the Cauchy problem for the second order nonlinear hyperbolic equation

$$\frac{\partial^2 v}{\partial t^2} = \frac{\partial}{\partial x_i}\Big(\rho\frac{\partial v}{\partial x_i}\Big), \qquad \text{in } \Omega \times (-T, T), \tag{3.4.45}$$

where

$$\rho = \frac{1}{\sqrt{1 + |\nabla v|^2}},$$

and Ω is the interior of an $n-$dimensional ball centered at the origin in $\mathbf{R}^n$. The solution of this equation is compared to a solution h of the linear wave equation

$$\frac{\partial^2 h}{\partial t^2} = \Delta h. \tag{3.4.46}$$

Under the appropriate constraints, it can be established that in suitable neighbourhoods of $\partial\Omega \times (-T, T)$, Ω_β, the estimate

$$\int_{\Omega_\beta} (v - \epsilon h)^2 dx\, dt = O\big(\epsilon^{(6-d)/3}\big), \tag{3.4.47}$$

for $d < 6$ holds.

3.5 Modeling Errors in Micropolar Fluid Dynamics and in Magnetohydrodynamics

Payne & Straughan (1989a) study modeling questions in initial - boundary value and final - boundary value problems for the micropolar fluid theory of Eringen (1966). The micropolar fluid equations provide additional structure to the Navier - Stokes equations by including a description of the microstructure in the fluid. Micropolar fluids have received immense attention in the literature where the novel effects of the microstructure are often pronounced. For example, Payne & Straughan (1989c) have shown they may account for a plausible explanation of the drastic reduction observed in the critical temperature gradient necessary for convection in a gas with added smoke particles. The work of Payne & Straughan (1989a) addresses the question of how the velocity field of the micropolar equations converges to the velocity field of the Navier-Stokes equations as an interaction parameter tends to zero.

The continuity, momentum, and moment of momentum equations for an incompressible, isotropic micropolar fluid can be written in the form

$$\begin{aligned}\frac{\partial v_i}{\partial x_i} &= 0,\\ \frac{\partial v_i}{\partial t} + v_j \frac{\partial v_i}{\partial x_j} &= -\frac{\partial p}{\partial x_i} + (\nu+\lambda)\Delta v_i + \lambda\epsilon_{ijk}\frac{\partial n_k}{\partial x_j},\\ j\Big(\frac{\partial n_i}{\partial t} + v_j\frac{\partial n_i}{\partial x_j}\Big) &= -2\lambda n_i + \lambda\epsilon_{ijk}\frac{\partial v_k}{\partial x_j}\\ &\quad + (\alpha+\beta)\frac{\partial^2 n_k}{\partial x_k \partial x_i} + \gamma\Delta n_i,\end{aligned} \tag{3.5.1}$$

where v_i denote the components of the fluid velocity and n_i is peculiar to the micropolar fluid theory. The vector n_i accounts for microstructure effects in that it is a particle spin vector. These equations are assumed to hold on $\Omega \times (0,T)$ where $\Omega \subset \mathbf{R}^3$ is a bounded domain and $T > 0$ (T may be infinite for the initial - boundary value problem). The constants j, λ, ν, and γ are nonnegative while α, β, and γ additionally satisfy

$$3\alpha + \beta + \gamma \geq 0, \qquad \gamma + \beta \geq 0, \qquad \gamma - \beta \geq 0.$$

The vectors $\mathbf{v}$ and $\mathbf{n}$ are assumed to satisfy the initial and boundary conditions

$$\begin{aligned} v_i = n_i &= 0, && \text{on} \quad \Gamma \times [0,T],\\ v_i(\mathbf{x},0) &= f_i(\mathbf{x}), && \mathbf{x}\in\Omega,\\ n_i(\mathbf{x},0) &= g_i(\mathbf{x}), && \mathbf{x}\in\Omega,\end{aligned} \tag{3.5.2}$$

where Γ is the boundary of Ω. Observing that when the constants $\lambda, \alpha, \beta, \gamma$, and j are all zero equations (3.5.1) reduce to the Navier-Stokes equations, Payne & Straughan (1989a) investigate the behavior of v_i as $\lambda \to 0$. In particular they study the relationship between v_i and the velocity field u_i of a Navier - Stokes fluid that is governed by the initial - boundary value problem

$$\left.\begin{aligned} &\frac{\partial u_i}{\partial x_i} = 0, \\ &\frac{\partial u_i}{\partial t} + u_j \frac{\partial u_i}{\partial x_j} = -\frac{\partial q}{\partial x_i} + \nu \Delta u_i, \end{aligned}\right\} \tag{3.5.3$_1$}$$

$$\begin{aligned} &u_i = 0, \qquad \text{on} \quad \Gamma \times [0,T], \\ &u_i(\mathbf{x},0) = f_i(\mathbf{x}), \qquad \mathbf{x} \in \Omega. \end{aligned} \tag{3.5.3$_2$}$$

Convergence in the Backward in Time Problem

To study the backward in time problem for (3.5.1) and (3.5.3), Payne & Straughan (1989a) replace t by $-t$ and deal with the resulting system of equations forward in time. Hence, let again Ω be a bounded, three-dimensional domain with boundary Γ. The idea is to compare the solutions v_i and n_i of the system

$$\begin{aligned} \frac{\partial v_i}{\partial x_i} =& 0, \\ \frac{\partial v_i}{\partial t} =& v_j \frac{\partial v_i}{\partial x_j} - (\nu + \lambda)\Delta v_i - \lambda \epsilon_{ijk} \frac{\partial n_k}{\partial x_j} + \frac{\partial p}{\partial x_i}, \\ j\frac{\partial n_i}{\partial t} =& j v_j \frac{\partial n_i}{\partial x_j} + 2\lambda n_i - \lambda \epsilon_{ijk} \frac{\partial v_k}{\partial x_j} \\ & - (\alpha + \beta) \frac{\partial^2 n_k}{\partial x_k \partial x_i} - \gamma \Delta n_i, \end{aligned} \tag{3.5.4}$$

in $\Omega \times (0,T)$, subject to the boundary and initial conditions

$$\begin{aligned} &v_i = 0, \qquad n_i = 0, \qquad \text{on} \quad \Gamma \times [0,T], \\ &v_i(\mathbf{x},0) = f_i(\mathbf{x}), \qquad \mathbf{x} \in \Omega, \\ &n_i(\mathbf{x},0) = h_i(\mathbf{x}), \qquad \mathbf{x} \in \Omega, \end{aligned} \tag{3.5.5}$$

with the solution u_i to the Navier-Stokes equations

$$\left.\begin{aligned} &\frac{\partial u_i}{\partial x_i} = 0, \\ &\frac{\partial u_i}{\partial t} = u_j \frac{\partial u_i}{\partial x_j} - \nu \Delta u_i + \frac{\partial q}{\partial x_i}, \end{aligned}\right\} \tag{3.5.6}$$

in $\Omega \times (0,T)$, which satisfies the boundary and initial conditions

$$\begin{aligned} u_i &= 0, \qquad \text{on} \quad \Gamma \times [0,T], \\ u_i(\mathbf{x},0) &= f_i(\mathbf{x}), \qquad \mathbf{x} \in \Omega. \end{aligned} \tag{3.5.7}$$

Both of these initial-boundary value problems are ill posed; hence, in the course of the analysis it is necessary to require that solutions v_i, n_i, and u_i belong to constraint sets which we now define. A function $\psi(x_k,t)$ belongs to the set $\mathcal{M}_1$ if

$$\sup_{\Omega\times[0,T]} \psi_i\psi_i \le M_1^2, \tag{3.5.8}$$

for a given constant M_1; it belongs to $\mathcal{M}_2$ if

$$\begin{aligned} &\sup_{\Omega\times[0,T]} \psi_i\psi_i + \sup_{\Omega\times[0,T]} (\psi_{i,j}-\psi_{j,i})(\psi_{i,j}-\psi_{j,i}) \\ &\qquad + \sup_{t\in[0,T]} \int_\Omega (\psi_{i,t}\psi_{i,t})^2 dx \le M_2^2, \end{aligned} \tag{3.5.9}$$

for a constant M_2; and it belongs to $\mathcal{M}_3$ if

$$\sup_{t\in[0,T]} \left[\int_\Omega (\psi_{i,j}-\psi_{j,i})(\psi_{i,j}-\psi_{j,i})dx + \int_\Omega \psi_{i,t}\psi_{i,t}dx \right] \le M_3^2, \tag{3.5.10}$$

for a constant M_3. We assume that $v_i \in \mathcal{M}_1$, $u_i \in \mathcal{M}_2$, and $n_i \in \mathcal{M}_3$, in what follows. The argument used by Payne & Straughan (1989a) to derive continuous dependence on λ is the logarithmic convexity method as outlined below.

To compare the velocity fields we begin by defining

$$w_i = v_i - u_i, \qquad \text{and} \qquad \pi = p - q,$$

and observe that w satisfies the equations

$$\begin{aligned} \frac{\partial w_i}{\partial x_i} =& 0, \\ \frac{\partial w_i}{\partial t} =& v_j \frac{\partial w_i}{\partial x_j} + w_j \frac{\partial u_i}{\partial x_j} + \frac{\partial \pi}{\partial x_i} - (\nu+\lambda)\Delta w_i \\ & - \lambda \epsilon_{ijk} \frac{\partial n_k}{\partial x_j} - \lambda \Delta u_i, \end{aligned} \tag{3.5.11}$$

in $\Omega \times (0,T)$, and the boundary and initial conditions

$$\begin{aligned} w_i &= 0, \qquad \text{on} \quad \Gamma \times [0,T], \\ w_i(\mathbf{x},0) &= 0, \qquad \mathbf{x} \in \Omega. \end{aligned} \tag{3.5.12}$$

The key function in the application of the logarithmic convexity method is

$$\phi(t) = \|w(t)\|^2 + C\lambda^2, \tag{3.5.13}$$

where C is a constant to be chosen. The next step is to calculate ϕ' and ϕ''. Hence, differentiating (3.5.13), substituting (3.5.11), (3.5.12) and integrating by parts, one finds

$$\begin{aligned}\phi'(t) =& 2\int_\Omega w_i w_j u_{i,j} dx + 2(\nu + \lambda)\|\nabla w\|^2 \\ &+ 2\lambda \int_\Omega u_{i,j} w_{i,j} dx + 2\lambda\epsilon_{ijk} \int_\Omega n_k w_{i,j} dx.\end{aligned} \tag{3.5.14}$$

An alternative way to write ϕ' is to define

$$\varphi_i = w_{i,t} - \frac{1}{4} w_j (u_{i,j} - u_{j,i}) - \frac{1}{2} v_j w_{i,j}, \tag{3.5.15}$$

so that

$$\phi'(t) = 2\int_\Omega w_i \varphi_i dx. \tag{3.5.16}$$

The function φ_i is introduced because it allows the removal of awkward terms. Equation (3.5.14) is differentiated, the differential equation (3.5.11) is reintroduced, followed by several integrations by parts to obtain an expression for $\phi''(t)$. An estimate for $\phi''(t)$ is then obtained by judiciously using inequalities such as the Cauchy - Schwarz inequality and the bounds (3.5.8) - (3.5.10). The result is

$$\phi'' \geq 4\|\varphi\|^2 - c_1\lambda\|\varphi\| - c_2\|\nabla w\|^2 - c_3\|w\|^2 - c_4\lambda^2, \tag{3.5.17}$$

for constants $c_1, \dots, c_4$ that depend on M_1, M_2, M_3 and Ω but are independent of λ.

We need a way to control the $\|\nabla w\|^2$ term in (3.5.17) and with this in mind we apply the arithmetic - geometric mean inequality to (3.5.14). It can be shown that

$$\begin{aligned}-2(\nu + \lambda)\|\nabla w\|^2 \geq& -\phi' - (a_1 + a_2 + a_3)\|\nabla w\|^2 - \frac{M_2^2}{a_3}\|w\|^2 \\ &- \lambda^2\Big[\frac{1}{a_1}\|\nabla u\|^2 + \frac{2}{a_2}\|n\|^2\Big],\end{aligned}$$

for positive constants a_i $(i = 1, 2, 3)$. The a_i must now be chosen to leave a negative piece of $-\|\nabla w\|^2$ on the left. A possible choice is

$$a_1 + a_2 + a_3 = \nu.$$

Then with the aid of (3.5.9) and (3.5.10) we can find constants K_1 and K_2 independent of λ such that

$$-\|\nabla w\|^2 \geq -\frac{1}{\nu}\phi' - K_1\|w\|^2 - K_2\lambda^2. \tag{3.5.18}$$

Upon use of this bound in (3.5.17) and an appropriate choice of C in (3.5.13) we may obtain

$$\phi'' \geq 4\|\varphi\|^2 - c_1\lambda\|\varphi\| - c_5\phi' - c_6\phi, \tag{3.5.19}$$

where c_5 and c_6 are constants independent of λ.

Now form the basic convexity expression $\phi\phi'' - (\phi')^2$ using (3.5.13), (3.5.16) and (3.5.19) to derive the inequality

$$\phi\phi'' - (\phi')^2 \geq S^2 - \frac{c_1}{C}\|\varphi\|\phi^{3/2} - c_5\phi\phi' - c_6\phi^2, \tag{3.5.20}$$

where we have defined

$$S^2 = 4\phi\|\varphi\|^2 - 4\left(\int_\Omega w_i\varphi_i dx\right)^2 = 4\phi\|\varphi\|^2 - (\phi')^2, \tag{3.5.21}$$

and note that $S^2 \geq 0$ by virtue of the Cauchy - Schwarz inequality. From (3.5.21) it follows that (3.5.20) may be rewritten as

$$\begin{aligned}\phi\phi'' - (\phi')^2 \geq& S^2 - \frac{c_1}{2\sqrt{C}}\phi\big[S^2 + (\phi')^2\big]^{1/2} - c_5\phi\phi' - c_6\phi^2,\\ \geq& -\left(\frac{c_1^2}{16C} + c_6\right)\phi^2 - c_5\phi\phi' - \frac{c_6}{2\sqrt{C}}\phi|\phi'|. \end{aligned}\tag{3.5.22}$$

Rewriting (3.5.14) and using (3.5.9) and (3.5.10), $|\phi'|$ is estimated to find

$$|\phi'| \leq \phi' + c_7\phi,$$

for a constant c_7 independent of λ. Hence, from (3.5.22) one may show that $\exists$ constants α_1, α_2 such that

$$\phi\phi'' - (\phi')^2 \geq -\alpha_1\phi\phi' - \alpha_2\phi^2.$$

Since this is the standard convexity inequality it may be integrated as in chapter 1 to obtain for $t \in [0, T)$,

$$\phi(t) \leq \hat{c}\big[\phi(0)\big]^{\delta(t)}\big[\phi(T)\big]^{1-\delta(t)},$$

where $0 < \delta(t) \leq 1$. It follows from this estimate that

$$\|u - v\|^2 \leq \hat{K}(\delta)\lambda^{2\delta}, \tag{3.5.23}$$

which is the desired inequality derived by Payne & Straughan (1989). Consequently we can conclude that w depends Hölder continuously on λ in $L^2(\Omega)$ on compact subintervals of $[0, T)$. For $t < T$, the order of convergence of v_i to u_i in $L^2(\Omega)$ is $O(\lambda^\delta)$, $0 < \delta \le 1$.

Modeling in Magnetohydrodynamics Incorporating Hall and Ion - Slip Effects

We now give brief details of work of Mulone *et al.* (1992) who show that the solution to the equations of magnetohydrodynamics depends continuously on the body force and the source term in the magnetic field equation. By this technique they are able to deduce that the solution depends continuously on the modeling process when the Hall effect or the ion-slip effect is present. The Hall and ion-slip effects are prominent in many physical situations, see e.g. Cowling (1976), Maiellaro *et al.* (1989), and thus it is important to establish rigorous quantitative estimates on how these effects contribute to the solution in magnetohydrodynamics.

Mulone *et al.* (1992) treat, without any loss of generality, the viscosity and other coefficients as unity (although they do indicate how to modify the analysis when this is not the case) and then they write down directly the equations for the backward in time problem, but by changing t to $-t$ and dealing with the modified problem forward in time. Thus their equations for an incompressible, linear viscous fluid permeated by a magnetic field, may be written

$$\begin{aligned} \frac{\partial v_i}{\partial t} &= v_j \frac{\partial v_i}{\partial x_j} - H_j \frac{\partial H_i}{\partial x_j} + \frac{\partial p}{\partial x_i} - \Delta v_i - F_i, \\ \frac{\partial H_i}{\partial t} &= v_j \frac{\partial H_i}{\partial x_j} - H_j \frac{\partial v_i}{\partial x_j} - \Delta H_i, \\ \frac{\partial v_i}{\partial x_i} &= 0, \qquad \frac{\partial H_i}{\partial x_i} = 0. \end{aligned} \tag{3.5.24}$$

Here v_i, H_i, p represent the velocity, magnetic field, and pressure, respectively, and $F_i(\mathbf{x}, t)$ is a body force which is additional to that associated with the Lorentz force. These equations are again defined on a bounded spatial domain $\Omega(\subset \mathbf{R}^3)$, and $t \in (0, T)$. On the boundary Γ for $\hat{v}_i, \hat{H}_i$ prescribed functions we have

$$v_i(x_m, t) = \hat{v}_i(x_m, t), \qquad H_i(x_m, t) = \hat{H}_i(x_m, t), \tag{3.5.25}$$

while the initial data are,

$$v_i(x_m, 0) = v_i^0(x_m), \qquad H_i(x_m, 0) = H_i^0(x_m), \tag{3.5.26}$$

with v_i^0, H_i^0 given.

When the Hall and ion-slip effects are included equations (3.5.24) are modified to (3.5.27), in which the β_1 term represents the Hall effect whereas the β_2 term is associated with the effect of ion-slip.

$$\begin{aligned}
\frac{\partial v_i^*}{\partial t} =& v_j^* \frac{\partial v_i^*}{\partial x_j} - H_j^* \frac{\partial H_i^*}{\partial x_j} + \frac{\partial p^*}{\partial x_i} - \Delta v_i^* - F_i^*, \\
\frac{\partial H_i^*}{\partial t} =& v_j^* \frac{\partial H_i^*}{\partial x_j} - H_j^* \frac{\partial v_i^*}{\partial x_j} - \Delta H_i^* - \beta_1 \big[\nabla \times (\mathbf{H}^* \times \nabla \times \mathbf{H}^*)\big]_i \\
& - \beta_2 \Big[\nabla \times \big\{\mathbf{H}^* \times (\mathbf{H}^* \times \nabla \times \mathbf{H}^*)\big\}\Big]_i, \\
& \frac{\partial v_i^*}{\partial x_i} = 0, \qquad \frac{\partial H_i^*}{\partial x_i} = 0.
\end{aligned} \tag{3.5.27}$$

Equations (3.5.27) are defined on the same domain as (3.5.24) and (v_i^*, H_i^*) satisfy the same boundary and initial data as (3.5.25) and (3.5.26).

Mulone *et al.* (1992) set

$$\begin{aligned}
f_i =& F_i^* - F_i, \\
g_i(x_m, t) =& \beta_1 \big[\nabla \times (\mathbf{H}^* \times \nabla \times \mathbf{H}^*)\big]_i \\
& + \beta_2 \Big[\nabla \times \big\{\mathbf{H}^* \times (\mathbf{H}^* \times \nabla \times \mathbf{H}^*)\big\}\Big]_i,
\end{aligned}$$

and then define u_i, h_i, π by

$$u_i = v_i^* - v_i, \quad h_i = H_i^* - H_i, \quad \pi = p^* - p.$$

These difference variables then satisfy the boundary initial value problem

$$\begin{aligned}
\frac{\partial u_i}{\partial t} &= v_j^* \frac{\partial u_i}{\partial x_j} + u_j \frac{\partial v_i}{\partial x_j} - H_j^* \frac{\partial h_i}{\partial x_j} - h_j \frac{\partial H_i}{\partial x_j} + \frac{\partial \pi}{\partial x_i} - \Delta u_i - f_i, \\
\frac{\partial h_i}{\partial t} &= v_j^* \frac{\partial h_i}{\partial x_j} + u_j \frac{\partial H_i}{\partial x_j} - H_j^* \frac{\partial u_i}{\partial x_j} - h_j \frac{\partial v_i}{\partial x_j} - \Delta h_i - g_i, \\
\frac{\partial u_i}{\partial x_i} &= 0, \qquad \frac{\partial h_i}{\partial x_i} = 0,
\end{aligned} \tag{3.5.28}$$

in $\Omega \times (0, T]$,

$$\begin{aligned}
u_i(x_m, t) &= 0, \qquad h_i(x_m, t) = 0, \qquad x_m \in \Gamma,\ t \in [0, T], \\
u_i(x_m, 0) &= 0, \qquad h_i(x_m, 0) = 0, \qquad x_m \in \Omega.
\end{aligned} \tag{3.5.29}$$

The constraint set adopted by Mulone *et al.* (1992) consists of those solutions satisfying

$$\begin{aligned}
& |\nabla \mathbf{v}|, |\nabla \mathbf{H}|, |\mathbf{v}^*|, |\mathbf{H}^*| \le M, \\
& \big|\nabla \times (\mathbf{H}^* \times \nabla \times \mathbf{H}^*)\big| \le N, \\
& \Big|\nabla \times \big\{\mathbf{H}^* \times (\mathbf{H}^* \times \nabla \times \mathbf{H}^*)\big\}\Big| \le N,
\end{aligned}$$

for M, N constants we know. The latter two bounds simply say that the Hall and ion-slip effects are bounded which is entirely reasonable.

To establish continuous dependence on the source terms f_i and g_i, and so establish continuous dependence on the Hall and ion-slip effects, Mulone *et al.* (1992) use the method of logarithmic convexity and utilize the functional

$$F(t) = \int_0^t (\|\mathbf{u}\|^2 + \|\mathbf{h}\|^2)ds + \sup_{t\in[0,T]} \|\mathbf{f}\|^2 + \sup_{t\in[0,T]} \|\mathbf{g}\|^2.$$

Omitting the details they are able to show F satisfies

$$FF'' - (F')^2 \geq -k_1FF' - k_2F^2,$$

this being the standard logarithmic convexity inequality, and then as shown in section 1.5 this leads to

$$F(t) \leq \big[F(0)\big]^{(\sigma-\sigma_2)/(1-\sigma_2)}\big[F(T)e^{\mu T}\big]^{(1-\sigma)/(1-\sigma_2)}e^{-\mu t},$$

where $\mu = k_2/k_1$, $\sigma_1 = 1$, $\sigma_2 = \exp(-k_1T)$ and $\sigma = \exp(-k_1t)$. A final imposition is made that

$$F(T)e^{\mu T} \leq K,$$

and then it is deduced that

$$\begin{aligned}\int_0^t (\|\mathbf{u}\|^2 + \|\mathbf{h}\|^2)ds \leq & e^{-\mu t}K^{(1-\sigma)/(1-\sigma_2)} \\ & \times \Big(\sup_{t\in[0,T]} \|\mathbf{f}\|^2 + \sup_{t\in[0,T]} \|\mathbf{g}\|^2\Big)^{(\sigma-\sigma_2)/(1-\sigma_2)}.\end{aligned}$$

This inequality, of course, shows the solution depends Hölder continuously on $\mathbf{f}$, $\mathbf{g}$ on compact subintervals of $[0, T)$, and consequently establishes continuous dependence on the modeling.

In particular, Mulone *et al.* (1992) deduce continuous dependence on the Hall and ion-slip effects in the sense that as $\beta_1, \beta_2 \to 0$, a quantitative estimate is known for the rate of approach of $v_i^* \to v_i$ and $H_i^* \to H_i$. They also observe that by using the dissipative nature of the ion-slip term one may directly establish continuous dependence on this effect via the β_2 coefficient.

Continuous Dependence on the Velocity for an Equation Arising from Dynamo Theory

We conclude this section with a brief account of recent work of Franchi & Straughan (1994) who study continuous dependence on the modeling in an equation which arises from the study of a dynamo model. The dynamo aspect may be found in Lortz & Meyer-Spasche (1982).

For the backward in time case Franchi & Straughan (1994) consider the following equation

$$-\frac{\partial H}{\partial t} = \frac{\partial}{\partial x_i}\Big(\eta\frac{\partial H}{\partial x_i}\Big) - \frac{\partial}{\partial x_i}\Big(v_i H\Big), \qquad \text{in} \quad \Omega \times (0,T), \tag{3.5.30}$$

where Ω is a bounded domain in $\mathbf{R}^3$. They suppose that on the boundary Γ of the domain Ω, H satisfies the data

$$H(\mathbf{x},t) = H_B(\mathbf{x},t), \tag{3.5.31}$$

for a prescribed function H_B, while H also satisfies the inital data

$$H(\mathbf{x},0) = H_0(\mathbf{x}), \qquad \mathbf{x} \in \Omega. \tag{3.5.32}$$

In the dynamo problem H is the toroidal part of the magnetic field, and v_i is a *known* velocity field.

Franchi & Straughan (1994) use a logarithmic convexity argument and so we here only outline the result.

If H and H^* are two solutions to (3.5.30) on $\Omega \times (0,T)$, corresponding to different velocity fields $\mathbf{v}$ and $\mathbf{v}^*$, but satisfying the *same* boundary and initial conditions, then the difference solution satisfies the boundary-initial value problem,

$$\frac{\partial h}{\partial t} = -\big(\eta(\mathbf{x})h_{,i}\big)_{,i} + \big(v_i^*(\mathbf{x})h\big)_{,i} + \big(u_i(\mathbf{x})H\big)_{,i}, \quad \text{in } \Omega \times (0,T), \tag{3.5.33}$$

$$h = 0, \quad \text{on } \Gamma \times [0,T], \tag{3.5.34}$$

$$h(\mathbf{x},0) = 0, \quad \mathbf{x} \in \Omega. \tag{3.5.35}$$

In addition it is necessary that on the boundary Γ,

$$v_i^* n_i = 0. \tag{3.5.36}$$

Franchi & Straughan (1994) when dealing with the improperly posed problem impose the bounds,

$$|v_i^*|, |v_{i,i}^*|, |v_{i,m}^*|, |\eta_{,i}|, |H|, |H_{,i}| \le N, \tag{3.5.37}$$

for a positive constant N.

To show that H depends continuously on $\mathbf{v}$ Franchi & Straughan (1994) utilize the functional

$$F(t) = \int_0^t \|h\|^2 ds + Q, \tag{3.5.38}$$

where $\|\cdot\|$ is the norm on $L^2(\Omega)$ and where Q is the data term

$$Q = \sup_{t\in[0,T]} \|\mathbf{u}\|^2 + \sup_{t\in[0,T]} \|u_{i,i}\|^2. \tag{3.5.40}$$

We omit the details of their calculation, but note that they arrive at the logarithmic convexity inequality

$$FF'' - (F')^2 \geq -k_1 FF' - k_2 F^2, \tag{3.5.41}$$

where k_1, k_2 are given explicitly in that work. Inequality (3.5.41) may now be integrated as shown in section 1.5 to find

$$F(t) \leq \Big[F(0)\Big]^{(\sigma-\sigma_1)/(1-\sigma_1)} \Big[F(T)e^{\mu T}\Big]^{(1-\sigma)/(1-\sigma_1)} e^{-\mu t}, \tag{3.5.42}$$

where

$$\sigma = e^{-k_1 t}, \qquad \sigma_1 = e^{-k_1 T}, \qquad \mu = \frac{k_2}{k_1}.$$

The continuous dependence estimate follows from (3.5.42) by supposing $F(T)$ satisfies a constraint such as the assumption that an *a priori* estimate is known for the quantity

$$L(T) = \Big[F(T)e^{\mu T}\Big]^{(1-\sigma)/(1-\sigma_1)}.$$

The continuous dependence inequality thus follows and is:

$$\int_0^t \|h\|^2 ds \leq L(T)e^{-\mu t}\big[\sup_{t\in[0,T]} \|\mathbf{u}\|^2 + \sup_{t\in[0,T]} \|u_{i,i}\|^2\big]^{(\sigma-\sigma_1)/(1-\sigma_1)}. \tag{3.5.43}$$

This inequality establishes continuous dependence on the velocity for the backward in time problem on compact subintervals of $[0, T)$.

3.6 Continuous Dependence on Modeling in Porous Convection Problems

We begin this section by considering an ill posed problem connected with the phenomenon of penetrative convection or non-Boussinesq convection, in a porous medium. The process of penetrative convection is examined in detail in a recent book by Straughan (1993). Here we review some results of Ames & Payne (1994b) who derive estimates which imply continuous dependence on modeling. More specifically, they show that solutions to the backward in time equations for a heat conducting incompressible fluid in a porous medium depend Hölder continuously on changes in the form of the buoyancy law.

The equations adopted by Ames & Payne (1994b) which have a linear buoyancy law are

$$\begin{aligned}\frac{\partial p}{\partial x_i} &= - v_i - g_i^1 T,\\ \frac{\partial v_i}{\partial x_i} &=0,\\ \frac{\partial T}{\partial t} &=v_i \frac{\partial T}{\partial x_i} - \Delta T.\end{aligned} \tag{3.6.1}$$

These equations are defined on $\Omega \times (0, t_0)$ where Ω is a bounded domain in $\mathbf{R}^3$ with a smooth boundary Γ. In (3.6.1), v_i is the velocity, T is the temperature, and p is the pressure. For simplicity, the viscosity, density, and thermal diffusivity have been set equal to unity and $g_i^1 = g_i^1(x_m, t)$ is a combination of the gravity vector and coefficient of thermal expansion. Associated with the equations (3.6.1) are boundary and initial conditions of the form

$$\begin{aligned}T(x_m, t) &=T_1(x_m, t),\\ v_i(x_m, t) &=\hat{v}_i(x_m, t),\end{aligned} \qquad \text{on} \quad \Gamma \times [0, t_0], \tag{3.6.2}$$

$$\begin{aligned}T(x_m, 0) &=T_2(x_m),\\ v_i(x_m, 0) &=\tilde{v}_i(x_m),\end{aligned} \qquad \mathbf{x} \in \Omega. \tag{3.6.3}$$

The problem which is compared to (3.6.1) - (3.6.3) has a quadratic buoyancy law, so we consider

$$\begin{aligned}\frac{\partial p^*}{\partial x_i} &= - v_i^* - g_i^1 T^* - \alpha g_i^2 (T^*)^2,\\ \frac{\partial v_i^*}{\partial x_i} &=0,\\ \frac{\partial T^*}{\partial t} &=v_i^* \frac{\partial T^*}{\partial x_i} - \Delta T^*,\end{aligned} \tag{3.6.4}$$

and require that (v_i^*, T^*, p^*) satisfy the same boundary and initial conditions as (v_i, T, p); in (3.6.4) g_i^2 is another function associated with gravity and the thermal expansion coefficient.

The goal in Ames & Payne (1994b) is to derive an estimate that indicates how solutions depend on the constant α. To this end set

$$u_i = v_i^* - v_i, \qquad \theta = T^* - T, \qquad P = p^* - p, \tag{3.6.5}$$

and observe that (u_i, θ, P) satisfy the initial - boundary value problem

$$\begin{aligned}\frac{\partial P}{\partial x_i} =& - u_i - g_i^1\theta - \alpha g_i^2(T^*)^2,\\ \frac{\partial u_i}{\partial x_i} =&0,\\ \frac{\partial \theta}{\partial t} =&v_i\frac{\partial \theta}{\partial x_i} + u_i\frac{\partial T^*}{\partial x_i} - \Delta\theta,\end{aligned} \tag{3.6.6}$$

in $\Omega \times (0, t_0)$ with

$$\begin{aligned}\theta(x_m, t) =&0, \quad u_i(x_m, t) = 0, \qquad \text{on } \Gamma \times [0, t_0],\\ \theta(x_m, 0) =&0, \quad u_i(x_m, 0) = 0, \qquad \mathbf{x} \in \Omega.\end{aligned} \tag{3.6.7}$$

It is henceforth assumed that

$$\sup_{\Omega\times[0,t_0]} g_i^\alpha g_i^\alpha \leq K^2, \tag{3.6.8}$$

for $\alpha = 1$ or 2 and some positive constant K, and we now introduce two classes of function that we will need in the course of the analysis. The function ψ belongs to $\mathcal{M}_1$ if

$$\sup_{\Omega\times[0,t_0]} \left(\psi^2 + |\nabla\psi|^2\right) \leq M_1^2, \tag{3.6.9}$$

for a prescribed constant M_1 and the vector w_i is of class $\mathcal{M}_2$ if

$$\sup_{\Omega\times[0,t_0]} w_iw_i \leq M_2^2, \tag{3.6.10}$$

for some prescribed constant M_2.

To establish continuous dependence on α Ames & Payne (1994b) use a logarithmic convexity method. They define the functional

$$\phi(t) = \int_0^t \|\theta\|^2 d\eta + \alpha^2, \tag{3.6.11}$$

and show that if $T^* \in \mathcal{M}_1$ and $v_i \in \mathcal{M}_2$, then

$$\phi\phi'' - (\phi')^2 \geq -c_1\phi\phi' - c_2\phi^2, \tag{3.6.12}$$

for computable, positive constants c_1 and c_2.

To review the analysis of Ames & Payne (1994b) we begin by obtaining an estimate for $\|u\|$ which will be needed to obtain (3.6.12). Multiplying the first equation in (3.6.6) by u_i and integrating over Ω, we have

$$\begin{aligned}\|u\|^2 =& - \int_\Omega g_i^1 u_i \theta \, dx - \alpha \int_\Omega g_i^2 u_i (T^*)^2 dx, \\ \leq& K\|\theta\| + \alpha K M_1^2 |\Omega|^{1/2}, \end{aligned} \tag{3.6.13}$$

where $|\Omega|$ represents the volume of the domain. The inequality follows from Schwarz's inequality, (3.6.8), and the constraint on T^*.

To apply the logarithmic convexity method it is necessary to calculate ϕ' and ϕ'' and so differentiating (3.6.11), introducing the third of equations (3.6.6) and integrating by parts, we find that

$$\phi'(t) = \|\theta\|^2 = 2\int_0^t \int_\Omega \theta \theta_{,\eta} dx \, d\eta, \tag{3.6.14}$$

$$= 2\int_0^t \int_\Omega \theta u_i T^*_{,i} dx \, d\eta + 2\int_0^t \int_\Omega \theta_{,i} \theta_{,i} dx \, d\eta. \tag{3.6.15}$$

If we differentiate (3.6.15), we then have

$$\begin{aligned}\phi''(t) =& 2\int_\Omega \theta u_i T^*_{,i} dx + 2\int_\Omega \theta_{,i}\theta_{,i} dx, \\ =& 4\int_0^t \int_\Omega \theta_{,i} \theta_{,i\eta} dx \, d\eta + 2\int_\Omega \theta u_i T^*_{,i} dx, \end{aligned} \tag{3.6.16}$$

which upon integration by parts and substitution of the differential equation, leads to

$$\begin{aligned}\phi''(t) =& 4\int_0^t \int_\Omega \theta^2_{,\eta} dx \, d\eta - 4\int_0^t \int_\Omega \theta_{,\eta}(v_i \theta_{,i} + u_i T^*_{,i}) dx \, d\eta \\ &+ 2\int_\Omega \theta u_i T^*_{,i} dx. \end{aligned} \tag{3.6.17}$$

The term involving $\theta_{,t} v_i \theta_{,i}$ causes problems and to remove it, it is convenient to write

$$\phi'(t) = 2\int_0^t \int_\Omega \theta \psi \, dx \, d\eta, \tag{3.6.18}$$

where we have set

$$\psi = \theta_{,\eta} - \frac{1}{2} v_i \theta_{,i}. \tag{3.6.19}$$

We may then obtain the expression

$$\begin{aligned}\phi\phi'' - (\phi')^2 =& 4S^2 - \left(\int_0^t \|\theta\|^2 d\eta\right)\left(\int_0^t \int_\Omega [v_i\theta_{,i}]^2 dx\, d\eta\right)\\ &+ 4\alpha^2 \int_0^t \|\theta_{,\eta}\|^2 d\eta - 4\phi \int_0^t \int_\Omega \theta_{,\eta} u_i T^*_{,i} dx\, d\eta \\ &+ 2\phi \int_\Omega \theta u_i T^*_{,i} dx,\end{aligned} \tag{3.6.20}$$

where

$$S^2 = \left(\int_0^t \|\theta\|^2 d\eta\right)\left(\int_0^t \int_\Omega \psi^2 dx\, d\eta\right) - \left(\int_0^t \int_\Omega \theta\psi\, dx\, d\eta\right)^2. \tag{3.6.21}$$

Note that $S^2 \geq 0$ by Schwarz's inequality. We now need to bound each of the terms on the right hand side of (3.6.20). Full details are not given here; however, we observe that the necessary analysis involves the inequalities (3.6.22) and (3.6.23). Since $T^* \in \mathcal{M}_1$, we derive from (3.6.15)

$$\int_0^t \int_\Omega \theta_{,i}\theta_{,i} dx\, d\eta \leq \frac{1}{2}\phi' + A\phi, \tag{3.6.22}$$

while from (3.6.13), it follows that

$$\int_0^t \|u\|^2 d\eta \leq B\phi, \tag{3.6.23}$$

where the constants A and B are given by

$$A = \frac{1}{2} M_1 K \{2 + M_1^4 |\Omega| t_0\}^{1/2}, \tag{3.6.24}$$

and

$$B = K^2\{1 + M_1^4 |\Omega| t_0\}. \tag{3.6.25}$$

The second last term in (3.6.20) is rearranged as

$$\begin{aligned}-4\phi \int_0^t \int_\Omega \theta_{,\eta} u_i T^*_{,i} dx\, d\eta =& -4\phi \int_0^t \int_\Omega \psi u_i T^*_{,i} dx\, d\eta \\ &- 2\phi \int_0^t \int_\Omega v_j \theta_{,j} u_i T^*_{,i} dx\, d\eta.\end{aligned} \tag{3.6.26}$$

With the observation that

$$\left[\phi \int_0^t \|\psi\|^2 d\eta\right]^{1/2} = \left[S^2 + (\phi')^2\right]^{1/2} \leq S + \phi, \tag{3.6.27}$$

and that

$$\|u\|\|\theta\| \leq K^{1/2}\big[\phi'\big]^{1/2}\big[(\phi')^{1/2} + c\phi^{1/2}\big], \tag{3.6.28}$$

where the constant c is

$$c = KM_1^2|\Omega|^{1/2},$$

one may show from (3.6.20), after employing (3.6.22) - (3.6.28), that ϕ satisfies the logarithmic convexity inequality

$$\phi\phi'' - (\phi')^2 \geq -2\gamma_1\phi\phi' - \Big(\gamma_3 + \frac{\gamma_2}{\gamma_1}\Big)\phi^2, \tag{3.6.29}$$

where

$$\begin{aligned} \gamma_1 =& \frac{1}{2}M_2^2 + 4M_1B^{1/2} + 2M_1K_1^2, \\ \gamma_2 =& (\sqrt{2}M_1M_2B^{1/2} + 2M_1c^{1/2})^2, \\ \gamma_3 =& M_1^2B + 2M_1M_2B^{1/2}A^{1/2}. \end{aligned} \tag{3.6.30}$$

Expression (3.6.29) is the desired logarithmic convexity inequality. Integration of this as in section 1.5 results in a continuous dependence on α estimate provided that solutions to the governing equations are suitably constrained. In this case the constraints are $T^* \in \mathcal{M}_1$, $v_i \in \mathcal{M}_2$, and θ satisfies the bound

$$\int_0^{t_0} \|\theta\|^2 d\eta + \alpha^2 \leq M$$

for a positive constant M independent of α. Then there exists a constant $\hat{K}(t_0)$ independent of α such that for $t \in [0, t_0)$

$$\int_0^t \|\theta\|^2 d\eta \leq \hat{K}(t_0)\alpha^{2[1-\delta(t)]}M^{\delta(t)}, \tag{3.6.31}$$

where $0 \leq \delta(t) \leq 1$. We have thus demonstrated continuous dependence of solutions on the buoyancy law as in Ames & Payne (1994b).

The Modeling Problem in a Viscous Fluid

Franchi & Straughan (1993b) obtain similar results in their study of an ill posed problem for thermal convection in a heat conducting viscous fluid. They establish that a solution to the backward in time Navier - Stokes equations coupled with a backward in time convective heat equation depends continuously on changes in the modeling of the buoyancy law under the requirements that the perturbed velocity and temperature fields as well as the gradients of base velocity and temperature are bounded.

The equations which Franchi & Straughan (1993b) consider with a linear buoyancy law are

$$
\begin{aligned}
\frac{\partial v_i}{\partial t} &= v_j \frac{\partial v_i}{\partial x_j} + \frac{\partial p}{\partial x_i} - \Delta v_i - b_i \Theta, \\
\frac{\partial v_i}{\partial x_i} &= 0, \\
\frac{\partial \Theta}{\partial t} &= v_j \frac{\partial \Theta}{\partial x_j} - \Delta \Theta,
\end{aligned}
\tag{3.6.32}
$$

where v_i, p, Θ are velocity, pressure, and temperature, respectively, and b_i is essentially gravity. The above equations are defined on a bounded spatial domain $\Omega\, (\subset \mathbf{R}^3)$, and on the bounded time interval $(0, T)$. On the boundary of Ω, Γ, it is assumed that

$$
v_i(\mathbf{x}, t) = \hat{v}_i(\mathbf{x}, t), \quad \Theta(\mathbf{x}, t) = \hat{\Theta}(\mathbf{x}, t), \tag{3.6.33}
$$

where $\hat{v}_i$, $\hat{\Theta}$ are given functions. The solutions are required to satisfy the initial data

$$
v_i(\mathbf{x}, 0) = h_i(\mathbf{x}), \quad \Theta(\mathbf{x}, 0) = \Phi(\mathbf{x}), \tag{3.6.34}
$$

for $\mathbf{x} \in \Omega$ and prescribed functions h_i, Φ.

The problem with which to compare (3.6.32) - (3.6.34) is that with a quadratic buoyancy law. The governing equations for this in $\Omega \times (0, T)$ are

$$
\begin{aligned}
\frac{\partial v_i^*}{\partial t} &= v_j^* \frac{\partial v_i^*}{\partial x_j} + \frac{\partial p^*}{\partial x_i} - \Delta v_i^* - b_i \Theta^* - \epsilon b_i \Theta^{*2}, \\
\frac{\partial v_i^*}{\partial x_i} &= 0, \\
\frac{\partial \Theta^*}{\partial t} &= v_j^* \frac{\partial \Theta^*}{\partial x_j} - \Delta \Theta^*.
\end{aligned}
\tag{3.6.35}
$$

Since the object is to study continuous dependence on changes in the model itself it is sufficient to take (v_i^*, Θ^*, p^*) to satisfy the *same* boundary data and the *same* initial data as that satisfied by (v_i, Θ, p). To investigate continuous dependence on ϵ, we define the variables u_i, π and θ by

$$
u_i = v_i^* - v_i, \quad \pi = p^* - p, \quad \theta = \Theta^* - \Theta,
$$

and observe that these variables satisfy the equations

$$
\begin{aligned}
\frac{\partial u_i}{\partial t} &= v_j^* \frac{\partial u_i}{\partial x_j} + u_j \frac{\partial v_i}{\partial x_j} + \frac{\partial \pi}{\partial x_i} - \Delta u_i - b_i \theta - \epsilon b_i \Theta^{*2}, \\
\frac{\partial u_i}{\partial x_i} &= 0, \\
\frac{\partial \theta}{\partial t} &= v_j^* \frac{\partial \theta}{\partial x_j} + u_j \frac{\partial \Theta}{\partial x_j} - \Delta \theta,
\end{aligned}
\tag{3.6.36}
$$

in $(\mathbf{x}, t) \in \Omega \times (0, T)$, and the appropriate boundary and initial conditions are

$$\begin{aligned} u_i(\mathbf{x}, t) &= 0, \quad \theta(\mathbf{x}, t) = 0, \qquad \mathbf{x} \in \Gamma, \quad t \in [0, T], \\ u_i(\mathbf{x}, 0) &= 0, \quad \theta(\mathbf{x}, 0) = 0, \qquad \mathbf{x} \in \Omega. \end{aligned} \tag{3.6.37}$$

Franchi & Straughan (1993b) assume the solutions $v_i, \Theta, v_i^*, \Theta^*$ satisfy the bounds:

$$|\nabla \mathbf{v}|, \ |\nabla \Theta|, \ |\mathbf{v}^*| \ |\Theta^*|^2 \le M,$$

where M is a known constant. The above restrictions thus define the constraint set. It is also supposed b_i is bounded.

Franchi & Straughan (1993b) employ the method of logarithmic convexity and define the functional $F(t)$ by

$$F(t) = \int_0^t (\|u\|^2 + \|\theta\|^2) ds + \epsilon^2. \tag{3.6.38}$$

After several manipulations they show this functional satisfies the logarithmic convexity inequality

$$FF'' - (F')^2 \ge -k_1 FF' - k_2 F^2, \tag{3.6.39}$$

and obtain the constants k_1, k_2 explicitly. This inequality is integrated as in section 1.5 by putting $\sigma = e^{-k_1 t}$, $\sigma_1 = 1$, $\sigma_2 = e^{-k_1 T}$, to find

$$F(t) \le \Big[F(0)\Big]^{(\sigma - \sigma_2)/(1 - \sigma_2)} \Big[F(T) e^{\mu T}\Big]^{(1 - \sigma)/(1 - \sigma_2)} e^{-\mu t}, \tag{3.6.40}$$

where $\mu = k_2/k_1$.

Inequality (3.6.40) is used to establish continuous dependence on the modeling, i.e. on ϵ, by supposing a bound is known for $F(T)$. Franchi & Straughan (1993b) assume

$$F(T) e^{\mu T} \le K,$$

where K is a constant independent of ϵ. Then (3.6.40) implies that

$$\int_0^t (\|\mathbf{u}\|^2 + \|\theta\|^2) ds \le e^{-\mu t} K^{(1-\sigma)/(1-\sigma_2)} \epsilon^{2(\sigma - \sigma_2)/(1-\sigma_2)}. \tag{3.6.41}$$

This inequality thus establishes Hölder continuous dependence of the solution on compact subintervals of $[0, T)$, and hence continuous dependence on the buoyancy.

Remark

We have only considered the question of changes in the buoyancy from a linear to a quadratic law. Straughan (1993) includes an exposition of various buoyancy laws which have been used in the literature, such as cubic, fifth and sixth order polynomials in temperature, and other functions. Franchi & Straughan (1993b) remark that it is not difficult to extend their analysis to the equivalent modeling problem of comparing a particular buoyancy law with another different one.

Continuous Dependence on Modeling Parameters

Several of the studies of thermal convection have been devoted to obtaining continuous dependence on parameters such as a heat source or a body force. Such parameters are subject to error during the formulation of the mathematical model. Strictly speaking these investigations would not be classified as "continuous dependence on modeling" studies in the sense that we have so far used the terminology. However, it can be argued that these parameters are part of the model and consequently, this work can be included in the present chapter. Examples of this kind of study can be found in Ames & Cobb (1993), Ames & Payne (1994a,1995b), Franchi (1995), Franchi & Straughan (1993a,1996), Song (1988) and Straughan (1993). Recently Franchi & Straughan (1993a) have established continuous dependence of the solution on changes in the heat supply in a model for double diffusive convective motion in a porous medium, as we now show.

For the problem considered in Franchi & Straughan (1993a), the equations for flow in a porous medium, backward in time, with a buoyancy law linear in solute and quadratic in temperature, may be written as

$$\begin{aligned} \frac{\partial p}{\partial x_i} &= b_i\big[1-\alpha(\Theta-\Theta_0)^2+\beta(C-C_0)\big]-v_i, \\ \frac{\partial v_i}{\partial x_i} &= 0, \\ \frac{\partial \Theta}{\partial t} &= v_i\frac{\partial \Theta}{\partial x_i}-\Delta\Theta-R, \\ \frac{\partial C}{\partial t} &= v_i\frac{\partial C}{\partial x_i}-\Delta C, \end{aligned} \tag{3.6.42}$$

on the domain $x\in\Omega$, $t\in(0,T)$. Here, v_i, p, Θ, C, R are velocity, pressure, temperature, concentration of solute, and heat supply. $\alpha, \beta, \Theta_0, C_0$ are constants and we suppose

$$|\mathbf{b}(\mathbf{x},t)|\le 1. \tag{3.6.43}$$

The boundary conditions we consider are

$$\begin{aligned} n_i v_i(\mathbf{x},t) &= n_i \hat{v}_i(\mathbf{x},t),\\ \Theta(\mathbf{x},t) &= \hat{\Theta}(\mathbf{x},t),\\ C(\mathbf{x},t) &= \hat{C}(\mathbf{x},t), \end{aligned} \tag{3.6.44}$$

where $\mathbf{x} \in \Gamma$, $t \in [0,T]$, $\mathbf{n}$ is the unit outward normal to Γ, $\hat{v}_i$, $\hat{\Theta}$, $\hat{C}$ are prescribed, and the initial conditions are

$$\begin{aligned} \Theta(\mathbf{x},0) &= \Theta_B(\mathbf{x}),\\ C(\mathbf{x},0) &= C_B(\mathbf{x}). \end{aligned} \tag{3.6.45}$$

(Initial data are given only for the temperature and concentration fields Θ and C, since this is all which is required in the ensuing analysis.)

Since we shall study continuous dependence on the heat supply R, we suppose

$$(v_i, \Theta, C, p) \qquad \text{and} \qquad (v_i^*, \Theta^*, C^*, p^*)$$

are two solutions to (3.6.42) - (3.6.45) for the same boundary and initial data, but for *different* heat sources R and R^*. In terms of the difference variables

$$\begin{aligned} u_i = & v_i^* - v_i, \quad \pi = p^* - p, \quad r = R^* - R,\\ &\theta = \Theta^* - \Theta, \quad c = C^* - C, \end{aligned}$$

the perturbation solution (u_i, θ, c, π) satisfies

$$\begin{aligned} \frac{\partial \pi}{\partial x_i} &= b_i\big[-\alpha(\Theta + \Theta^*)\theta + 2\alpha\Theta_0\theta + \beta c\big] - u_i,\\ \frac{\partial u_i}{\partial x_i} &= 0,\\ \frac{\partial \theta}{\partial t} &= v_i^* \frac{\partial \theta}{\partial x_i} + u_i \frac{\partial \Theta}{\partial x_i} - \Delta\theta - r,\\ \frac{\partial c}{\partial t} &= v_i^* \frac{\partial c}{\partial x_i} + u_i \frac{\partial C}{\partial x_i} - \Delta C, \end{aligned} \tag{3.6.46}$$

on the domain $\mathbf{x} \in \Omega$, $t \in (0,T)$, together with the boundary and initial conditions

$$\begin{aligned} & n_i u_i(\mathbf{x},t) = \theta(\mathbf{x},t) = c(\mathbf{x},t) = 0, \qquad \mathbf{x} \in \Gamma, \quad t \in [0,T],\\ & \theta(\mathbf{x},0) = c(\mathbf{x},0) = 0, \qquad \mathbf{x} \in \Omega. \end{aligned} \tag{3.6.47}$$

The bounds we impose are that

$$|v_i^*|,\ |\Theta|,\ |\Theta^*|,\ |\Theta_{,i}|,\ |C_{,i}| \le M, \tag{3.6.48}$$

for a prescribed constant M.

To apply the method of logarithmic convexity we define the functional $F(t)$ by

$$F(t) = \int_0^t (\|\theta\|^2 + \|c\|^2) ds + Q, \tag{3.6.49}$$

where Q is now the data term

$$Q = \sup_{t\in[0,T]} \|r\|^2, \tag{3.6.50}$$

and where $\|\cdot\|$ denotes the norm on $L^2(\Omega)$.

By differentiating F we obtain

$$\begin{aligned}
F'(t) =&\|\theta\|^2 + \|c\|^2, && (3.6.51)\\
=&2\int_0^t\int_\Omega \theta\theta_{,s} dx\, ds + 2\int_0^t\int_\Omega cc_{,s} dx\, ds, && (3.6.52)\\
=&2\int_0^t\int_\Omega \theta u_i\Theta_{,i} dx\, ds + 2\int_0^t\int_\Omega cu_iC_{,i} dx\, ds - 2\int_0^t\int_\Omega \theta r dx\, ds\\
&+2\int_0^t \|\nabla\theta\|^2 ds + 2\int_0^t \|\nabla c\|^2 ds, && (3.6.53)
\end{aligned}$$

where use has been made of (3.6.46) and (3.6.47). Next, differentiate F' and after integration by parts and further use of (3.6.46) we find

$$\begin{aligned}
F'' =&2\int_\Omega \theta u_j\Theta_{,j} dx - 2\int_\Omega \theta r\, dx + 2\int_\Omega cu_iC_{,i} dx - 4\int_0^t\int_\Omega \theta_{,s}u_i\Theta_{,i} dx\, ds\\
&+4\int_0^t\int_\Omega \theta_{,s} r\, dx\, ds - 4\int_0^t\int_\Omega c_{,s}u_iC_{,i} dx\, ds + 4\int_0^t \|\phi\|^2 ds\\
&+4\int_0^t \|\psi\|^2 ds - \int_0^t\int_\Omega (v_i^*\theta_{,i})^2 dx\, ds\\
&-\int_0^t\int_\Omega (v_i^* c_{,i})^2 dx\, ds, && (3.6.54)
\end{aligned}$$

where ϕ and ψ have been defined as

$$\phi = \theta_{,t} - \frac{1}{2}v_j^*\theta_{,j}, \qquad \psi = c_{,t} - \frac{1}{2}v_j^* c_{,j}. \tag{3.6.55}$$

The next step is to multiply the first equation of (3.6.46) by u_i and integrate over Ω to find,

$$\|\mathbf{u}\|^2 = \alpha\int_\Omega b_i(\Theta+\Theta^*)\theta u_i dx + 2\Theta_0\alpha\int_\Omega b_i\theta u_i dx + \beta\int_\Omega b_i c u_i dx.$$

We use the arithmetic-geometric mean inequality on this equation together with the bounds (3.6.43), (3.6.48), and hence obtain

$$\|\mathbf{u}\|^2 \leq k(\|\theta\|^2 + \|c\|^2), \tag{3.6.56}$$

where k is the constant

$$k = 12 \max\{\alpha^2(M^2 + \Theta_0^2), \beta^2\}.$$

Now use (3.6.52), (3.6.54) and form the expression $FF'' - (F')^2$, to find

$$\begin{aligned} FF'' - (F')^2 \geq & 4S^2 + 4Q\int_0^t (\|\phi\|^2 + \|\psi\|^2)ds + 2F\int_\Omega \theta u_i \Theta_{,i} dx \\ & - 2F\int_\Omega \theta r\, dx + 2F\int_\Omega c u_i C_{,i} dx - 4F\int_0^t\int_\Omega \theta_{,s} u_i \Theta_{,i} dx\, ds \\ & + 4F\int_0^t\int_\Omega \theta_{,s} r\, dx\, ds - 4F\int_0^t\int_\Omega c_{,s} u_i C_{,i} dx\, ds \\ & - F\int_0^t\int_\Omega (v_i^* \theta_{,i})^2 dx\, ds - F\int_0^t\int_\Omega (v_i^* c_{,i})^2 dx\, ds, \end{aligned} \tag{3.6.57}$$

where S^2 has been defined by

$$\begin{aligned} S^2 = & \int_0^t (\|\theta\|^2 + \|c\|^2)ds \int_0^t (\|\phi\|^2 + \|\psi\|^2)ds \\ & - \left(\int_0^t \left[\int_\Omega \theta\phi\, dx + \int_\Omega c\psi\, dx\right] ds\right)^2, \end{aligned} \tag{3.6.58}$$

and is non-negative by virtue of the Cauchy-Schwarz inequality.

Next we derive from (3.6.58) the result

$$\begin{aligned} \sqrt{\int_0^t (\|\theta\|^2 + \|c\|^2)ds \int_0^t (\|\phi\|^2 + \|\psi\|^2)ds} & = \sqrt{S^2 + \frac{(F')^2}{4}} \\ & \leq S + \frac{F'}{2}. \end{aligned} \tag{3.6.59}$$

The sixth, seventh and eighth terms on the right of (3.6.57) are bounded as follows, using (3.6.56),

$$\begin{aligned} & -\int_0^t\int_\Omega \theta_{,s} u_i \Theta_{,i} dx\, ds - \int_0^t\int_\Omega c_{,s} u_i C_{,i} dx\, ds \\ & = -\int_0^t\int_\Omega \phi u_i \Theta_{,i} dx\, ds - \int_0^t\int_\Omega \psi u_i C_{,i} dx\, ds \end{aligned}$$

$$
\begin{aligned}
& -\frac{1}{2}\int_0^t\int_\Omega v_m^*\theta_{,m}u_i\Theta_{,i}dx\,ds-\frac{1}{2}\int_0^t\int_\Omega v_m^*c_{,m}u_iC_{,i}dx\,ds\\
&\geq -M\sqrt{\int_0^t\|\mathbf{u}\|^2ds\int_0^t\|\phi\|^2ds}-M\sqrt{\int_0^t\|\mathbf{u}\|^2ds\int_0^t\|\psi\|^2ds}\\
&\quad-\frac{M^2}{2}\sqrt{\int_0^t\|\mathbf{u}\|^2ds\int_0^t\|\nabla\theta\|^2ds}-\frac{M^2}{2}\sqrt{\int_0^t\|\mathbf{u}\|^2ds\int_0^t\|\nabla c\|^2ds}\\
&\geq -M\sqrt{k}\sqrt{\int_0^t\|\phi\|^2ds\int_0^t(\|\theta\|^2+\|c\|^2)ds}\\
&\quad-M\sqrt{k}\sqrt{\int_0^t\|\psi\|^2ds\int_0^t(\|\theta\|^2+\|c\|^2)ds}\\
&\quad-\frac{M^2}{4}\int_0^t\|\nabla\theta\|^2ds-\frac{M^2}{4}\int_0^t\|\nabla c\|^2ds\\
&\quad-\frac{kM^2}{2}\int_0^t(\|\theta\|^2+\|c\|^2)ds,
\end{aligned}
\tag{3.6.60}
$$

$$
\begin{aligned}
\int_0^t\int_\Omega\theta_{,s}r\,dx\,ds&=\int_0^t\int_\Omega\phi r\,dx\,ds+\frac{1}{2}\int_0^t\int_\Omega rv_m^*\theta_{,m}dx\,ds\\
&\geq-\sqrt{\int_0^t\|r\|^2ds\int_0^t\|\phi\|^2ds}\\
&\quad-\frac{TQ}{4}-\frac{M^2}{4}\int_0^t\|\nabla\theta\|^2ds.
\end{aligned}
\tag{3.6.61}
$$

Inequalities (3.6.60) and (3.6.61) are now used in (3.6.57), and the third, fourth, fifth, ninth and tenth terms are bounded using also (3.6.51), to see that:

$$
\begin{aligned}
FF''-(F')^2\geq&\,4S^2+4Q\int_0^t(\|\phi\|^2+\|\psi\|^2)ds\\
&-(1+3\sqrt{k}M)FF'-(1+T)FQ\\
&-4\sqrt{k}MF\bigg\{\sqrt{\int_0^t\|\phi\|^2ds\int_0^t(\|\theta\|^2+\|c\|^2)ds}\\
&\qquad+\sqrt{\int_0^t\|\psi\|^2ds\int_0^t(\|\theta\|^2+\|c\|^2)ds}\bigg\}\\
&-2M^2kF\int_0^t(\|\theta\|^2+\|c\|^2)ds-4F\sqrt{\int_0^t\|\phi\|^2ds\int_0^t\|r\|^2ds}
\end{aligned}
$$

$$- 3M^2F\int_0^t(\|\nabla\theta\|^2+\|\nabla c\|^2)ds. \tag{3.6.62}$$

From equation (3.6.53) we now derive an estimate for the gradient terms

$$-\int_0^t(\|\nabla\theta\|^2+\|\nabla c\|^2)ds \geq -\frac{1}{2}F' - KF, \tag{3.6.63}$$

where

$$K=\frac{M(2k+1)+\sqrt{M^2(2k+1)^2+4TM}}{2T}.$$

Inequality (3.6.63) is next utilized in (3.6.62) to obtain

$$\begin{aligned}FF''-(F')^2 \geq & 4S^2+4Q\int_0^t(\|\phi\|^2+\|\psi\|^2)ds - 3M^2KF^2\\ & -(\frac{3}{2}M^2+1+3\sqrt{k}M)FF' - (1+T)FQ\\ & - 2M^2kF\int_0^t(\|\theta\|^2+\|c\|^2)ds - 4\sqrt{T}F\sqrt{Q\int_0^t\|\phi\|^2ds}\\ & - 4\sqrt{k}MF\Bigg[\sqrt{\int_0^t\|\phi\|^2ds\int_0^t(\|\theta\|^2+\|c\|^2)ds}\\ & \qquad + \sqrt{\int_0^t\|\psi\|^2ds\int_0^t(\|\theta\|^2+\|c\|^2)ds}\Bigg].\end{aligned} \tag{3.6.64}$$

The seventh term on the right in (3.6.64) is bounded as follows

$$-4\sqrt{T}F\sqrt{Q\int_0^t\|\phi\|^2ds} \geq -TF^2 - 4Q\int_0^t\|\phi\|^2ds. \tag{3.6.65}$$

Inequality (3.6.65) is employed in (3.6.64), and the last term in (3.6.64) is estimated with the aid of (3.6.59) to obtain

$$\begin{aligned}FF''-(F')^2 \geq & 4S^2-4M\sqrt{2k}FS-(3M^2K+1+2T+2kM^2)F^2\\ & -(\frac{3}{2}M^2+1+(3+2\sqrt{2})\sqrt{k}M)FF'.\end{aligned} \tag{3.6.66}$$

Upon completion of the square we obtain the fundamental inequality

$$FF''-(F')^2 \geq -k_1FF' - k_2F^2, \tag{3.6.67}$$

where now

$$k_1 = \frac{3}{2}M^2 + 1 + (3 + 2\sqrt{2})\sqrt{k}M,$$
$$k_2 = M^2(3K + 4k) + 1 + 2T.$$

Inequality (3.6.67) is integrated exactly as in section 1.5, to obtain in this case

$$\int_0^t (\|\theta\|^2 + \|c\|^2)ds \leq e^{-\mu t} K^{(1-\sigma)/(1-\sigma_2)} \left[\sup_{t\in[0,T]} \|r\|^2\right]^{(\sigma-\sigma_2)/(1-\sigma_2)}, \qquad (3.6.68)$$

which clearly establishes continuous dependence of θ, c on r, on compact subintervals of $[0, T)$. Furthermore, we may then utilize (3.6.56) to deduce a similar inequality to (3.6.68) for $\|\mathbf{u}\|^2$, and hence derive continuous dependence of $\mathbf{u}$ on r, again, on any compact subinterval of $[0, T)$.

3.7 Modeling Errors in Theories of Heat Conduction with Finite Propagation Speed

There has been an enormous explosion of activity in the field of heat propagation at low temperatures; this is also known as the phenomenon of second sound. Extensive reviews of second sound theories are available, see e.g. Chandrasekharaiah (1986), Dreyer & Struchtrup (1993), Fichera (1992), Jou *et al.* (1988), Morro & Ruggeri (1984), and Müller (1966,1967,1985). Experimental work is briefly reviewed in Caviglia *et al.* (1992) and explains the need for a theoretical model. Theories have often centered on ramifications of the heat flux relaxation model of Cattaneo (1948), usually also associated with the work of Maxwell (1867), or the Müller approach, see e.g. Müller (1985) which includes the rate of change of temperature, $\dot{T}$, as an independent variable in the constitutive theory. Fichera (1992) gives a critical appraisal of the work of Cattaneo, and especially others, and puts it firmly into context with what Maxwell did. His excellent article is a pointed account of this topic. Since it has became fashionable to associate Maxwell's name with second sound theories we shall, therefore, follow this convention.

Our goal here is to follow Franchi & Straughan (1993d) and describe their analysis of properties of solutions to the classical Maxwell - Cattaneo

(MC) model:

$$\begin{aligned} \tau\frac{\partial Q_i}{\partial t}+Q_i&=-\kappa\frac{\partial T}{\partial x_i},\\ c\frac{\partial T}{\partial t}&=-\frac{\partial Q_i}{\partial x_i}, \end{aligned} \tag{3.7.1}$$

where Q_i, T are heat flux and temperature, τ is the relaxation time, κ thermal conductivity, and c specific heat. They also investigate a generalization of (3.7.1) which accounts for space correlation at low temperatures. This is the generalized Maxwell - Cattaneo (GMC) model, which is described in Bampi *et al.* (1981) and Jou *et al.* (1988) §4,

$$\begin{aligned} \tau\frac{\partial Q_i}{\partial t}+Q_i&=-\kappa\frac{\partial T}{\partial x_i}+\mu\Delta Q_i+\nu\frac{\partial^2 Q_j}{\partial x_j\partial x_i},\\ c\frac{\partial T}{\partial t}&=-\frac{\partial Q_i}{\partial x_i}. \end{aligned} \tag{3.7.2}$$

A thermodynamic derivation of (3.7.2) may be found in Morro *et al.* (1990).

To account for the experimentally observed wavespeed behavior with temperature over a sufficiently large temperature range, the MC model has been found inadequate, in that the form of (3.7.1) holds good, but the coefficients must depend on temperature and the rate of change of temperature, thereby converting (3.7.1) into a nonlinear model. In fact, a thermodynamically consistent theory was developed by Morro & Ruggeri (1988) who derive the equations

$$\begin{aligned} F(T)\frac{\partial Q_i}{\partial t}+\kappa\frac{\partial T}{\partial x_i}+\Big(1+\Gamma\frac{\partial T}{\partial t}\Big)Q_i&=0,\\ c_0(T)\frac{\partial T}{\partial t}+\frac{\partial Q_i}{\partial x_i}&=0, \end{aligned} \tag{3.7.3}$$

where

$$\begin{aligned} \Gamma&=-\kappa\big[5\tilde{A}T^{-4}+(5-n)\tilde{B}T^{n-4}\big],\\ F&=\kappa[\tilde{A}T^{-3}+\tilde{B}T^{n-3}], \end{aligned}$$

$\tilde{A}, \tilde{B}, n$ being determined from experiments. The function $c_0(T)$ is the equilibrium specific heat, and fitting to experimental data suggests

$$c_0(T)=\epsilon T^3,$$

ϵ constant.

Franchi & Straughan (1993d) concentrate on the linearized version of (3.7.3), namely (3.7.1); however, it is important to realize that for a description of nonlinear wave motion, the highly nonlinear form (3.7.3) is necessary.

Continuous Dependence on the Relaxation Time for the MC Model Backward in Time

In this section we study improperly posed problems associated with the heat propagation theories referred to above. In particular, we study the continuous dependence of the solution to (3.7.1) and (3.7.2), upon changes in the relaxation time τ, for the backward in time problem.

Without loss of generality, for the study envisaged here, we may take $c = \kappa = 1$ in (3.7.1) and (3.7.2), and we treat τ, μ, ν as constants.

To study continuous dependence of solutions to (3.7.1) on τ, backward in time, Franchi & Straughan (1993d) obtain weaker restrictions on solutions by first integrating (3.7.1) with respect to time. Hence, consider (3.7.1) with $t < 0$ by reversing the sign of the time derivative terms, and let the modified (3.7.1) be defined on the Cartesian product of a bounded spatial domain Ω with a time interval $(0, \tilde{\theta})$. Then define

$$S(\mathbf{x}, t) = \int_0^t T(\mathbf{x}, s)ds, \qquad R_i(\mathbf{x}, t) = \int_0^t Q_i(\mathbf{x}, s)ds, \tag{3.7.4}$$

and note S, R_i satisfy

$$\begin{aligned} \tau\frac{\partial R_i}{\partial t} &= R_i + \frac{\partial S}{\partial x_i} + \tau Q_i(\mathbf{x}, 0), \\ \frac{\partial S}{\partial t} &= \frac{\partial R_i}{\partial x_i} + T(\mathbf{x}, 0). \end{aligned} \tag{3.7.5}$$

We suppose (Q_i, T) and (Q_i^*, T^*) [equivalently (R_i, S), (R_i^*, S^*)] are solutions to (3.7.5) with the *same* boundary and initial data but with (R_i, S) satisfying (3.7.5) for a constant τ while (R_i^*, S^*) satisfies (3.7.5) for another constant τ^*, with $0 < \tau < \tau^*$.

Define the difference variables r_i, s, β by

$$r_i = R_i^* - R_i, \quad s = S^* - S, \quad \beta = \tau^* - \tau, \tag{3.7.6}$$

and then we observe (r_i, s) satisfies:

$$\begin{aligned} \tau^*\frac{\partial r_i}{\partial t} &= -\beta\frac{\partial R_i}{\partial t} + r_i + \frac{\partial s}{\partial x_i} + \beta Q_i(\mathbf{x}, 0), \\ \frac{\partial s}{\partial t} &= \frac{\partial r_i}{\partial x_i}, \end{aligned} \tag{3.7.7}$$

on $\Omega \times (0, \tilde{\theta})$, with

$$s = r_i = 0, \tag{3.7.8}$$

on the boundary Γ, of Ω, and when $t = 0$.

Even though this is a backward in time problem, the structure of the MC system allows us to obtain continuous dependence on τ by means of a relatively straightforward energy method. To do this we let $\|\cdot\|$ denote the norm on $L^2(\Omega)$, and then we multiply $(3.7.7)_1$ by r_i, $(3.7.7)_2$ by s, integrate over Ω and use (3.7.8) to see that

$$\begin{aligned}\frac{d}{dt}\Big(\frac{1}{2}\|s\|^2+\frac{1}{2}\tau^*\|\mathbf{r}\|^2\Big) &=\|\mathbf{r}\|^2-\beta\int_\Omega Q_i r_i dx+\beta\int_\Omega Q_i^0 r_i dx,\\ &\le\|\mathbf{r}\|^2+\beta\|\mathbf{r}\|(\|\mathbf{Q}\|+\|\mathbf{Q}^0\|), \end{aligned} \tag{3.7.9}$$

where we have used the Cauchy-Schwarz inequality, and where

$$Q_i^0\equiv Q_i(\mathbf{x},0).$$

Suppose now the constraint set is defined by

$$\|\mathbf{Q}(t)\|\le M, \qquad \forall t\in[0,\tilde\theta]. \tag{3.7.10}$$

Then, put

$$F(t)=\frac{1}{2}\|s\|^2+\frac{1}{2}\tau^*\|\mathbf{r}\|^2, \tag{3.7.11}$$

to see that from (3.7.9) and (3.7.10),

$$\frac{dF}{dt}\le\frac{2}{\tau^*}F+\frac{2^{3/2}\beta M}{\sqrt{\tau^*}}\sqrt{F}. \tag{3.7.12}$$

By using an integrating factor we may integrate this to obtain

$$\sqrt{F(t)}\le(e^{t/\tau^*}-1)\sqrt{2\tau^*}\,\beta M.$$

Thus,

$$\frac{1}{2}\|s(t)\|^2+\frac{1}{2}\tau^*\|\mathbf{r}(t)\|^2\le 2\tau^* M^2(e^{\tilde\theta/\tau^*}-1)^2\,\beta^2, \tag{3.7.13}$$

$\forall t\in(0,\tilde\theta)$, which shows s, r_i depend continuously on β, i.e. continuous dependence on τ is established.

Continuous Dependence on the Relaxation Time for the GMC System Backward in Time

We now give an exposition of the work of Franchi & Straughan (1993d) which establishes a result analogous to that above, but for the GMC system (3.7.2). An energy method does not work because of the presence of the μ,ν terms. Instead they employ a Lagrange identity argument.

Take now $c = \kappa = 1$ in (3.7.2), treat τ, μ, ν constants, and define S, R_i as in (3.7.4). From the basic equations (3.7.2) for $t < 0$ we may then deduce that R_i and S satisfy the partial differential equations

$$\begin{aligned}\tau\frac{\partial R_i}{\partial t} &= R_i + \frac{\partial S}{\partial x_i} - \mu\Delta R_i - \nu\frac{\partial^2 R_j}{\partial x_j \partial x_i} + \tau Q_i(\mathbf{x},0),\\ \frac{\partial S}{\partial t} &= \frac{\partial R_i}{\partial x_i} + T(\mathbf{x},0),\end{aligned} \tag{3.7.14}$$

where (3.7.14) are defined on $\Omega \times (0, \tilde{\theta})$.

We again define the difference variables r_i, s, β as in (3.7.6), and they here satisfy the equations

$$\tau^*\frac{\partial r_i}{\partial t} = -\beta\frac{\partial R_i}{\partial t} + r_i + \frac{\partial s}{\partial x_i} - \mu\Delta r_i - \nu\frac{\partial^2 r_j}{\partial x_j \partial x_i} + \beta Q_i(\mathbf{x},0), \tag{3.7.15}$$

$$\frac{\partial s}{\partial t} = \frac{\partial r_i}{\partial x_i}. \tag{3.7.16}$$

The boundary and initial data for s and r_i are again zero as in (3.7.8).

To form the Lagrange identity we take (3.7.15) at $t = u$ and multiply by $r_i(2t-u)$ and integrate over Ω, then integrate over $(0,t)$, and then multiply (3.7.16) by $s(2t-u)$ and do the same integrations, to obtain

$$\int_0^t\int_\Omega \dot{s}(u)s(2t-u)dx\,du = \int_0^t\int_\Omega r_{i,i}(u)s(2t-u)dx\,du, \tag{3.7.17}$$

$$\begin{aligned}\tau^*\int_0^t\int_\Omega \dot{r}_i(u)r_i(2t-u)dx\,du &= -\beta\int_0^t\int_\Omega Q_i(u)r_i(2t-u)dx\,du\\ &+\int_0^t\int_\Omega r_i(u)r_i(2t-u)dx\,du + \int_0^t\int_\Omega s_{,i}(u)r_i(2t-u)dx\,du\\ &+\mu\int_0^t\int_\Omega r_{i,j}(u)r_{i,j}(2t-u)dx\,du + \nu\int_0^t\int_\Omega r_{i,i}(u)r_{j,j}(2t-u)dx\,du\\ &+\beta\int_0^t\int_\Omega Q_i(0)r_i(2t-u)dx\,du,\end{aligned} \tag{3.7.18}$$

where the superposed dot denotes partial differentiation with respect to the time argument, e.g. in (3.7.17),

$$\dot{s}(u) \equiv \frac{\partial s}{\partial u}(\mathbf{x},u).$$

We now reverse the roles of $u, 2t-u$ in (3.7.17), (3.7.18) to find

$$\int_0^t\int_\Omega \dot{s}(2t-u)s(u)dx\,du = \int_0^t\int_\Omega r_{i,i}(2t-u)s(u)dx\,du, \tag{3.7.19}$$

$$\begin{aligned}\tau^* \int_0^t \int_\Omega \dot{r}_i(2t-u)r_i(u)dx\,du = &-\beta \int_0^t \int_\Omega Q_i(2t-u)r_i(u)dx\,du \\ &+ \int_0^t \int_\Omega r_i(2t-u)r_i(u)dx\,du + \int_0^t \int_\Omega s_{,i}(2t-u)r_i(u)dx\,du \\ &+ \mu \int_0^t \int_\Omega r_{i,j}(2t-u)r_{i,j}(u)dx\,du + \nu \int_0^t \int_\Omega r_{i,i}(2t-u)r_{j,j}(u)dx\,du \\ &+ \beta \int_0^t \int_\Omega Q_i(0)r_i(u)dx\,du. \end{aligned} \tag{3.7.20}$$

The next step involves subtracting equation (3.7.20) from (3.7.18), and then subtracting equation (3.7.17) from (3.7.19). The two resulting equations are added. After integration, recalling the initial data, we deduce

$$\tau^* \|\mathbf{r}(t)\|^2 - \|s(t)\|^2 = \beta \int_0^t H(u)du, \tag{3.7.21}$$

where

$$\begin{aligned} H(u) = &\int_\Omega Q_i(2t-u)r_i(u)dx - \int_\Omega Q_i(u)r_i(2t-u)dx \\ &+ \int_\Omega Q_i(0)r_i(2t-u)dx - \int_\Omega Q_i(0)r_i(u)dx. \end{aligned} \tag{3.7.22}$$

In this section we suppose Q_i, R_i and R_i^* satisfy the constraint (3.7.10) on $[0, 2\tilde{\theta}]$. Then, note that from (3.7.21) we derive immediately

$$\tau^* \|\mathbf{r}(t)\|^2 - \|s(t)\|^2 \le 4\tilde{\theta} M^2 \beta. \tag{3.7.23}$$

At this stage Franchi & Straughan (1993d) employ a device introduced by Professor Lawrence E. Payne. Observe that from (3.7.21),

$$\|s(t)\|^2 = 2\|s(t)\|^2 - \tau^* \|\mathbf{r}(t)\|^2 + \beta \int_0^t H(u)du.$$

Differentiate this and substitute fom (3.7.15), (3.7.16) to see that

$$\begin{aligned} \frac{d}{dt}\|s(t)\|^2 = &6 \int_\Omega s r_{i,i} dx - 2\|\mathbf{r}\|^2 - 2\mu \int_\Omega r_{i,j} r_{i,j} dx - 2\nu \|r_{i,i}\|^2 \\ &+ \beta H(t) + 2\beta \int_\Omega Q_i r_i dx - 2\beta \int_\Omega Q_i(0) r_i dx. \end{aligned} \tag{3.7.24}$$

The last three terms are immediately bounded using the constraint, by $8M^2\beta$. Furthermore, using the arithmetic-geometric mean inequality

$$6 \int_\Omega s r_{i,i} dx \le \frac{9}{2\nu} \|s\|^2 + 2\nu \|r_{i,i}\|^2. \tag{3.7.25}$$

Thus, use of (3.7.25) in (3.7.24) leads to

$$\frac{d}{dt}\|s(t)\|^2 \le \frac{9}{2\nu}\|s\|^2 + 8M^2\beta. \tag{3.7.26}$$

This inequality is easily integrated to yield

$$\|s(t)\|^2 \le K\beta, \tag{3.7.27}$$

where

$$K = \frac{16\nu M^2}{9}\big(e^{9\tilde{\theta}/2\nu} - 1\big).$$

This inequality is the required continuous dependence estimate for s. A similar continuous dependence estimate for r_i then follows from (3.7.23). This has form

$$\tau^*\|\mathbf{r}(t)\|^2 \le (K + 4\tilde{\theta}M^2)\beta. \tag{3.7.28}$$

Thus continuous dependence has been established on the parameter τ for the GMC system.

Continuous Dependence on Modeling for the MC and GMC Systems

In this part we review the work of Franchi & Straughan (1993d) which studies continuous dependence on the model itself. We commence with a result for the GMC system. Thus, let (R_i, S) be a solution to (3.7.14) for $\tau(> 0)$ and let (R_i^*, S^*) be a solution to (3.7.14) with $\tau = 0$. We wish to derive conditions which ensure (R_i, S) remains close to (R_i^*, S^*) for τ small. Suppose (R_i, S) and (R_i^*, S^*) satisfy the same boundary and initial data. Define now the variables (r_i, s) by

$$r_i = R_i^* - R_i, \qquad s_i = S_i^* - S_i,$$

and then (r_i, s) satisfy the system of equations:

$$\begin{aligned} r_i + \frac{\partial s}{\partial x_i} - \mu\Delta r_i - \nu\frac{\partial^2 r_j}{\partial x_j \partial x_i} &= -\,\tau Q_i + \tau Q_i(0), \\ \frac{\partial s}{\partial t} &= \frac{\partial r_i}{\partial x_i}, \end{aligned} \tag{3.7.29}$$

in $\Omega \times (0, \tilde{\theta})$, with $r_i = s = 0$ on Γ and at $t = 0$.

We may first of all use a Lagrange identity method. To do this we follow the steps leading to (3.7.21). In this case we find

$$\begin{aligned} \|s(t)\|^2 =& \tau\int_0^t\int_\Omega r_i(u)\big[Q_i(2t-u) - Q_i(0)\big]\,dx\,du \\ & - \tau\int_0^t\int_\Omega r_i(2t-u)\big[Q_i(u) - Q_i(0)\big]\,dx\,du. \end{aligned} \tag{3.7.30}$$

If we request that Q_i, R_i, R_i^* satisfy the constraint (3.7.10) on $[0, 2\tilde{\theta}]$, then from (3.7.29) one easily derives

$$\|s(t)\|^2 \le 4M^2\tilde{\theta}\,\tau\,. \qquad (3.7.31)$$

This establishes continuous dependence on modeling for s, for $t \in [0, \tilde{\theta}]$. To continue note that multiplication of $(3.7.29)_1$ by r_i and integration over Ω leads to

$$\begin{aligned}\|\mathbf{r}\|^2 = &\int_\Omega r_{i,i}s\,dx - \mu\int_\Omega r_{i,j}r_{i,j}dx - \nu\|r_{i,i}\|^2\\ &+\tau\int_\Omega \big[Q_i(0) - Q_i\big]r_i dx.\end{aligned} \qquad (3.7.32)$$

To proceed, one employs the arithmetic-geometric mean inequality on the first term on the right of (3.7.32). This is followed by use of the Cauchy-Schwarz inequality and the bound imposed by the constraint set. Finally utilizing (3.7.31) leads to

$$\|\mathbf{r}\|^2 \le (2M^2\tilde{\theta} + 4M^2\tau)\tau. \qquad (3.7.33)$$

This inequality together with the bound (3.7.31) clearly establishes continuous dependence on modeling for $t \in [0, \tilde{\theta}]$.

To demonstrate continuous dependence on modeling for the MC system Franchi & Straughan (1993d) use a logarithmic convexity method. This has the advantage over the Lagrange identity method in that the constraint is imposed only on $[0, \tilde{\theta}]$; however, the constraint set is stronger. We give their general proof for a system which encompasses both the MC and GMC systems.

Observe that if we take the x_i derivative of $(3.7.29)_1$ and use $(3.7.29)_2$, we find s satisfies the equation

$$\big[1 - (\mu+\nu)\Delta\big]\frac{\partial s}{\partial t} = -\Delta s - f, \qquad (3.7.34)$$

where

$$f = \tau\big[Q_{i,i}(\mathbf{x}, t) - Q_{i,i}(\mathbf{x}, 0)\big]. \qquad (3.7.35)$$

Thus, if we restrict attention to the MC system, (3.7.34) may be regarded as a forced diffusion equation whereas the GMC version is a forced modified diffusion equation.

A generalization which includes equation (3.7.34) may be obtained by considering the abstract equation in a Hilbert space, H,

$$Au_t = Lu - f, \qquad (3.7.36)$$

with $u(0) = 0$, where A, L are densely defined symmetric, linear operators, with A positive definite, i.e.

$$(Ax, x) \geq \zeta \|x\|^2, \qquad \forall x \in D,$$

for some $\zeta > 0$, where D is the domain of A and where $\|\cdot\|$ and $(\,,\,)$ now denote the norm and inner product on H.

To establish continuous dependence on modeling we prove u depends continuously on f. Hence, let $F(t)$ be defined by

$$F(t) = \int_0^t (u, Au)ds + \sup_{t\in[0,\tilde{\theta}]} \|f(t)\|^2. \tag{3.7.37}$$

Differentiate F,

$$\begin{aligned} F' =&(u, Au),\\ =&2\int_0^t (u, Au_s)ds,\\ =&2\int_0^t (u, Lu)ds - 2\int_0^t (u, f)ds, \end{aligned}$$

and differentiate again,

$$\begin{aligned} F'' =&4\int_0^t (u_s, Lu)ds - 2(u, f),\\ =&4\int_0^t (u_s, Au_s)ds + 4\int_0^t (u_s, f)ds - 2(u, f). \end{aligned}$$

We now form $FF'' - (F')^2$ to find

$$\begin{aligned} FF'' - (F')^2 =&4S^2 + 4Q\int_0^t (u_s, Au_s)ds + 4F\int_0^t (u_s, f)ds\\ &- 2F(u, f), \end{aligned} \tag{3.7.38}$$

where

$$Q = \sup_{t\in[0,\tilde{\theta}]} \|f(t)\|^2,$$

and

$$S^2 = \int_0^t (u, Au)ds \int_0^t (u_s, Au_s)ds - \left(\int_0^t (u, Au_s)ds\right)^2 \ (\geq 0).$$

By using the properties of F and A, we may obtain

$$-2F(u,f) \geq -\frac{1}{\zeta}FF' - F^2, \tag{3.7.39}$$

and by use of the arithmetic - geometric mean inequality,

$$4F\int_0^t (u_s, f)ds \geq -\frac{\tilde{\theta}}{\zeta}F^2 - 4Q\int_0^t (u_s, Au_s)ds. \tag{3.7.40}$$

Upon combining (3.7.39) and (3.7.40) in (3.7.38), we may show that

$$FF'' - (F')^2 \geq -k_1 FF' - k_2 F^2, \tag{3.7.41}$$

where

$$k_1 = \frac{1}{\zeta}, \quad k_2 = 1 + \frac{\tilde{\theta}}{\zeta}.$$

We may integrate inequality (3.7.41) as in section 1.5 by putting $\sigma = e^{-k_1 t}$, $\sigma_1 = 1$, $\sigma_2 = e^{-k_1\tilde{\theta}}$, to find

$$F(t) \leq \Big[F(0)\Big]^{(\sigma-\sigma_2)/(1-\sigma_2)}\Big[F(\tilde{\theta})e^{\alpha\tilde{\theta}}\Big]^{(1-\sigma)/(1-\sigma_2)} e^{-\alpha t}, \tag{3.7.42}$$

$t \in [0, \tilde{\theta})$, where $\alpha = k_2/k_1$.

To employ (3.7.42) to establish continuous dependence on the modeling, i.e. to derive continuous dependence on τ, suppose now a bound is known for $F(\tilde{\theta})$. Hence, we assume

$$F(\tilde{\theta})e^{\alpha\tilde{\theta}} \leq K.$$

Then, (3.7.42) allows us to see that

$$\int_0^t (u, Au)ds \leq e^{-\alpha t}K^{(1-\sigma)/(1-\sigma_2)}\Big(\sup_{t\in[0,\tilde{\theta}]} \|f(t)\|^2\Big)^{(\sigma-\sigma_2)/(1-\sigma_2)}, \tag{3.7.43}$$

for $t \in [0, \tilde{\theta})$. Inequality (3.7.43) is the basic estimate which establishes Hölder continuous dependence of the solution on compact subintervals of $[0, \tilde{\theta})$.

Remark

As Franchi & Straughan observe, inequality (3.7.43) interpreted in terms of the MC theory takes the form

$$\int_0^t \|s(v)\|^2 dv \leq e^{-\alpha t}K^{(1-\sigma)/(1-\sigma_2)}K^*\tau^{2(\sigma-\sigma_2)/(1-\sigma_2)},$$

where

$$K^* = \left[\sup_{t \in [0,\tilde{\theta}]} \|Q_{i,i}(t) - Q_{i,i}(0)\|^2 \right]^{(\sigma-\sigma_2)/(1-\sigma_2)},$$

while the corresponding inequality for the GMC system reads,

$$\int_0^t \left(\|s(\eta)\|^2 + (\nu+\mu)\|\nabla s(\eta)\|^2 \right) d\eta \leq e^{-\alpha t} K^{(1-\sigma)/(1-\sigma_2)} K^* \tau^{2(\sigma-\sigma_2)/(1-\sigma_2)}.$$

An analogous inequality for $\|\mathbf{r}\|^2$ may be found from $(3.7.29)_1$ for the MC theory and from (3.7.32) for the GMC theory. These estimates clearly demonstrate how the solution difference depends on τ.

4

Continuous Dependence on Modeling Forward in Time

4.1 Modeling Errors for the Navier-Stokes Equations Forward in Time

A beautiful and practical illustration of the modeling problem may be obtained by studying how a solution to the Navier-Stokes equations will converge to a solution to the Stokes equations as the nonlinearities are "neglected" in a certain sense. Our consideration of the effects of modeling errors for the Navier-Stokes equations will begin with a comparison of solutions to the well posed forward in time problem. Not only is this highly illustrative and important, it preceded the improperly posed backward in time study, and was first tackled by Payne (1987b).

The Stokes, Navier - Stokes Problem Forward in Time

The study of Payne (1987b) begins by looking for a vector u_i, $(i = 1,2,3)$ of the problem:

$$\left.\begin{aligned} &\frac{\partial u_i}{\partial t} + u_j \frac{\partial u_i}{\partial x_j} = \nu \Delta u_i - \frac{\partial p}{\partial x_i}, \\ &\frac{\partial u_i}{\partial x_i} = 0, \end{aligned}\right\} \quad \text{in} \quad \Omega \times (0,T), \tag{4.1.1$_1$}$$

$$\begin{aligned} &u_i = 0, \qquad \text{on} \quad \partial\Omega \times [0,T], \\ &u_i(x,0) = \epsilon f_i(x), \qquad x \in \Omega, \end{aligned} \tag{4.1.1$_2$}$$

where Ω is a bounded region in $\mathbf{R}^3$ with a smooth boundary $\partial\Omega$. We shall say problem (4.1.1) is composed of the partial differential equations $(4.1.1)_1$ solved subject to the boundary and initial conditions $(4.1.1)_2$, and continue this usage of notation when the numbering of a boundary-initial value problem is split. Problem (4.1.1) is a Dirichlet boundary - initial value problem for the Navier-Stokes equations defined on a bounded domain Ω with p the unknown pressure and u_i the components of velocity. One question that arises is if ϵ is very small, can the solution to (4.1.1) be approximated adequately by ϵv_i, where v_i is a solution to the Stokes boundary -initial value problem:

$$\left.\begin{aligned} \frac{\partial v_i}{\partial t} &= \nu\Delta v_i - \frac{\partial q}{\partial x_i}, \\ \frac{\partial v_i}{\partial x_i} &= 0, \end{aligned}\right\} \quad \text{in} \quad \Omega\times(0,T), \tag{4.1.2$_1$}$$

$$\begin{aligned} v_i &= 0, \qquad \text{on} \quad \partial\Omega\times[0,T], \\ v_i(x,0) &= f_i(x), \qquad x\in\Omega. \end{aligned} \tag{4.1.2$_2$}$$

Payne (1987b) compares u_i and ϵv_i, and in the process derives a continuous dependence on ϵ result.

We here give a slightly different treatment from that of Payne (1987b), and so set

$$w_i = u_i - \epsilon v_i, \tag{4.1.3}$$

and observe that w_i satisfies:

$$\left.\begin{aligned} \frac{\partial w_i}{\partial t} + u_j\frac{\partial w_i}{\partial x_j} + \epsilon u_j\frac{\partial v_i}{\partial x_j} &= \nu\Delta w_i - \frac{\partial P}{\partial x_i}, \\ \frac{\partial w_i}{\partial x_i} &= 0, \end{aligned}\right\} \quad \text{in} \quad \Omega\times(0,T), \tag{4.1.4$_1$}$$

$$\begin{aligned} w_i &= 0, \qquad \text{on} \quad \partial\Omega\times[0,T], \\ w_i(x,0) &= 0, \qquad x\in\Omega, \end{aligned} \tag{4.1.4$_2$}$$

where $P = p - \epsilon q$. Define the function $\phi(t)$ by

$$\phi(t) = \int_\Omega w_iw_i dx \equiv \|w\|^2, \tag{4.1.5}$$

and we derive an estimate for $\phi(t)$ in terms of ϵ. Upon differentiation of (4.1.5) we find

$$\begin{aligned} \frac{d\phi}{dt} &= 2\int_\Omega w_iw_{i,t}dx, \\ &= 2\int_\Omega w_i\big[\nu\Delta w_i - P_{,i} - u_jw_{i,j} - \epsilon u_jv_{i,j}\big]dx, \\ &= -2\nu\int_\Omega w_{i,j}w_{i,j}dx - 2\epsilon\int_\Omega w_iu_jv_{i,j}dx, \end{aligned} \tag{4.1.6}$$

after substituting the differential equation and integrating by parts. Rewriting (4.1.6) using (4.1.3), we have

$$\frac{d\phi}{dt} = -2\nu\int_\Omega w_{i,j}w_{i,j}dx - 2\epsilon\int_\Omega w_iw_jv_{i,j}dx + 2\epsilon^2\int_\Omega w_iv_jv_{i,j}dx, \tag{4.1.7}$$

and then the last term may be integrated by parts to obtain

$$\frac{d\phi}{dt} = -2\nu \int_\Omega w_{i,j}w_{i,j}dx - 2\epsilon \int_\Omega w_iw_jv_{i,j}dx - 2\epsilon^2 \int_\Omega w_{i,j}v_iv_j\, dx. \quad (4.1.8)$$

The Cauchy - Schwarz inequality is used on the last two terms to derive

$$\begin{aligned}\frac{d\phi}{dt} \leq & -2\nu \int_\Omega w_{i,j}w_{i,j}dx + 2\epsilon\Big(\int_\Omega (w_iw_i)^2dx\Big)^{1/2}\|\nabla v\| \\ & + 2\epsilon^2\|\nabla w\|\Big(\int_\Omega (v_jv_j)^2dx\Big)^{1/2}. \end{aligned} \quad (4.1.9)$$

The next step is to employ the Sobolev inequality (in $\Omega \subset \mathbf{R}^3$),

$$\gamma \int_\Omega (\psi_i\psi_i)^2dx \leq \|\psi\|\|\nabla\psi\|^3, \quad (4.1.10)$$

where the Sobolev constant γ is defined as,

$$\gamma = \inf_{\psi_i \in H_0^1(\Omega)} \frac{\|\psi\|\,\|\nabla\psi\|^3}{\int_\Omega (\psi_i\psi_i)^2dx}. \quad (4.1.11)$$

This then allows us to see that

$$\begin{aligned}\frac{d\phi}{dt} \leq & -2\nu\|\nabla w\|^2 + 2\frac{\epsilon}{\sqrt{\gamma}}\|w\|^{1/2}\|\nabla w\|^{3/2}\|\nabla v\| \\ & + 2\frac{\epsilon^2}{\sqrt{\gamma}}\|\nabla w\|\|v\|^{1/2}\|\nabla v\|^{3/2}. \end{aligned} \quad (4.1.12)$$

A bound for $\|\nabla v\|$ is found by observing that

$$\begin{aligned}\frac{d}{dt}\int_\Omega v_{i,j}v_{i,j}dx &= 2\int_\Omega v_{i,j}v_{i,jt}dx, \\ &= -2\int_\Omega (\Delta v_i)v_{i,t}dx, \\ &= -\frac{2}{\nu}\int_\Omega v_{i,t}v_{i,t}dx, \end{aligned} \quad (4.1.13)$$

where integration by parts has been employed and the differential equation has been used in the last step. Using the Cauchy-Schwarz inequality, we may see that

$$\int_\Omega v_iv_{i,t}\, dx \leq \|v\|\sqrt{\int_\Omega v_{i,t}v_{i,t}\, dx}\,,$$

and upon use of this in (4.1.13) there follows

$$\frac{d}{dt}\int_\Omega v_{i,j}v_{i,j}dx \le -\frac{2}{\nu}\left[\int_\Omega v_iv_{i,t}dx\right]^2 \Big/ \|v\|^2 . \tag{4.1.14}$$

Let us now introduce the constant λ_1 which is the first eigenvalue of the membrane problem:

$$\begin{aligned} \Delta\psi + \lambda\psi &= 0, \qquad \text{in} \quad \Omega,\\ \psi &= 0, \qquad \text{on} \quad \partial\Omega. \end{aligned} \tag{4.1.15}$$

Inequality (4.1.14) may then be rewritten after resubstituting the differential equation, and then after using Poincaré's inequality it follows that

$$\begin{aligned} \frac{d}{dt}\int_\Omega v_{i,j}v_{i,j}dx \le& -2\nu\left[\int_\Omega v_{i,j}v_{i,j}dx\right]^2 \Big/ \|v\|^2 ,\\ \le& -2\nu\lambda_1\int_\Omega v_{i,j}v_{i,j}dx. \end{aligned} \tag{4.1.16}$$

If we let

$$F(t) = \int_\Omega v_{i,j}v_{i,j}dx,$$

then (4.1.16) becomes

$$\frac{dF}{dt} \le -2\nu\lambda_1 F. \tag{4.1.17}$$

Integration of (4.1.17) using the fact that

$$F(0) = \int_\Omega f_{i,j}f_{i,j}dx,$$

yields the bound

$$\|\nabla v(t)\|^2 \le \|\nabla f\|^2 e^{-2\nu\lambda_1 t}. \tag{4.1.18}$$

A not unrelated calculation leads to a similar bound for $\|v\|^2$, namely

$$\|v(t)\|^2 \le \|f\|^2 e^{-2\nu\lambda_1 t}. \tag{4.1.19}$$

Making use of (4.1.18) and (4.1.19) in (4.1.12), we then have

$$\begin{aligned} \frac{d\phi}{dt} \le& -2\nu\|\nabla w\|^2 + 2\frac{\epsilon}{\lambda_1^{1/4}\sqrt{\gamma}}\|\nabla f\|\,\|\nabla w\|^2 e^{-\nu\lambda_1 t}\\ &+ 2\frac{\epsilon^2}{\sqrt{\gamma}}\|f\|^{1/2}\|\nabla f\|^{3/2}\|\nabla w\| e^{-2\nu\lambda_1 t}. \end{aligned} \tag{4.1.20}$$

Application of the arithmetic-geometric mean inequality to the last term on the right hand side of (4.1.20) gives

$$\frac{d\phi}{dt} \leq -2\Big(\nu - \frac{\epsilon}{\lambda_1^{1/4}\sqrt{\gamma}}\|\nabla f\|e^{-\nu\lambda_1 t} - \alpha\Big)\|\nabla w\|^2 + \frac{\epsilon^4}{2\alpha\gamma}\|f\|\|\nabla f\|^3 e^{-4\nu\lambda_1 t}, \tag{4.1.21}$$

where α is a positive constant. We now assume that the initial data and viscosity ν are such that

$$\epsilon < \frac{\nu\lambda_1^{1/4}\gamma^{1/2}}{\|\nabla f\|}, \tag{4.1.22}$$

and choose α as

$$\alpha = \frac{1}{2}\Big(\nu - \frac{\epsilon\|\nabla f\|}{\lambda_1^{1/4}\sqrt{\gamma}}\Big). \tag{4.1.23}$$

It then follows from (4.1.21) that

$$\frac{d\phi}{dt} \leq -2\alpha\|\nabla w\|^2 + \left(\frac{\epsilon^4\|f\|\|\nabla f\|^3}{2\gamma\alpha}\right)e^{-4\nu\lambda_1 t}. \tag{4.1.24}$$

Thus, with the aid of Poincaré's inequality we find

$$\frac{d\phi}{dt} \leq -2\alpha\lambda_1\phi + \left(\frac{\epsilon^4\|f\|\|\nabla f\|^3}{2\gamma\alpha}\right)e^{-4\nu\lambda_1 t}. \tag{4.1.25}$$

Integration of the differential inequality (4.1.25) leads us to

$$\|u(t) - \epsilon v(t)\|^2 \leq K\epsilon^4(e^{-2\alpha\lambda_1 t} - e^{-4\nu\lambda_1 t}), \tag{4.1.26}$$

where the positive constant K is given by

$$K = \frac{\|f\|\|\nabla f\|^3}{(3\lambda_1^{3/4}\nu\gamma^{1/2} + 2\lambda^{1/2}\epsilon\|\nabla f\|)(\lambda_1^{1/4}\gamma^{1/2}\nu - \epsilon\|\nabla f\|)}.$$

Inequality (4.1.26) shows the modeling error to be bounded by the product of a decaying exponential and a term of $O(\epsilon^4)$. This inequality gives an estimate on the size of the error incurred in approximating the solution u_i of the nonlinear problem by the solution ϵv_i of the linear one.

4.2 Modeling Errors in Micropolar Fluid Dynamics

The theory of micropolar fluids developed by Eringen (1966) has been introduced in section 3.5. As we pointed out there the micropolar fluid equations provide additional structure to the Navier - Stokes equations by including a description of the microstructure in the fluid. Also micropolar fluids have received immense attention in the literature where the novel effects of the microstructure are often pronounced. In section 3.5 we reviewed the part of the work of Payne & Straughan (1989a) which addresses the question of how the velocity field of the micropolar equations converges to the velocity field of the Navier-Stokes equations as an interaction parameter tends to zero, in the situation where final data are prescribed. We here wish to review that part of the work of Payne & Straughan (1989a) which studies the analogous question of the manner in which the velocity field of the micropolar equations converges to the velocity field of the Navier-Stokes equations as an interaction parameter tends to zero, but for the forward in time problem.

Although the equations for a micropolar fluid are given in section 3.5, we recall them here in the interests of clarity. The continuity, momentum, and moment of momentum equations for an incompressible, isotropic micropolar fluid can be written in the form

$$\begin{aligned}\frac{\partial v_i}{\partial x_i} &= 0,\\ \frac{\partial v_i}{\partial t} + v_j\frac{\partial v_i}{\partial x_j} &= -\frac{\partial p}{\partial x_i} + (\nu+\lambda)\Delta v_i + \lambda\epsilon_{ijk}\frac{\partial n_k}{\partial x_j},\\ j\Big(\frac{\partial n_i}{\partial t} + v_j\frac{\partial n_i}{\partial x_j}\Big) &= -2\lambda n_i + \lambda\epsilon_{ijk}\frac{\partial v_k}{\partial x_j}\\ &\qquad + (\alpha+\beta)\frac{\partial^2 n_k}{\partial x_k\partial x_i} + \gamma\Delta n_i,\end{aligned} \tag{4.2.1}$$

where v_i denote the components of the fluid velocity while n_i is a particle spin vector. These equations are assumed to hold on $\Omega\times(0,T)$ where $\Omega\subset\mathbf{R}^3$ is a bounded domain and $T>0$ (T may be infinite for the initial - boundary value problem). The constants j,λ,ν, and γ are nonnegative, and satisfaction of the second law of thermodynamics requires, see Eringen (1966)

$$3\alpha+\beta+\gamma\geq 0, \qquad \gamma+\beta\geq 0, \qquad \gamma-\beta\geq 0.$$

The vectors $\mathbf{v}$ and $\mathbf{n}$ are subject to the initial and boundary conditions

$$
\begin{aligned}
v_i = n_i &= 0, \qquad \text{on} \quad \Gamma \times [0,T], \\
v_i(\mathbf{x},0) &= f_i(\mathbf{x}), \qquad \mathbf{x} \in \Omega, \\
n_i(\mathbf{x},0) &= g_i(\mathbf{x}), \qquad \mathbf{x} \in \Omega,
\end{aligned} \tag{4.2.2}
$$

where Γ is the boundary of Ω. As we noted in section 3.5, when the constants $\lambda, \alpha, \beta, \gamma$, and j are all zero equations (4.2.1) reduce to the Navier-Stokes equations. Payne & Straughan (1989a) investigate the behaviour of v_i as $\lambda \to 0$. In particular they study the relationship between v_i and the velocity field u_i of a Navier - Stokes fluid that is governed by the initial - boundary value problem

$$
\left.
\begin{aligned}
&\frac{\partial u_i}{\partial x_i} = 0, \\
&\frac{\partial u_i}{\partial t} + u_j \frac{\partial u_i}{\partial x_j} = - \frac{\partial q}{\partial x_i} + \nu \Delta u_i,
\end{aligned}
\right\} \tag{4.2.3$_1$}
$$

$$
\begin{aligned}
u_i &= 0, \qquad \text{on} \quad \Gamma \times [0,T], \\
u_i(\mathbf{x},0) &= f_i(\mathbf{x}), \qquad \mathbf{x} \in \Omega.
\end{aligned} \tag{4.2.3$_2$}
$$

Convergence in the Forward in Time Problem

Since one recovers the Navier - Stokes equations (4.2.3) from (4.2.1) in the limit $\lambda \to 0$ it is reasonable to expect $v_i \to u_i$ in some sense.

To compare the velocity fields set

$$
w_i = v_i - u_i, \qquad \text{and} \qquad \pi = p - q.
$$

Then w satisfies the equations

$$
\begin{aligned}
\frac{\partial w_i}{\partial x_i} &= 0, \\
\frac{\partial w_i}{\partial t} + v_j \frac{\partial w_i}{\partial x_j} &= - \frac{\partial \pi}{\partial x_i} + (\nu + \lambda) \Delta w_i \\
&\quad - w_j \frac{\partial u_i}{\partial x_j} + \lambda \epsilon_{ijk} \frac{\partial n_k}{\partial x_j} + \lambda \Delta u_i,
\end{aligned} \tag{4.2.4}
$$

which hold on $\Omega \times (0,T)$, and the boundary and initial conditions

$$
\begin{aligned}
w_i &= 0, \qquad \text{on} \quad \Gamma \times [0,T), \\
w_i(\mathbf{x},0) &= 0, \qquad \mathbf{x} \in \Omega.
\end{aligned} \tag{4.2.5}
$$

The goal is to derive an estimate for $\|w\|$ in terms of λ, where as usual $\|\cdot\|$ denotes the norm on $L^2(\Omega)$. We stress that we are here including an exposition of Payne & Straughan's (1989a) analysis when Ω is a bounded domain in $\mathbf{R}^3$. If $\Omega \subset \mathbf{R}^2$ the results achieved in Payne & Straughan (1989a) are much sharper.

Let us now define $\phi(t)$ as

$$\phi(t) = \|w(t)\|^2. \tag{4.2.6}$$

Differentiating and substituting (4.2.4) in the result, we obtain

$$\begin{aligned}\phi' =& 2\int_\Omega w_i w_{i,t} dx \\ =& -2(\nu+\lambda)\|\nabla w\|^2 - 2\int_\Omega u_{i,j} w_i w_j dx \\ & - 2\lambda\epsilon_{ijk}\int_\Omega w_{i,j} n_k dx - 2\lambda\int_\Omega u_{i,j} w_{i,j} dx.\end{aligned} \tag{4.2.7}$$

To bound $\int_\Omega u_{i,j} w_i w_j dx$ we make use of the Cauchy - Schwarz inequality and the following Sobolev inequality,

$$\int_\Omega \phi^4 dx \leq \Lambda\|\phi\|\,\|\nabla\phi\|^3. \tag{4.2.8}$$

(In section 3.2 we also employed (4.2.8) but Λ was there denoted by $1/\gamma$. However, γ is here reserved for the micropolar coefficient in $(4.2.1)_3$. The analogous inequality in two dimensions has form

$$\int_\Omega \phi^4 dx \leq \Lambda\|\phi\|^2\,\|\nabla\phi\|^2,$$

and it is this fact which allows Payne & Straughan (1989a) to sharpen considerably the convergence results in that case.) The result of using (4.2.8) is

$$-\int_\Omega u_{i,j} w_i w_j dx \leq \Lambda^{1/2}\phi^{1/4}\|\nabla w\|^{3/2}\|\nabla u\|,$$

which leads upon application of Young's inequality to

$$-\int_\Omega u_{i,j} w_i w_j dx \leq \frac{1}{4}\Lambda^{1/2}\big[a^3\phi\|\nabla u\|^4 + \frac{3}{a}\|\nabla w\|^2\big], \tag{4.2.9}$$

for a positive constant a. The term $\|\nabla u\|^4$ is next estimated using arguments developed by Song (1988) and Payne (1992). Observe that

$$\begin{aligned}\nu^2\|\nabla u\|^4 =& \nu^2\Big(\int_\Omega u_i \Delta u_i dx\Big)^2 \\ =& \Big(\int_\Omega u_i u_{i,t} dx\Big)^2,\end{aligned} \tag{4.2.10}$$

where the second equality follows from substitution of the differential equations in (4.2.3). Use of the Cauchy - Schwarz inequality in expression (4.2.10) then allows us to deduce that

$$\nu^2\|\nabla u\|^4 \leq \|u\|^2\|u_{,t}\|^2. \tag{4.2.11}$$

If the Navier - Stokes equations, i.e. $(4.2.3)_1$, are now multiplied by u_i and the result integrated over space and time, it follows from (4.2.3) that

$$\|u(t)\|^2 \leq e^{-2\lambda_1\nu t}\|f\|^2, \tag{4.2.12}$$

where $\lambda_1 > 0$ is the lowest eigenvalue in the membrane problem for Ω. Substituting the above two inequalities in (4.2.9) leads to

$$\begin{aligned}
&-\int_\Omega u_{i,j}w_iw_j dx \\
&\qquad \leq \frac{1}{4}\Lambda^{1/2}\Big[\frac{a^3}{\nu^2}e^{-2\lambda_1\nu t}\phi\|f\|^2\int_\Omega u_{i,t}u_{i,t}dx + \frac{3}{a}\|\nabla w\|^2\Big].
\end{aligned}\tag{4.2.13}$$

The above estimate together with the arithmetic - geometric mean inequality is next used in (4.2.7) to find that

$$\begin{aligned}
\phi' \leq &- \Big[2(\nu+\lambda) - \frac{3}{2a}\Lambda^{1/2} - b - c\Big]\|\nabla w\|^2 + \frac{2\lambda^2}{c}\|n\|^2 \\
&+ \frac{\lambda^2}{b}\|\nabla u\|^2 + \frac{a^3\Lambda^{1/2}}{2\nu^2e^{2\lambda_1\nu t}}\|f\|^2\,\phi\,\|u_{i,t}\|^2,
\end{aligned}\tag{4.2.14}$$

for positive constants b and c. The term in the Dirichlet integral of w may be discarded if we use the arbitrariness of the numbers a, b, c to ensure it is non-positive as a member of the right hand side of inequality (4.2.14), and to this end the constants a, b and c are now chosen so that

$$-2\nu + \frac{3}{2a}\Lambda^{1/2} + b + c = 0.$$

For example, we may pick

$$b = c = \frac{1}{2}\nu, \qquad a = \frac{3\Lambda^{1/2}}{2\nu}.$$

If we drop the $-2\lambda\|\nabla w\|^2$ term which remains (4.2.14) leads to

$$\phi' \leq A(t)\phi + \lambda^2\Big[\frac{4}{\nu}\|n\|^2 + \frac{2}{\nu}\|\nabla u\|^2\Big], \tag{4.2.15}$$

where

$$A(t) = \frac{27\Lambda^2}{16\nu^5 e^{2\lambda_1 \nu t}} \|f\|^2 \|u_{i,t}\|^2. \tag{4.2.16}$$

After integration (4.2.15) yields

$$\phi(t) \le \lambda^2 \int_0^t \exp\left[\int_\tau^t A(\eta)d\eta\right]\left[\frac{4}{\nu}\|n\|^2 + \frac{2}{\nu}\|\nabla u\|^2\right] d\tau. \tag{4.2.17}$$

The $\|n\|^2$ and $\|\nabla u\|^2$ terms will be bounded directly from the differential equations and the real difficulty is to bound $\|u_{i,t}\|^2$ in the exponential term in (4.2.17). To achieve this define

$$J(t) = \int_\Omega u_{i,t} u_{i,t} dx,$$

and upon differentiation with respect to t and use of (4.2.3) one obtains

$$\frac{dJ}{dt} = -2\nu \int_\Omega u_{i,jt} u_{i,jt} dx - 2\nu \int_\Omega u_{i,t} u_{j,t} u_{i,j} dx. \tag{4.2.18}$$

Applying the Cauchy - Schwarz inequality and Sobolev inequality (4.2.8) on the second term on the right hand side of (4.2.18) and using (4.2.12), we are led to

$$\frac{dJ}{dt} \le -2\nu \int_\Omega u_{i,jt} u_{i,jt} dx + 2\Lambda^{1/2}\left[\frac{\|f\|^2 J^2}{\nu^2 e^{2\lambda_1 \nu t}}\right]^{1/4}\left(\int_\Omega u_{i,jt} u_{i,jt} dx\right)^{3/4}.$$

It then follows from Young's inequality that

$$\frac{dJ}{dt} \le \left[-2\nu + \frac{3}{2\beta_1}\Lambda^{1/2}\right] \int_\Omega u_{i,jt} u_{i,jt} dx + \frac{1}{2}\beta_1^3 \Lambda^{1/2}\left[\frac{\|f\|^2 J^2}{\nu^2 e^{2\lambda_1 \nu t}}\right],$$

for an arbitrary positive constant β_1. The method is to select β_1 such that the first term on the right is non-positive. We follow Payne & Straughan (1989a) and pick

$$\beta_1 = \frac{3\Lambda^{1/2}}{4\nu},$$

which leads to the differential inequality

$$\frac{dJ}{dt} \le \frac{1}{2}\beta_1^3 \Lambda^{1/2}\left[\frac{\|f\|^2 J^2}{\nu^2 e^{2\lambda_1 \nu t}}\right]. \tag{4.2.19}$$

Next separate variables and integrate (4.2.19) from 0 to t, and we then obtain

$$\frac{1}{J(0)} - \frac{\beta_1^3 \Lambda^{1/2} \|f\|^2}{4\lambda_1 \nu^3}\left[1 - e^{-2\lambda_1 \nu t}\right] \le \frac{1}{J(t)}. \tag{4.2.20}$$

By rearranging (4.2.20) we may then derive a bound for $J(t)$ in terms of $J(0)$. However, the term $J(0) = \int_{\Omega(0)} u_{i,t}u_{i,t}dx$ is not data, and so we replace $u_{i,t}$ using the differential equation in (4.2.3) to observe that

$$J(0) = \int_{\Omega} u_{i,t}u_{i,t}dx\bigg|_{t=0} \leq \int_{\Omega} (\nu\Delta f_i - f_j f_{i,j})(\nu\Delta f_i - f_k f_{i,k})dx \equiv J_1(0).$$

The idea is to replace $J(0)$ in (4.2.20) by $J_1(0)$. At this point Payne & Straughan (1989a) consider two cases.

Case 1

If

$$\frac{1}{J_1(0)} > K = \frac{\beta_1^3\Lambda^{1/2}\|f\|^2}{4\lambda_1\nu^3},$$

then (4.2.20) implies that

$$J(t) \leq \frac{J(0)}{1 - KJ_1(0)} = K_1. \tag{4.2.21}$$

Hence,

$$A(t) \leq K_2 e^{-2\lambda_1\nu t}, \tag{4.2.22}$$

where

$$K_2 = \frac{27\Lambda^2 K_1\|f\|^2}{16\nu^5}.$$

Since A is now bounded we may use (4.2.22) in (4.2.17) to see that

$$\phi(t) \leq \lambda^2\int_0^t \exp\left[K_2\int_\tau^t e^{-2\lambda_1\nu\eta}d\eta\right]\left[\frac{4}{\nu}\|n\|^2 + \frac{2}{\nu}\|\nabla u\|^2\right]d\tau. \tag{4.2.23}$$

It is now shown how to estimate the integrals of $\|n\|^2$ and $\|\nabla u\|^2$. From the Navier - Stokes equations (4.2.3), we may derive the equality

$$\|u\|^2 + 2\nu\int_0^t \|\nabla u\|^2 d\eta = \|f\|^2. \tag{4.2.24}$$

Also, using the two systems of equations in (4.2.1), i.e. the micropolar equations, multiplication of $(4.2.1)_2$ by v_i and $(4.2.1)_3$ by n_i leads to

$$\begin{aligned}\frac{1}{2}\frac{d}{dt}(\|v\|^2 + j\|n\|^2) = &-(\nu\|\nabla v\|^2 + \gamma\|\nabla n\|^2) - \lambda\|\mathrm{curl}\,\mathbf{v} - \mathbf{n}\|^2\\ &- \lambda\|n\|^2 - (\alpha+\beta)\|n_{i,i}\|^2.\end{aligned} \tag{4.2.25}$$

Poincaré's inequality and the sign restrictions on the constants then allow us to deduce that there exists a constant μ independent of λ such that

$$\frac{d}{dt}\big(\|v\|^2 + j\|n\|^2\big) \le -\mu\big(\|v\|^2 + j\|n\|^2\big). \qquad (4.2.26)$$

Integration of this inequality yields

$$\|v\|^2 + j\|n\|^2 \le e^{-\mu t}\big(\|f\|^2 + j\|g\|^2\big). \qquad (4.2.27)$$

It follows from (4.2.27) that

$$\int_0^t \|n\|^2 d\tau \le \frac{1}{\mu}\Big(\frac{1}{j}\|f\|^2 + \|g\|^2\Big) = K_3, \qquad (4.2.28)$$

where K_3 is a data term independent of λ. In view of (4.2.23), one may then show

$$\phi(t) \le \lambda^2 e^{K_2/2\lambda_1\nu}\bigg[\frac{4}{\nu}\int_0^t \|n\|^2 d\tau + \frac{2}{\nu}\int_0^t \|\nabla u\|^2 d\tau\bigg],$$

and then using (4.2.24) and (4.2.28) we establish

$$\phi(t) \le K_4\lambda^2, \qquad (4.2.29)$$

where K_4 is a another constant given by

$$K_4 = \Big(\frac{4}{\nu}K_3 + \frac{1}{\nu^2}\|f\|^2\Big)e^{K_2/2\lambda_1\nu}.$$

Inequality (4.2.29) is the desired continuous dependence estimate in case 1.

Case 2

In this case,

$$\frac{1}{J_1(0)} \le K,$$

and there follows instead of (4.2.21) the bound

$$J(t) \le \frac{J_1(0)}{1 - KJ_1(0)(1 - e^{-2\lambda_1\nu t})}, \qquad (4.2.30)$$

provided the denominator in (4.2.30) *remains positive.* This will be true for t small enough. Assuming this is the case, Payne & Straughan (1989a) obtain

$$\exp\int_\tau^t A(\eta)d\eta \le \bigg[\frac{B(\tau)}{B(t)}\bigg]^{a^3/\beta_1^3}, \qquad (4.2.31)$$

where $B(t)$ is defined by

$$B(t) = 1 - KJ_1(0)\big(1 - e^{-2\lambda_1 \nu t}\big).$$

Substituting (4.2.31) as well as (4.2.24) and (4.2.28) in (4.2.17), yields

$$\phi(t) \le \lambda^2 \Big(\frac{4}{\nu} K_3 + \frac{1}{\nu^2}\|f\|^2\Big) \Big/ \big[B(t)\big]^{a^3/\beta_1^3}, \tag{4.2.32}$$

from which we may again infer continuous dependence on λ provided that t is small enough, i.e. such that the inequality below is valid,

$$1 - KJ_1(0)(1 - e^{-2\lambda_1 \nu t}) > 0.$$

We observe that in both cases one obtains the result that $\|w\| = \|u - v\|$ is $O(\lambda)$ for the forward in time problem when Ω is a three-dimensional bounded region. Payne & Straughan (1989a) are able to extend their analysis to the situation in which Ω is unbounded in all directions, obtaining the same order of convergence. They additionally treat the forward in time problem in two space dimensions. Again the order of convergence in $L^2(\Omega)$ is $O(\lambda)$ but the restrictions are much weaker. The analysis is not dissimilar to that expounded above except that instead of the Sobolev inequality (4.2.8) the equivalent inequality in two dimensions is

$$\int_\Omega \phi^4 dx \le \Lambda \|\phi\|^2 \|\nabla \phi\|^2.$$

This leads to the bound

$$-\int_\Omega u_{i,j} w_i w_j dx \le \frac{1}{2} a \|w\|^2 \|\nabla u\|^2 + \frac{1}{2a} \Lambda \|\nabla w\|^2,$$

instead of (4.2.9). The resulting analysis does not place a restriction on the time interval for convergence, or on the initial data. We omit the details of the argument in the two - dimensional case.

4.3 Continuous Dependence on the Velocity for an Equation Arising from Dynamo Theory

We conclude this chapter with an account of recent work of Franchi & Straughan (1994) who study continuous dependence on the modeling in an equation which arises from the study of a dynamo model. The dynamo aspect may be found in Lortz & Meyer-Spasche (1982).

Franchi & Straughan (1994) consider the following equation

$$\frac{\partial H}{\partial t} = \frac{\partial}{\partial x_i}\Big(\eta \frac{\partial H}{\partial x_i}\Big) - \frac{\partial}{\partial x_i}\Big(v_i H\Big), \qquad \text{in} \quad \Omega \times (0,T), \tag{4.3.1}$$

where Ω is a bounded domain in $\mathbf{R}^3$. They suppose that on the boundary Γ of the domain Ω, H satisfies the data

$$H(\mathbf{x},t) = H_B(\mathbf{x},t), \tag{4.3.2}$$

for a prescribed function H_B, while H also satisfies the inital data

$$H(\mathbf{x},0) = H_0(\mathbf{x}), \qquad \mathbf{x} \in \Omega. \tag{4.3.3}$$

In the dynamo problem H is the toroidal part of the magnetic field, and v_i is a *known* velocity field.

First we recount the analysis of Franchi & Straughan (1994) which studies continuous dependence on the velocity field v_i. They split the problem into two parts, one being continuous dependence on the prescribed velocity field for the forward in time problem, while the other part treats the improperly posed analogous backward in time problem.

Continuous Dependence on the Velocity, Forward in Time

Franchi & Straughan (1994) point out that the forward in time problem is well posed, at least when the velocity $\mathbf{v}$ is not too large, and hence they are able to obtain truly *a priori* estimates without the need to restrict H whatsoever. To achieve this they let H and H^* be solutions to (4.3.1), H being a solution for a prescribed velocity v_i whereas H^* is a solution for another prescribed velocity v_i^*. The functions H and H^* are supposed to satisfy the (same) inhomogeneous boundary and initial data H_B, H_0. Define the difference variables $h, \mathbf{u}$ by

$$h = H^* - H, \qquad u_i = v_i^* - v_i,$$

and then these variables satisfy the initial - boundary value problem:

$$\frac{\partial h}{\partial t} = \big(\eta(\mathbf{x})h_{,i}\big)_{,i} - \big(v_i^*(\mathbf{x})h\big)_{,i} - \big(u_i(\mathbf{x})H\big)_{,i}, \quad \text{in } \Omega \times (0,T], \tag{4.3.4}$$

$$h = 0, \quad \text{on } \Gamma \times [0,T], \tag{4.3.5}$$

$$h(\mathbf{x},0) = 0, \quad \mathbf{x} \in \Omega. \tag{4.3.6}$$

To achieve an *a priori* estimate multiply (4.3.4) by h, integrate over Ω, and then make use of the boundary conditions to find

$$\frac{d}{dt}\frac{1}{2}\|h\|^2 = -\int_\Omega \eta h_{,i}h_{,i}dx + \int_\Omega v_i^* hh_{,i}dx + \int_\Omega u_i h_{,i}H\,dx. \tag{4.3.7}$$

The velocity field is $C^1(\bar{\Omega})$ and is prescribed and so there is a known bound, M say, for $|v_i^*|$. Then the second term on the right of (4.3.7) may be estimated by employing the arithmetic-geometric mean inequality as follows

$$\int_\Omega v_i^* hh_{,i}dx \le \frac{M^2}{2\eta_0}\|h\|^2 + \frac{1}{2}\int_\Omega \eta h_{,i}h_{,i}dx. \tag{4.3.8}$$

Further use of the arithmetic-geometric mean inequality leads to a bound for the third term on the right of (4.3.7) in the following manner

$$\int_\Omega u_i h_{,i}H\,dx \le \frac{1}{2}\int_\Omega \eta h_{,i}h_{,i}dx + \frac{1}{2\eta_0}\|H\|^2 \sup_{\Omega\times[0,T]} |\mathbf{u}|^2. \tag{4.3.9}$$

Upon using (4.3.8) and (4.3.9) in (4.3.7) the weighted Dirichlet integrals add out and we may obtain:

$$\frac{d}{dt}\frac{1}{2}\|h\|^2 \le \frac{M^2}{2\eta_0}\|h\|^2 + \frac{1}{2\eta_0}\|H\|^2 \sup_{\Omega\times[0,T]} |\mathbf{u}|^2. \tag{4.3.10}$$

An *a priori* estimate follows easily from (4.3.10) if one can obtain an *a priori* estimate for the L^2 norm of H. To derive such an estimate Franchi & Straughan (1994) introduce the function $\phi(\mathbf{x},t)$ to be that function which satisfies the boundary value problem

$$\begin{aligned} \Delta\phi = 0 \qquad &\text{in} \qquad \Omega\times[0,T],\\ \phi = H(\mathbf{x},t) \qquad &\text{on} \qquad \Gamma\times[0,T]. \end{aligned} \tag{4.3.11}$$

Since H satisfies equation (4.3.1) they begin with the identity

$$0 = \int_0^t \int_\Omega (H-\phi)\big[H_{,s} - (\eta H_{,i})_{,i} + (v_i H)_{,i}\big]dx\,ds.$$

This identity may be rearranged with several integrations by parts to lead to

$$\begin{aligned} &\frac{1}{2}\|H(t)\|^2 + \int_0^t\int_\Omega \phi_{,s}H\,dx\,ds - \int_\Omega \phi H\,dx\Big|_t \\ &+ \int_0^t\int_\Omega \eta H_{,i}H_{,i}dx\,ds - \int_0^t\int_\Omega \eta H_{,i}\phi_{,i}dx\,ds - \int_0^t\int_\Omega v_i HH_{,i}dx\,ds \\ &+ \int_0^t\int_\Omega \phi_{,i}v_i H\,dx\,ds = \frac{1}{2}\|H_0\|^2 - \int_\Omega \phi_0 H_0 dx, \end{aligned} \tag{4.3.12}$$

where ϕ_0 denotes the value ϕ takes when $t = 0$, i.e. $\phi_0(\mathbf{x}) = \phi(\mathbf{x}, 0)$. Estimates of the various terms in (4.3.12) are needed and these are

$$-\int_0^t \int_\Omega \phi_{,s} H dx\, ds \leq \frac{1}{2} \int_0^t \int_\Omega \phi_{,s}^2 dx\, ds + \int_0^t \int_\Omega H^2 dx\, ds, \tag{4.3.13}$$

$$\int_\Omega \phi H dx \Big|_t \leq \frac{1}{4} \|H(t)\|^2 + \|\phi(t)\|^2, \tag{4.3.14}$$

$$\int_0^t \int_\Omega \eta H_{,i} \phi_{,i} dx\, ds \leq \int_0^t \int_\Omega \eta H_{,i} H_{,i} dx\, ds + \frac{\bar{\eta}}{4} \int_0^t \int_\Omega \phi_{,i} \phi_{,i} dx\, ds, \tag{4.3.15}$$

where $\bar{\eta}$ is a constant defined by

$$\bar{\eta} = \max_{\mathbf{x} \in \Omega} |\eta|,$$

$$\begin{aligned} \int_0^t \int_\Omega v_i H H_{,i} dx\, ds = & -\frac{1}{2} \int_0^t \int_\Omega v_{i,i} H^2 dx\, ds \\ & + \int_0^t \oint_\Gamma \frac{1}{2} v_i n_i H^2 dS\, ds, \end{aligned} \tag{4.3.16}$$

and,

$$\begin{aligned} -\int_0^t \int_\Omega v_i H \phi_{,i} dx\, ds \leq & \frac{1}{2} m \int_0^t \int_\Omega H^2 dx\, ds \\ & + \frac{1}{2} m \int_0^t \int_\Omega \phi_{,i} \phi_{,i} dx\, ds, \end{aligned} \tag{4.3.17}$$

where m is the constant

$$m = \max_{\Omega \times [0,T]} |v_i|.$$

Upon using estimates (4.3.13) - (4.3.17) in expression (4.3.12), Franchi & Straughan (1994) derive the following estimate for computable constants $k_1 - k_4$ which depend on the known value of max $|v_{i,i}|$.

$$\begin{aligned} \|H(t)\|^2 \leq & 2 \int_0^t \oint_\Gamma v_i n_i H^2 dS\, ds + k_1 \int_0^t \|H\|^2 ds \\ & + k_2 \|\phi\|^2 + k_3 \int_0^t \|\phi_{,s}\|^2 ds + k_4 \int_0^t \|\nabla \phi\|^2 ds \\ & + 2\|H_0\|^2 - 4 \int_\Omega \phi_0 H_0 dx. \end{aligned} \tag{4.3.18}$$

To continue with the *a priori* estimate it is necessary to show that the ϕ terms are bounded by data. Franchi & Straughan (1994) do this by introducing a Rellich identity, following Payne & Weinberger (1958). The relevant identity is

$$-\int_\Omega x^i_{,j}\phi_{,i}\phi_{,j}\,dx + \frac{1}{2}\int_\Omega x^i_{,i}\phi_{,j}\phi_{,j}\,dx$$
$$-\frac{1}{2}\oint_\Gamma n_i x^i\phi_{,j}\phi_{,j}dS + \oint_\Gamma x^i\phi_{,i}n_j\phi_{,j}dS = 0,$$

which arises from (4.3.11), since

$$\int_\Omega x^i\phi_{,i}\Delta\phi\,dx = 0.$$

Upon simplification the Rellich identity yields

$$\frac{1}{2}\|\nabla\phi\|^2 = \frac{1}{2}\oint_\Gamma n_i x^i\phi_{,j}\phi_{,j}dS - \oint_\Gamma x^i\phi_{,i}\frac{\partial\phi}{\partial n}\,dS. \tag{4.3.19}$$

To proceed from this the unit normal and tangential vectors to Γ are introduced namely, $\mathbf{n}$ and $\mathbf{s}$, and then on Γ the derivative may be written as

$$\frac{\partial\phi}{\partial x_i} = \frac{\partial\phi}{\partial n}n^i + \nabla_s\phi\, s^i,$$

where $\nabla_s\phi$ is the tangential derivative to Γ. Then (4.3.19) may be rewritten as

$$\frac{1}{2}\|\nabla\phi\|^2 + \frac{1}{2}\oint_\Gamma x^i n_i\Big(\frac{\partial\phi}{\partial n}\Big)^2 dS = \frac{1}{2}\oint_\Gamma x^i n_i|\nabla_s\phi|^2 dS - \oint_\Gamma x^i s_i\frac{\partial\phi}{\partial n}\nabla_s\phi\,dS. \tag{4.3.20}$$

At this stage Franchi & Straughan (1994) assume that the domain Ω is star shaped with respect to the origin and put

$$m_1 = \min_\Gamma |x^i n_i|,$$

so from (4.3.20) one may derive constants c_1, c_2 such that

$$\|\nabla\phi\|^2 + c_1\oint_\Gamma\Big(\frac{\partial\phi}{\partial n}\Big)^2 dS \le c_2\oint_\Gamma|\nabla_s H|^2 dS, \tag{4.3.21}$$

where the fact that $H = \phi$ on Γ has been used. Then upon use of Poincaré's inequality on $\|\nabla\phi\|^2$ in (4.3.21) it may be shown for c_3, c_4 computable constants

$$\|\phi\|^2 \le c_3\oint_\Gamma H^2 dS + c_4\oint_\Gamma|\nabla_s H|^2 dS. \tag{4.3.22}$$

An integration of (4.3.21) leads to

$$\int_0^t \|\nabla\phi\|^2 ds \le c_2 \int_0^t \oint_\Gamma |\nabla_s H|^2 dS. \tag{4.3.23}$$

The observation that $\phi_{,t}$ satisfies

$$\Delta\phi_{,t} = 0, \qquad \text{in} \qquad \Omega \times (0,T),$$

and the boundary data $\phi_{,t} = H_{,t}$ on Γ, now allows a process similar to (4.3.21) and (4.3.23) to derive for computable constants c_5, c_6

$$\int_0^t \|\phi_{,s}\|^2 ds \le c_5 \int_0^t \oint_\Gamma H_{,\tau}^2 dS\, d\tau + c_6 \int_0^t \oint_\Gamma |\nabla_s H_{,\tau}|^2 dS\, d\tau. \tag{4.3.24}$$

The procedure of Franchi & Straughan (1994) is to employ (4.3.22) - (4.3.24) in inequality (4.3.18) and this gives

$$\|H(t)\|^2 \le k_1 \int_0^t \|H\|^2 ds + \mathcal{D}. \tag{4.3.25}$$

The term $\mathcal{D}$ is comprised solely of data and is given by

$$\begin{aligned}\mathcal{D} = & 2\int_0^t \oint_\Gamma v_i n_i H^2 dS\, ds + q_1 \oint_\Gamma H^2 dS + q_2 \oint_\Gamma |\nabla_s H|^2 dS \\ & + q_3 \int_0^t \oint_\Gamma |\nabla_s H|^2 dS\, d\tau + q_4 \int_0^t \oint_\Gamma H_{,\tau}^2 dS\, d\tau \\ & + q_5 \int_0^t \oint_\Gamma |\nabla_s H_{,\tau}|^2 dS\, d\tau + 4\|H_0\|^2. \end{aligned} \tag{4.3.26}$$

The constants $q_1 - q_5$ may be computed explicitly for a given domain D.

Inequality (4.3.25) is now integrated to see that

$$\int_0^t \|H\|^2 ds \le \int_0^t e^{k_1(t-\tau)} \mathcal{D}(\tau) d\tau,$$

and upon use of this in (4.3.25), one deduces

$$\|H(t)\|^2 \le \mathcal{L}(t), \tag{4.3.27}$$

where $\mathcal{L}(t)$ depends solely on data and is given by

$$\mathcal{L}(t) = \mathcal{D} + k_1 \int_0^t e^{k_1(t-\tau)} \mathcal{D}(\tau) d\tau. \tag{4.3.28}$$

The estimate (4.3.27) may be now used in inequality (4.3.10) to obtain

$$\|h(t)\|^2 \leq \frac{N}{k} e^{kT} \sup_{\Omega \times [0,T]} |\mathbf{u}|^2, \tag{4.3.29}$$

where the constants k and N are defined by

$$k = \frac{M^2}{\eta_0}, \qquad N = \frac{1}{\eta_0} \max_{t \in [0,T]} \mathcal{L}(t).$$

Inequality (4.3.29) is the one derived by Franchi & Straughan (1994) and establishes the continuous dependence of H on $\mathbf{v}$. It is important to note that it is an *a priori* inequality in the sense that the right hand side is dependent only on given data.

4.4 Structural Stability for Infinite Prandtl Number Thermal Convection

As we explained in section 1.1, Bellomo & Preziosi (1994), pp. 85 - 86, define the concept of structural stability to be stability with respect to the model itself. For example, continuous dependence on the coefficients in the partial differential equations governing the model, continuous dependence on the boundary data, continuous dependence on the coefficients in the boundary conditions, or continuous dependence with respect to the partial differential equations themselves. We expect that for a model to be reasonable we should have some control over its structural stability. In this section we investigate some aspects of structural stability for a model of thermal convection in a viscous fluid, in the limit of infinite Prandtl number. The Prandtl number is the ratio of the fluid kinematic viscosity to its thermal diffusivity, and the infinite Prandtl number situation is a mathematical idealization which represents the physical situation where the kinematic viscosity is much larger than the thermal diffusivity. Such a situation is commonplace in terrestial convection situations. The infinite Prandtl number model is used in many applied mathematical situations, generally leads to a simplified mathematical procedure, and can yield very useful results, see e.g. Thess & Bestehorn (1995).

Precisely, in this section we study three questions of structural stability for flow in a linear viscous fluid, concentrating on forward in time problems. We are investigating problems forward in time and hence we derive *a priori* results. As pointed out by Franchi & Straughan (1996), when dealing with the Navier-Stokes equations truly *a priori* results have previously proved extremely difficult, cf. Payne (1967), Ames & Payne (1996), Song

(1988). We here firstly investigate continuous dependence on the cooling coefficient for thermal convection with Newton's law of cooling holding on the boundary. We derive an a *priori* result in three-dimensions and do not need to restrict the size of the interval or the size of the initial data. Such a restriction would presently appear necessary for a similar result involving the cooling coefficient in Navier-Stokes theory (see Ames & Payne (1996)), although for the Brinkman equations of porous media this restriction is not needed. The second investigation of this section studies continuous dependence on the model by comparing the theory which employs an equation of state linear in temperature to that with the theory for a quadratic density law. We might point out that theories with nonlinear equations of state are important in astrophysical and geophysical penetrative convection contexts, cf. Straughan (1992,1993). We close the chapter by examining continuous dependence on the heat supply for the infinite Prandtl number theory. Similar structural stability results for the Brinkman equations for flow in a porous material are established in Franchi & Straughan (1996). The work of Ames & Payne (1996) studies continuous dependence on the cooling coefficient in the full equations for thermal convection in both the forward and backward time cases.

Continuous Dependence on the Cooling Coefficient

If we adopt a linear density temperature relationship of form

$$\rho = \rho_0\big(1 - \alpha(T - T_0)\big), \tag{4.4.1}$$

then the equations of thermal convection have form

$$\begin{aligned} \frac{\partial v_i}{\partial t} + v_m \frac{\partial v_i}{\partial x_m} &= -\frac{\partial p}{\partial x_i} + \nu\Delta v_i + \alpha g T k_i, \\ \frac{\partial v_i}{\partial x_i} &= 0, \\ \frac{\partial T}{\partial t} + v_i \frac{\partial T}{\partial x_i} &= \kappa\Delta T, \end{aligned} \tag{4.4.2}$$

see e.g. Straughan (1992), chapter 3. Now non-dimensionalize the above equations with the scalings of *time* $= L^2/\kappa$, and speed $U = \kappa/L$, where L is a typical length. The temperature is rescaled by a typical temperature factor $T^\sharp$ and we introduce the Rayleigh number, $Ra = L^3\alpha g T^\sharp/\kappa\nu$. The resulting non - dimensional system has form

$$\begin{aligned} \frac{1}{Pr}\left[\frac{\partial v_i}{\partial t} + v_m \frac{\partial v_i}{\partial x_m}\right] &= -\frac{\partial p}{\partial x_i} + \Delta v_i + Ra T k_i, \\ \frac{\partial v_i}{\partial x_i} &= 0, \\ \frac{\partial T}{\partial t} + v_i \frac{\partial T}{\partial x_i} &= \Delta T. \end{aligned} \tag{4.4.3}$$

From (4.4.3) we see that the equations for thermal convection in the limit of infinite Prandtl number may be taken to be, cf. Thess & Bestehorn (1995),

$$-\frac{\partial p}{\partial x_i} + \Delta v_i + RaTk_i = 0, \tag{4.4.4}$$

$$\frac{\partial v_i}{\partial x_i} = 0, \tag{4.4.5}$$

$$\frac{\partial T}{\partial t} + v_i \frac{\partial T}{\partial x_i} = \Delta T, \tag{4.4.6}$$

in which v_i, p, T and Ra are, respectively, the non-dimensionalized velocity, pressure, temperature, and the Rayleigh number. The vector $\mathbf{k}$ is (0,0,1). These equations hold on a bounded domain Ω with boundary Γ sufficiently smooth to allow applications of the divergence theorem. On the boundary Γ the no slip condition holds for the velocity field so

$$v_i = 0, \qquad \text{on } \Gamma,$$

and we assume the temperature field satisfies the condition

$$\frac{\partial T}{\partial n} + \kappa T = F(\mathbf{x}, t), \qquad \text{on } \Gamma, \tag{4.4.7}$$

in which κ is a positive constant and F is a prescribed function. Equation (4.4.7) is essentially Newton's law of cooling with inhomogeneous outside temperature, i.e.

$$\frac{\partial T}{\partial n} = -\kappa(T - T_a), \tag{4.4.8}$$

in which T_a is the ambient outside temperature.

Ames & Payne (1996) derive results for continuous dependence on the cooling coefficient κ for the full equations for a heat conducting viscous fluid, by employing (4.4.8) with T_a constant. Thus, they consider instead of equation (4.4.4) the equation $(4.4.2)_3$. The continuous dependence on κ result given here is for a more general outside temperature than that studied in Ames & Payne (1996), but the major impact of the result here is of interest in its own right due to practical applications. In the three-dimensional situation Ames & Payne (1996) have to restrict the size of the initial data, or alternatively the size of the continuous dependence interval. This is because of the presence of fluid acceleration and the convective nonlinearity, i.e. the $v_{i,t}$ and $v_j v_{i,j}$ terms in $(4.4.2)_3$. In the limit case $Pr \to \infty$ under investigation here we are able to proceed without such restrictions, as are Franchi & Straughan (1996) who deal with the Brinkman system in a porous medium.

To examine continuous dependence on κ let (v_i, T, p) and (v_i^*, T^*, p^*) be solutions to (4.4.4)-(4.4.7) subject to the same initial data and same prescribed function F. However, the cooling coefficient for (v_i, T, p) is κ in (4.4.7) whereas that for (v_i^*, T^*, p^*) is κ^*.

Define α as the difference

$$\alpha = \kappa^* - \kappa, \tag{4.4.9}$$

and then the difference variables $u_i = v_i^* - v_i$, $\theta = T^* - T$, $\pi = p^* - p$, satisfy the partial differential equations

$$-\frac{\partial \pi}{\partial x_i} + \Delta u_i + Rak_i\theta = 0, \tag{4.4.10}$$

$$\frac{\partial u_i}{\partial x_i} = 0, \tag{4.4.11}$$

$$\frac{\partial \theta}{\partial t} + v_i^* \frac{\partial \theta}{\partial x_i} + u_i \frac{\partial T}{\partial x_i} = \Delta\theta. \tag{4.4.12}$$

The appropriate initial conditions are

$$u_i = 0, \qquad \theta = 0, \qquad \text{at } t = 0, \tag{4.4.13}$$

while the boundary conditions become

$$u_i = 0, \qquad \frac{\partial \theta}{\partial n} + \kappa\theta = -\alpha T^*, \qquad \text{on } \Gamma. \tag{4.4.14}$$

In our usual notation $\|.\|$ denotes the norm on $L^2(\Omega)$. Then multiply (4.4.10) by u_i and integrate over Ω to obtain

$$\begin{aligned}\|\nabla \mathbf{u}\|^2 &= k_i Ra \int_\Omega \theta u_i \, dx \\ &\leq Ra\|\theta\|\|\mathbf{u}\|,\end{aligned} \tag{4.4.15}$$

where the Cauchy-Schwarz inequality has also been employed. Now use Poincaré's inequality,

$$\lambda_1 \|\mathbf{u}\|^2 \leq \|\nabla \mathbf{u}\|^2,$$

in (4.4.15) to find

$$\|\nabla \mathbf{u}\| \leq c\|\theta\|, \tag{4.4.16}$$

where the constant c is given by $c = Ra/\sqrt{\lambda_1}$.

By multiplication of (4.4.12) by θ, integration over Ω and use of (4.4.14) we now derive

$$\frac{1}{2}\frac{d}{dt}\|\theta\|^2 = -\int_\Omega u_i T_{,i} \theta \, dx - \|\nabla\theta\|^2 - \kappa \oint_\Gamma \theta^2 dS - \alpha \oint_\Gamma T^* \theta \, dS. \tag{4.4.17}$$

To deal with the last term use the arithmetic - geometric mean inequality to see that

$$\alpha \oint_\Gamma T^*\theta \, dS \le \frac{\alpha^2}{4\kappa} \oint_\Gamma T^{*2} dS + \kappa \oint_\Gamma \theta^2 dS. \tag{4.4.18}$$

Then the first term in (4.4.17) may be manipulated by integration by parts followed by use of the Cauchy - Schwarz inequality to obtain

$$\begin{aligned} -\int_\Omega u_i T_{,i}\theta \, dx &= \int_\Omega u_i T\theta_{,i} \, dx, \\ &\le \|\nabla\theta\| \|\mathbf{u}\|_4 \|T\|_4, \end{aligned} \tag{4.4.19}$$

where again $\|\cdot\|_4$ denotes the $L^4(\Omega)$ norm. We then employ the Sobolev inequality

$$\|\mathbf{u}\|_4 \le \gamma \|\nabla \mathbf{u}\|,$$

in (4.4.19) and follow this with use of estimate (4.4.16) to derive

$$\begin{aligned} -\int_\Omega u_i T_{,i}\theta \, dx \le& \gamma \|\nabla\theta\| \|\nabla\mathbf{u}\| \|T\|_4, \\ \le& \gamma c \|\nabla\theta\| \|\theta\| \|T\|_4. \end{aligned} \tag{4.4.20}$$

Inequalities (4.4.18) and (4.4.20) are collected together in (4.4.17) to now see that

$$\frac{1}{2}\frac{d}{dt}\|\theta\|^2 \le -\|\nabla\theta\|^2 + \gamma c \|\nabla\theta\| \|\theta\| \|T\|_4 + \frac{\alpha^2}{4\kappa} \oint_\Gamma T^{*2} \, dS. \tag{4.4.21}$$

The next step is to estimate the $L^2(\Gamma)$ integral of T^*. To do this multiply the form of equation (4.4.6) which holds for T^* by T^* itself and integrate over Ω to derive

$$\frac{1}{2}\frac{d}{dt}\|T^*\|^2 = -\|\nabla T^*\|^2 - \kappa \oint_\Gamma T^{*2} dS + \oint_\Gamma T^* F \, dS.$$

The arithmetic-geometric mean inequality may be used on the last term, then the Dirichlet integral discarded. After integration in t we obtain

$$\frac{1}{2}\kappa \int_0^t \oint_\Gamma T^{*2} dS \, ds \le D_2(t), \tag{4.4.22}$$

where $D_2(t)$ is a data term defined by

$$D_2(t) = \frac{1}{2\kappa} \int_0^t \oint_\Gamma F^2 dS \, ds + \frac{1}{2}\|T_0^*\|^2. \tag{4.4.23}$$

It remains to bound $\|T\|_4$ in terms of data. To this end multiply (4.4.6) by T^3 to find

$$\frac{1}{4}\frac{d}{dt}\|T\|_4^4 = -\frac{3}{4}\|\nabla T^2\|^2 - \kappa \oint_\Gamma T^4 dS + \oint_\Gamma T^3 F\, dS. \tag{4.4.24}$$

Employ Young's inequality on the last term and we may derive

$$\frac{d}{dt}\|T\|_4^4 + 2\kappa \oint_\Gamma T^4 dS + 3\|\nabla T^2\|^2 \leq \frac{27}{8\kappa^3} \oint_\Gamma F^4\, dS. \tag{4.4.25}$$

The general Poincaré inequality for a function ϕ with non-zero boundary data can be written for a positive constant μ as

$$\mu\|\phi\|^2 \leq 2\kappa \oint_\Gamma \phi^2 dS + 3\|\nabla\phi\|^2.$$

Hence, use this in inequality (4.4.25) with $\phi = T^2$ to derive

$$\frac{d}{dt}\|T\|_4^4 + \mu\|T\|_4^4 \leq D_1(t), \tag{4.4.26}$$

where $D_1(t)$ is the data term

$$D_1(t) = \frac{27}{8\kappa^3} \oint_\Gamma F^4\, dS. \tag{4.4.27}$$

Inequality (4.4.26) is now integrated with the aid of an integrating factor to produce

$$\|T(t)\|_4^4 \leq D(t), \tag{4.4.28}$$

in which $D(t)$ is the following data term,

$$D(t) = \|T_0\|_4^4 e^{-\mu t} + \int_0^t e^{-\mu(t-s)} D_1(s) ds. \tag{4.4.29}$$

We observe that the data term $D(t)$ actually has a decay effect due to the exponentials.

The estimate (4.4.28) is now employed in inequality (4.4.21) to derive, with the aid of the arithmetic-geometric mean inequality,

$$\frac{d}{dt}\|\theta\|^2 - \frac{1}{2}\gamma^2 c^2 D^{1/2}(t)\|\theta\|^2 \leq \frac{\alpha^2}{2\kappa} \oint_\Gamma T^{*2} dS. \tag{4.4.30}$$

Inequality (4.4.30) is integrated in time and then estimate (4.4.22) is used to bound the right hand side. The result is

$$\|\theta(t)\|^2 \leq \frac{1}{2}\gamma^2 c^2 \int_0^t D^{1/2}(s)\|\theta(s)\|^2 ds + \frac{\alpha^2}{\kappa^2} D_2(t). \tag{4.4.31}$$

In the interests of clarity we define the data functions $A(t)$ and $B(t)$ by

$$A(t) = \frac{1}{2}\gamma^2 c^2 D^{1/2}(t), \qquad B(t) = \frac{D_2(t)}{\kappa^2}\,.$$

Finally, Gronwall's inequality is used in (4.4.31) and we obtain

$$\|\theta(t)\|^2 \le C(t)\alpha^2, \tag{4.4.32}$$

where the function $C(t)$ is given by

$$C(t) = B(t) + \int_0^t B(s)A(s)\Big[\exp\int_s^t A(u)\,du\Big]\,ds. \tag{4.4.33}$$

From inequality (4.4.16) we know that

$$\|\nabla\mathbf{u}\| \le c\|\theta\|$$

and so this together with (4.4.32) furnishes a continuous dependence estimate for $\|\nabla\mathbf{u}\|$, namely

$$\|\nabla\mathbf{u}(t)\| \le c\,C^{1/2}(t)|\alpha|, \tag{4.4.34}$$

and a further application of Poincaré's inequality allows us to deduce

$$\|\mathbf{u}(t)\| \le \frac{Ra}{\lambda_1}\,C^{1/2}(t)|\alpha|. \tag{4.4.35}$$

Inequalities (4.4.32), (4.4.34), and (4.4.35) demonstrate continuous dependence of the solution $(\mathbf{v}, T)$ on the cooling coefficient κ. Of course, they are truly *a priori* because $C(t)$ is a function of prescribed data.

Continuous Dependence on the Model in non - Boussinesq Convection

In section 3.6 we have seen studies of improperly posed problems in convection when the equation of state, i.e. the density - temperature relationship, is a nonlinear one. We here continue this study by comparing the solution to the equations of thermal convection for the linear law (4.4.1) to the solution of the analogous system when the equation of state is

$$\rho = \rho_0(1 - \alpha_1 T - \alpha_2 T^2). \tag{4.4.36}$$

Note that we are here dealing with a forward in time problem. The equation of state (4.4.36) is frequently used in penetrative convection which is the phenomenon which may arise when a convectively unstable layer of fluid is bounded by one (or more) stable fluid layers. The motions in the unstable layer may penetrate deeply into the stable layer, and this process has wide application in fields such as astrophysics, geophysics, and oceanography.

Several models of penetrative convection are reviewed in the monograph by Straughan (1993) and a commonly employed one involves a layer of thermally conducting viscous fluid with a buoyancy law in which the density ρ is a nonlinear function of temperature T. Since there are many possibilities for the functional form of $\rho(T)$, one may raise the question as to what effect the change in the model imparts to the solution.

Thus we now address the problem of examining the difference in solution behavior between that for a model for convection in a viscous fluid with a linear buoyancy law and that for a model with a quadratic law, in the limit of infinite Prandtl number. We shall continue to employ boundary condition (4.4.7) on the temperature field.

The equations of motion when a linear buoyancy law is adopted are, in the limit of infinite Prandtl number,

$$\begin{aligned} -\frac{\partial p}{\partial x_i} + \Delta v_i + RaTk_i &= 0, \\ \frac{\partial v_i}{\partial x_i} &= 0, \\ \frac{\partial T}{\partial t} + v_i \frac{\partial T}{\partial x_i} &= \Delta T. \end{aligned} \tag{4.4.37}$$

These equations hold on the bounded spatial domain $\Omega \in \mathbf{R}^3$, for positive time. On the boundary, Γ, of Ω we take

$$v_i(\mathbf{x}, t) = 0, \qquad \frac{\partial T}{\partial n} + \kappa T = F(\mathbf{x}, t), \tag{4.4.38}$$

for a prescribed function F, and for initial data we assume

$$v_i(\mathbf{x}, 0) = v_i^I(\mathbf{x}), \qquad T(\mathbf{x}, 0) = T_I(\mathbf{x}), \tag{4.4.39}$$

for prescribed functions v_i^I, T_I.

When we employ the quadratic law (4.4.36) then the system which arises from the procedure leading to (4.4.3) - (4.4.6) is

$$\begin{aligned} -\Delta v_i^* &= -\frac{\partial p^*}{\partial x_i} + Ra(1+\xi)k_i T^* + \epsilon Rak_i T^{*2}, \\ v_{i,i}^* &= 0, \\ \frac{\partial T^*}{\partial t} + v_i^* \frac{\partial T^*}{\partial x_i} &= \Delta T^*, \end{aligned} \tag{4.4.40}$$

where ϵ is a measure of the deviation from the linear equation (4.4.1) and ξ represents the variation in α in replacing (4.4.1) by (4.4.36).

We are studying continuous dependence on the model itself and hence suppose (v_i, T, p) and (v_i^*, T^*, p^*) are solutions to (4.4.37) and (4.4.40), respectively. These solutions satisfy the *same* boundary data in (4.4.38) and also satisfy the *same* initial data in (4.4.39). The idea is to now produce an *a priori* estimate, thereby demonstrating how the difference in solutions to (4.4.37) and (4.4.40) depends continuously on the coefficients ξ and ϵ.

We define the difference variables (u_i, θ, π) as

$$u_i = v_i^* - v_i, \qquad \theta = T^* - T, \qquad \pi = p^* - p.$$

By subtraction one may show (u_i, θ, π) satisfies the equations

$$\begin{aligned} &-\frac{\partial \pi}{\partial x_i} + \Delta u_i + Ra\theta k_i + \xi Ra T^* k_i + \epsilon Ra T^{*2} k_i = 0, \\ &u_{i,i} = 0, \\ &\frac{\partial \theta}{\partial t} + u_i \frac{\partial T}{\partial x_i} + v_i^* \frac{\partial \theta}{\partial x_i} = \Delta\theta. \end{aligned} \tag{4.4.41}$$

The boundary and initial conditions become

$$\begin{aligned} &u_i = 0, \qquad \frac{\partial \theta}{\partial n} + \kappa\theta = 0, \qquad \text{on} \quad \Gamma \times \{t > 0\}, \\ &u_i = 0, \qquad \theta = 0, \qquad \mathbf{x} \in \Omega, \ t = 0. \end{aligned} \tag{4.4.42}$$

By multiplying $(4.4.41)_1$ by u_i and integrating over Ω we obtain

$$\begin{aligned} \|\nabla \mathbf{u}\|^2 =& Rak_i \int_\Omega \theta u_i dx + \xi Rak_i \int_\Omega T^* u_i dx + \epsilon Rak_i \int_\Omega T^{*2} u_i dx, \\ \le& Ra\|\theta\| \|\mathbf{u}\| + \xi Ra\|T^*\| \|\mathbf{u}\| + \epsilon Ra \|T^*\|_4^2 \|\mathbf{u}\|, \end{aligned}$$

where in the last line the Cauchy - Schwarz inequality has been used. Thus, using Poincaré's inequality allows us to deduce

$$\|\nabla \mathbf{u}\| \le \frac{Ra}{\sqrt{\lambda_1}} \Big(\|\theta\| + \xi \|T^*\| + \epsilon \|T^*\|_4^2 \Big). \tag{4.4.43}$$

Next, multiply $(4.4.41)_3$ by θ and integrate over Ω to derive

$$\frac{1}{2}\frac{d}{dt}\|\theta\|^2 = -\int_\Omega u_i T_{,i} \theta \, dx - \|\nabla\theta\|^2 - \kappa \oint_\Gamma \theta^2 dA. \tag{4.4.44}$$

We integrate the cubic nonlinear term by parts and employ the Cauchy - Schwarz and Sobolev inequalities to see that

$$\begin{aligned} -\int_\Omega u_i T_{,i} \theta \, dx \le& \|\nabla\theta\| \|\mathbf{u}\|_4 \|T\|_4, \\ \le& \gamma \|\nabla\theta\| \|\nabla \mathbf{u}\| \|T\|_4. \end{aligned}$$

Then (4.4.43) is used to estimate the Dirichlet integral of **u** from which we derive the result

$$-\int_\Omega u_i T_{,i}\theta\, dx \le \frac{\gamma Ra}{\sqrt{\lambda_1}}\|\nabla\theta\|\|T\|_4\big(\|\theta\| + \xi\|T^*\| + \epsilon\|T^*\|_4^2\big).$$

The arithmetic - geometric mean inequality then allows us to deduce

$$\begin{aligned}-\int_\Omega u_i T_{,i}\theta\, dx \le &\|\nabla\theta\|^2\\ &+\frac{\gamma^2 Ra^2}{\lambda_1}\Big(\|T\|_4^2\|\theta\|^2 + \xi^2\|T\|_4^2\|T^*\|^2 + \epsilon^2\|T\|_4^2\|T^*\|_4^4\Big).\end{aligned}$$

Finally, we use this estimate in (4.4.44) to obtain

$$\begin{aligned}\frac{1}{2}\frac{d}{dt}\|\theta\|^2 \le &-\kappa\oint_\Gamma \theta^2 dS + \frac{\gamma^2 Ra^2}{4\lambda_1}\|T\|_4^2\|\theta\|^2\\ &+\frac{\gamma^2\xi^2 Ra^2}{4\lambda_1}\|T\|_4^2\|T^*\|^2 + \frac{\gamma^2\epsilon^2 Ra^2}{4\lambda_1}\|T\|_4^2\|T^*\|_4^4. \qquad (4.4.45)\end{aligned}$$

To proceed, multiply $(4.4.41)_3$ by T^3 and integrate over Ω to derive

$$\frac{1}{4}\frac{d}{dt}\|T\|_4^4 = -\frac{3}{4}\|\nabla T^2\|^2 - \kappa\oint_\Gamma T^4 dS + \oint_\Gamma FT^3 dS. \qquad (4.4.46)$$

After use of Young's inequality and integration we may show

$$\|T(t)\|_4^4 \le D(t), \qquad (4.4.47)$$

where $D(t)$ is defined as in (4.4.29) and (4.4.27). Since T^* satisfies the same equation and boundary conditions as T, we obtain an identical estimate for T^*.

Furthermore, from $(4.4.40)_3$ we derive

$$\frac{1}{2}\frac{d}{dt}\|T^*\|^2 = -\|\nabla T^*\|^2 - \kappa^*\oint_\Gamma T^{*2} dS + \oint_\Gamma T^* F\, dS,$$

and this leads to, after use of the Cauchy - Schwarz inequality

$$\frac{1}{2}\frac{d}{dt}\|T^*\|^2 \le -\|\nabla T^*\|^2 - \frac{1}{2}\kappa^*\oint_\Gamma T^{*2} dS + \frac{1}{2}\kappa^*\oint_\Gamma F^2 dS.$$

If we employ the Poincaré inequality in the form

$$\|\nabla T^*\|^2 + \frac{1}{2}\kappa^*\oint_\Gamma T^{*2} dS \ge c_1\|T^*\|^2,$$

for a computable constant c_1, we then find

$$\frac{d}{dt}\|T^*\|^2 + 2c_1\|T^*\|^2 \le \kappa^* \oint_\Gamma F^2 dS.$$

We may integrate this to produce

$$\|T^*(t)\|^2 \le D_3(t), \tag{4.4.48}$$

where the data term D_3 is defined by

$$D_3(t) = \|T_0^*\|^2 e^{-2c_1 t} + \kappa^* \int_0^t e^{-2c_1(t-s)} \oint_\Gamma F^2(\mathbf{x}, s)\, dx\, ds. \tag{4.4.49}$$

We then employ (4.4.7) and its T^* equivalent together with (4.4.48) in (4.4.45) to find

$$\frac{d}{dt}\|\theta\|^2 \le A(t)\|\theta\|^2 + \epsilon^2 B(t) + \xi^2 C(t), \tag{4.4.50}$$

where $A(t), B(t)$ and $C(t)$ are data terms given by

$$A(t) = \frac{\gamma^2 Ra^2 D^{1/2}(t)}{2\lambda_1}, \qquad B(t) = \frac{\gamma^2 D^{3/2}(t)}{2\lambda^2\lambda_1},$$

$$C(t) = \frac{\gamma^2 Ra^2}{2\lambda_1} D^{1/2}(t) D_3(t).$$

Finally we integrate (4.4.50) to derive

$$\|\theta(t)\|^2 \le \epsilon^2 D_4(t) + \xi^2 D_5(t). \tag{4.4.51}$$

The data terms D_4 and D_5 are given by

$$D_4(t) = \int_0^t \left(\exp \int_s^t A(\eta)\, d\eta\right) B(s)\, ds,$$

$$D_5(t) = \int_0^t \left(\exp \int_s^t A(\eta)\, d\eta\right) C(s)\, ds.$$

We also observe that due to (4.4.43) we may produce a similar inequality for $\|\nabla\mathbf{u}\|$, and then by employing Poincaré's inequality, for $\|\mathbf{u}\|$.

Inequality (4.4.51) and the analogous ones for $\|\nabla\mathbf{u}\|$ and $\|\mathbf{u}\|$ represent continuous dependence on the model results, since as $\xi, \epsilon \to 0$ the solutions converge in L^2 sense.

Continuous Dependence on the Heat Supply for Infinite Prandtl Number Convection

In section 3.6 we included a study of continuous dependence on the body force and heat supply for flow in porous media for the backward in time problem. Franchi & Straughan (1996) consider continuous dependence on the body force and heat supply for flow in a porous body governed by Brinkman's equations. In this section we develop an analysis for continuous dependence on the heat supply in infinite Prandtl number convection in a viscous fluid.

The relevant equations may be taken as, cf. (4.4.4) - (4.4.6),

$$\begin{aligned} &\Delta v_i - \frac{\partial p}{\partial x_i} + Rak_i T = 0, \\ &v_{i,i} = 0, \\ &\frac{\partial T}{\partial t} + v_i \frac{\partial T}{\partial x_i} = \Delta T + r, \end{aligned} \tag{4.4.53}$$

where r denotes the heat supply. The boundary conditions we consider are no slip on the velocity and the temperature prescribed, so

$$v_i = 0, \quad T = T_B \qquad \text{on} \quad \Gamma, \tag{4.4.54}$$

where T_B is a prescribed function. In addition the initial values of v_i and T are assumed given.

To develop an analysis of continuous dependence on the heat supply we suppose (v_i, T, p) and (v_i^*, T^*, p^*) are solutions to (4.4.53) for r and r^*, respectively. The functions (v_i, T, p) and (v_i^*, T^*, p^*) satisfy the same boundary and initial data. Let ω be the difference in heat supply, i.e.

$$\omega = r^* - r.$$

From (4.4.53) the difference solution (u_i, θ, π) is found to satisfy

$$\begin{aligned} \Delta u_i =& - \frac{\partial \pi}{\partial x_i} + k_i Ra\theta, \\ u_{i,i} =& 0, \\ \frac{\partial \theta}{\partial t} + u_i \frac{\partial T}{\partial x_i} + v_i^* \frac{\partial \theta}{\partial x_i} =& \Delta\theta + \omega, \end{aligned} \tag{4.4.55}$$

with the boundary and initial conditions being,

$$\begin{aligned} u_i = 0, \qquad \theta = 0, \qquad & \text{on} \quad \Gamma, \\ u_i = 0, \qquad \theta = 0, \qquad & \mathbf{x} \in \Omega, \ t = 0. \end{aligned} \tag{4.4.56}$$

To begin we multiply $(4.4.55)_1$ by u_i and integrate to find

$$\|\nabla \mathbf{u}\|^2 = Rak_i(\theta, u_i),$$

and so employing Poincaré's inequality we obtain,

$$\|\nabla\mathbf{u}\| \le \frac{Ra}{\sqrt{\lambda_1}}\|\theta\|. \tag{4.4.57}$$

Next, multiply $(4.4.53)_3$ by θ and integrate over Ω to see that

$$\frac{d}{dt}\|\theta\|^2 = -2\int_\Omega u_i T_{,i}\theta\,dx - 2\|\nabla\theta\|^2 + 2(\omega,\theta). \tag{4.4.58}$$

The cubic term is dealt with as in (4.4.20) so that

$$\begin{aligned} -\int_\Omega u_i T_{,i}\theta\,dx =& \int_\Omega u_i T\theta_{,i}\,dx,\\ \le&\|\mathbf{u}\|_4\|T\|_4\|\nabla\theta\|,\\ \le&\gamma\|\nabla\theta\|\|\nabla\mathbf{u}\|\|T\|_4,\\ \le&\frac{\gamma Ra}{\sqrt{\lambda_1}}\|\nabla\theta\|\|T\|_4\|\theta\|, \end{aligned}$$

where (4.4.57) has been used. Further, using the arithmetic - geometric mean inequality we find

$$-\int_\Omega u_i T_{,i}\theta\,dx \le \|\nabla\theta\|^2 + \frac{\gamma^2 Ra^2}{4\lambda_1}\|T\|_4^2\|\theta\|^2. \tag{4.4.59}$$

Employ (4.4.59) in (4.4.58) to find

$$\frac{d}{dt}\|\theta\|^2 \le \|\theta\|^2\Big(1 + \frac{Ra^2\gamma^2}{4\lambda_1}\|T\|_4^2\Big) + \|\omega\|^2. \tag{4.4.60}$$

The completion of the continuous dependence proof requires a bound for $\|T\|_4$. This is achieved as in Franchi & Straughan (1996) by introducing the functions ϕ and ψ which satisfy

$$\Delta\phi = 0 \quad \text{in} \quad \Omega, \qquad \phi = T_B \quad \text{on} \quad \Gamma, \tag{4.4.61}$$

$$\Delta\psi = 0 \quad \text{in} \quad \Omega, \qquad \psi = T_B^3 \quad \text{on} \quad \Gamma. \tag{4.4.62}$$

The idea is to form the identity

$$\int_0^t\int_\Omega (\psi - T^3)(T_{,t} + v_i T_{,i} - \Delta T - r)dx\,ds = 0. \tag{4.4.63}$$

This may be written after several integrations as

$$\begin{aligned}\frac{1}{4}\|T\|_4^4+\frac{3}{4}\int_0^t\|\nabla T^2\|^2ds=&\frac{1}{4}\|T_0\|_4^4+\int_0^t\int_\Omega\psi v_iT_{,i}dx\,ds\\&-\int_0^t\int_\Omega\psi_{,s}T dx\,ds\\&+\int_\Omega\psi(\mathbf{x},t)T(\mathbf{x},t)dx-\int_\Omega\psi(\mathbf{x},0)T(\mathbf{x},0)dx\\&+\int_0^t\int_\Omega\nabla\psi\cdot\nabla T dx\,ds-\int_0^t\int_\Omega\psi r\,dx\,ds\\&+\int_0^t\int_\Omega T^3r\,dx\,ds=0.\end{aligned}\tag{4.4.64}$$

A bound for $\|T\|_4$ is sought from this. We may use Young's inequality and estimate the cubic terms. However, this introduces L^2 quantities which we must in turn estimate in terms of data. This involves the function ϕ. Hence, we form the identity

$$\int_0^t\int_\Omega(\phi-T)(T_{,t}+v_iT_{,i}-\Delta T-r)dx\,ds=0.\tag{4.4.65}$$

After several integrations (4.4.65) may be shown to yield

$$\begin{aligned}\frac{1}{2}\|T\|^2+\int_0^t\|\nabla T\|^2ds=&\frac{1}{2}\|T_0\|^2+\int_0^t\int_\Omega\phi v_iT_{,i}dx\,ds\\&-\int_0^t\int_\Omega\phi_{,s}T dx\,ds\\&+\int_\Omega\phi(\mathbf{x},t)T(\mathbf{x},t)dx-\int_\Omega\phi(\mathbf{x},0)T(\mathbf{x},0)dx\\&+\int_0^t\int_\Omega\nabla\phi\cdot\nabla T\,dx\,ds-\int_0^t\int_\Omega\phi r\,dx\,ds\\&+\int_0^t\int_\Omega Tr\,dx\,ds=0.\end{aligned}\tag{4.4.66}$$

The cubic terms may be dealt with in the following manner,

$$\begin{aligned}\int_0^t\int_\Omega\phi v_iT_{,i}dx\,ds=&-\int_0^t\int_\Omega\phi_{,i}v_iT\,dx\,ds,\\\leq&\int_0^t\|\nabla\phi\|\|\mathbf{v}\|_4\|T\|_4ds,\\\leq&\gamma\int_0^t\|\nabla\phi\|\|\nabla\mathbf{v}\|\|T\|_4ds,\\\leq&\frac{\gamma Ra}{\sqrt{\lambda_1}}\int_0^t\|\nabla\phi\|\|T\|\|T\|_4ds,\end{aligned}$$

where from $(4.4.53)_1$ we note that

$$\|\nabla\mathbf{v}\| \leq \frac{Ra}{\sqrt{\lambda_1}}\,\|T\|.$$

Thus, by application of the arithmetic - geometric mean inequality,

$$\begin{aligned}\int_0^t\int_\Omega \phi v_i T_{,i} dx\, ds \leq & \frac{\gamma\xi Ra}{2\sqrt{\lambda_1}}\int_0^t \|T\|^2 ds + \frac{\gamma\xi_1 Ra}{4\xi\sqrt{\lambda_1}}\int_0^t \|T\|_4^4 ds\\ & + \frac{\gamma Ra}{4\xi\xi_1\sqrt{\lambda_1}}\int_0^t \|\nabla\phi\|^4 ds,\end{aligned} \tag{4.4.67}$$

where ξ, ξ_1 are positive constants at our disposal.

For other constants ξ_2, ξ_3 we develop an inequality analogous to (4.4.67) for the ψ term.

Then, expressions (4.4.64) and (4.4.66) together with the appropriate inequalities for the ϕ and ψ cubic terms lead to an inequality of form

$$\begin{aligned}\frac{1}{2}\|T\|^2 + & \frac{1}{4}\|T\|^4 + \int_0^t \|\nabla T\|^2 ds + \frac{3}{4}\int_0^t \|\nabla T^2\|^2 ds\\ \leq & \frac{1}{2}\|T_0\|^2 + \frac{1}{4}\|T_0\|_4^4 - \int_\Omega \phi(\mathbf{x},0)T(\mathbf{x},0)dx - \int_\Omega \psi(\mathbf{x},0)T(\mathbf{x},0)dx\\ & - \int_0^t\int_\Omega \phi r\, dx\, ds - \int_0^t\int_\Omega \psi r\, dx\, ds\\ & + k_1\int_0^t \|T\|^2 ds + k_2\int_0^t \|T\|_4^4 ds + k_3\|T\|^2 + k_4\int_0^t \|\nabla T\|^2 ds\\ & + c_1\int_0^t \|\nabla\phi\|^4 ds + c_2\int_0^t \|\nabla\psi\|^4 ds + c_3\int_0^t \|\phi_{,s}\|^2 ds\\ & + c_4\int_0^t \|\psi_{,s}\|^2 ds + c_5\|\phi\|^2 + c_6\|\psi\|^2\\ & + c_7\int_0^t \|\nabla\phi\|^2 ds + c_8\int_0^t \|\nabla\psi\|^2 ds + c_9\int_0^t \|r\|_4^4 ds,\end{aligned} \tag{4.4.68}$$

for known constants k_i, c_i. The idea is to choose k_3 and k_4 so that

$$k_3 < \frac{1}{2}, \quad k_4 < 1.$$

The next step involves estimating the terms $\|\nabla\phi\|^2, \|\nabla\psi\|^2$, $\int_0^t \|\phi_{,s}\|^2 ds$, $\int_0^t \|\psi_{,s}\|^2 ds$, $\|\phi\|^2, \|\psi\|^2$, in terms of the data function T_B. This may be done by using a Rellich identity in a manner similar to that after (4.3.17). The upshot is that one derives an inequality of form

$$\|T\|^2 + \mu_1 \|T\|_4^4 \leq d_1(t) + \mu_2 \left(\int_0^t \|T\|^2 ds + \mu_1 \int_0^t \|T\|_4^4 ds \right), \tag{4.4.69}$$

for $d_1(t)$ a specific data term. The last inequality may be integrated to find

$$\|T(t)\|^2 + \mu_1 \|T(t)\|_4^4 \leq d_2(t), \tag{4.4.70}$$

where the data term $d_2(t)$ is given by

$$d_2(t) = d_1(t) + \mu_2 \int_0^t e^{\mu_2(t-s)} d_1(s) ds. \tag{4.4.71}$$

If estimate (4.4.70) is used in (4.4.60) one may derive the inequality, for $f(t)$ a data function,

$$\frac{d}{dt}\|\theta\|^2 \leq f(t)\|\theta\|^2 + \|\omega\|^2. \tag{4.4.72}$$

Inequality (4.4.72) may be integrated to establish

$$\|\theta(t)\|^2 \leq \int_0^t \left(\exp \int_s^t f(\xi) d\xi \right) \|\omega(s)\|^2 ds. \tag{4.4.73}$$

Inequality (4.4.73) establishes continuous dependence on the heat supply r. By using equation (4.4.57) and the Poincaré inequality we may also similarly estimate $\|\nabla \mathbf{u}(t)\|$ and $\|\mathbf{u}(t)\|$. It is worth observing that our continuous dependence on the heat supply result is truly *a priori* because the coefficients are all expressed in terms of known data.

5

Non-Standard and Non-Characteristic Problems

5.1 The "Sideways" Problem for the Heat Equation

In this chapter we focus on various nonstandard problems for familiar differential equations, the nature of which are ill-posed. We begin with the non-characteristic (or sideways) Cauchy problem for the heat equation and related parabolic equations. By this we mean the heat equation with the solution and its spatial derivative simultaneously given on *only part* of the spatial boundary. The first studies of this problem by Cannon & Douglas (1967) and Ginsberg (1963) established Hölder continuous dependence on the Cauchy data for solutions of the heat equation but their methods do not generalize to related parabolic equations because they rely heavily on special representations and analyticity properties of the solution. Ewing & Falk (1979) derived a procedure for the numerical approximation of a more general class of parabolic equations by supplementing the non-characteristic Cauchy data with additional data. Bell (1981) studies a more general class of equations in one space dimension, obtaining uniqueness and a logarithmic continuous dependence on the data without requiring additional data or relying on analyticity properties of the solution. For this work, a weighted energy technique was used to derive the estimates. Payne (1985) later showed how the weighted energy method could be modified in the problems treated by Bell (1981) to yield Hölder continuous dependence on the data. Here we shall describe one such problem to illustrate the methodology that Payne developed.

Suppose Ω is an open region in the first quadrant of the (x,t) plane with a portion of its boundary, Γ, that is a segment of the t-axis, $0 \leq t \leq T$. The set Γ is thus a non-characteristic segment of the boundary of Ω. Consider the problem,

$$Lu \equiv a(x,t)\frac{\partial^2 u}{\partial x^2} - \frac{\partial u}{\partial t} = F\Big(x,t,u,\frac{\partial u}{\partial x}\Big), \qquad \text{in } \Omega, \tag{5.1.1}$$

$$u(0,t) = g(t), \qquad \frac{\partial u}{\partial x}(0,t) = h(t), \qquad \text{on } \Gamma. \tag{5.1.2}$$

The function $F(x, t, u, u_x)$ is assumed to be Lipschitz continuous in its last two arguments so that

$$|F(x,t,u_1,v_1) - F(x,t,u_2,v_2)| \le C\big(|u_1 - u_2| + |v_1 - v_2|\big) \tag{5.1.3}$$

for a positive constant C. In addition, the coefficient $a(x,t)$ is continuous and positive for each $(x,t) \in \Omega$ and $\partial a/\partial x$ is a bounded function.

Bell (1981) and Payne (1985) approach this problem by introducing the level sets of a function $f(x,t)$. They define

$$D_\alpha = \big\{(x,t)\big| f(x,t) < \alpha\big\} \cap \Omega, \tag{5.1.4}$$
$$S_\alpha = \partial D_\alpha \cap \Omega, \tag{5.1.5}$$
$$\Gamma_\alpha = \partial D_\alpha \cap \partial\Omega. \tag{5.1.6}$$

As we shall see in the section after next, Payne (1970) first used surfaces of this type in conjunction with logarithmic convexity arguments to study the Cauchy problem for elliptic equations.

For the problem of this section, the function $f(x,t)$ is assumed to satisfy the following conditions:

$$D_\alpha \subset D_\beta \text{ and } \Gamma_\alpha \subset \Gamma_\beta \quad \text{for } \alpha < \beta, \quad \alpha, \beta \in [0, \gamma], \tag{5.1.7}$$
$$\frac{\partial f}{\partial x} > c_1 > 0, \quad \text{in } D_\gamma, \tag{5.1.8}$$
$$\frac{\partial}{\partial x}\left(a\frac{\partial f}{\partial x}\right) \le -c_2 < 0, \quad \text{in } D_\gamma, \tag{5.1.9}$$
$$\frac{\partial}{\partial x}\left[a\left(\frac{\partial f}{\partial x}\right)^3\right] \le -c_3 < 0, \quad \text{in } D_\gamma. \tag{5.1.10}$$

Figure 3 illustrates the level surfaces of $f(x,t)$. An example of such a family of curves is given by

$$f(x,t) = A_1(t - t_0)^2 + A_2 \ln\big[A_3(x - x_0)\big] = \alpha,$$

for any point $(x_0, t_0) \in \mathbf{R}^2 - \bar{\Omega}$ near Γ, any positive constants A_1, A_2, A_3, and $0 < \alpha < \gamma$.

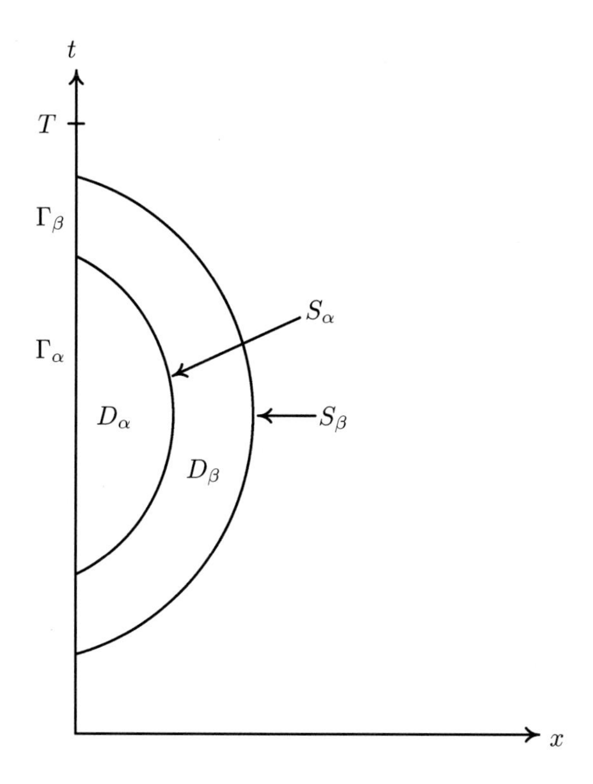

Figure 3. The geometry for problem (5.1.1),(5.1.2).

Using a weighted energy method Payne (1985) obtains a Hölder continuous dependence inequality in D_α $(\alpha < \gamma)$ for classical solutions of (5.1.1), (5.1.2) that are suitably constrained. We outline his argument here. Note that we can assume that $F(x,t,0,0) = 0$ since F is assumed to satisfy (5.1.3) and we are interested in continuous dependence on the Cauchy data. Let us now set

$$u(x,t) = e^{\lambda f} v(x,t) \tag{5.1.11}$$

for a constant λ to be determined and substitute into (5.1.1) so that

$$Kv \equiv Lv + 2\lambda \frac{\partial f}{\partial x}\frac{\partial v}{\partial x} + \left[\lambda^2 \left(\frac{\partial f}{\partial x}\right)^2 + \lambda Lf\right] v = e^{-\lambda f} F. \tag{5.1.12}$$

We form the expression

$$\int_{D_\alpha} \left[2\lambda a \frac{\partial f}{\partial x}\frac{\partial v}{\partial x} - \frac{\partial v}{\partial t}\right]\left[Kv - e^{-\lambda f}F\right] dx = 0. \tag{5.1.13}$$

Then using (5.1.8), (5.1.9), integrating by parts, and denoting the x and t components of the unit normal vector on S_α by

$$n_x = \frac{f_x}{[f_x^2 + f_t^2]^{1/2}}, \qquad n_t = \frac{f_t}{[f_x^2 + f_t^2]^{1/2}}, \tag{5.1.14}$$

it may be seen that for sufficiently large λ and computable, positive constants ρ_1 and ρ_2

$$\begin{aligned}
\lambda \int_{\partial D_\alpha} a \frac{\partial f}{\partial x}\left(\frac{\partial v}{\partial x}\right)^2 n_x d\sigma &+ \frac{1}{2}\int_{\partial D_\alpha}\left(\frac{\partial v}{\partial x}\right)^2 n_t d\sigma - \int_{\partial D_\alpha} \frac{\partial v}{\partial x}\frac{\partial v}{\partial t} n_x d\sigma \\
&+ \frac{1}{2}\lambda \int_{\partial D_\alpha}\left[2\lambda a \frac{\partial f}{\partial x} n_x - n_t\right]\left[\lambda\left(\frac{\partial f}{\partial x}\right)^2 - \frac{1}{a} L f\right] v^2 d\sigma \\
&\leq -\rho_1 \lambda \int_{D_\alpha}\left(\frac{\partial v}{\partial x}\right)^2 dx - \rho_2 \lambda^3 \int_{D_\alpha} v^2\, dx \\
&+ \frac{1}{4}\int_{D_\alpha} e^{-2\lambda f}\left[b_1 u^2 + b_2\left(\frac{\partial u}{\partial x}\right)^2\right] dx.
\end{aligned} \tag{5.1.15}$$

Observing that the right hand side of (5.1.15) can be made negative since

$$\frac{\partial u}{\partial x} = \left[\frac{\partial v}{\partial x} + \lambda v\right] e^{\lambda f},$$

we discard these terms to obtain the inequality

$$\begin{aligned}
\int_{S_\alpha} \left[\lambda a \left(\frac{\partial f}{\partial x}\right)^2 + \frac{\partial f}{\partial t}\right]\left(\frac{\partial v}{\partial x}\right)^2 \frac{1}{K_0} d\sigma &+ \int_{S_\alpha}\left[\lambda^3 a \left(\frac{\partial f}{\partial x}\right)^4 \frac{1}{K_0} + O(\lambda^2)\right] v^2 d\sigma \\
&- \int_{S_\alpha} \frac{\partial v}{\partial t}\frac{\partial v}{\partial x}\frac{\partial f}{\partial x}\frac{1}{K_0} d\sigma \\
&\leq \int_\Gamma \left[b_3 \lambda^3 v^2 + b_4 \lambda \left(\frac{\partial v}{\partial x}\right)^2 + b_5 \left(\frac{\partial v}{\partial t}\right)^2\right] d\sigma.
\end{aligned} \tag{5.1.16}$$

The right hand side is a computable data term of $O(\lambda^3)$ that we designate by Q^2 and

$$K_0 = \left[\left(\frac{\partial f}{\partial x}\right)^2 + \left(\frac{\partial f}{\partial t}\right)^2\right]^{1/2}.$$

Substituting $e^{-\lambda f}u(x,t)$ for $v(x,t)$, applying the arithmetic - geometric mean inequality and integrating with respect to α, one obtains the result

$$\begin{aligned}(\lambda - K_1)\int_{D_\alpha} a\Big(\frac{\partial f}{\partial x}\Big)^2\Big(\frac{\partial u}{\partial x}\Big)\,dx + 2\lambda^2(\lambda - K_1)\int_{D_\alpha} a\Big(\frac{\partial f}{\partial x}\Big)^4 u^2\,dx\\ -\frac{1}{2}\int_{S_\alpha} a\Big(\frac{\partial f}{\partial x}\Big)^2\Big(\frac{\partial u}{\partial x}\Big)^2\frac{1}{K_0}\,d\sigma - \lambda^2\int_{S_\alpha} a\Big(\frac{\partial f}{\partial x}\Big)^4 u^2\frac{1}{K_0}\,d\sigma\\ \leq K_2\lambda^2 e^{2\lambda\alpha}Q^2,\end{aligned} \qquad (5.1.17)$$

where K_1 and K_2 are computable constants and the differential equation (5.1.1) has been used to replace $\partial u/\partial t$. In addition, we have imposed the restriction (5.1.10).

Following Payne (1985), let us define

$$G(\alpha) = \int_{D_\alpha}\Big[a\Big(\frac{\partial f}{\partial x}\Big)^2\Big(\frac{\partial u}{\partial x}\Big)^2 + \lambda^2 a\Big(\frac{\partial f}{\partial x}\Big)^4 u^2\Big]\,dx \qquad (5.1.18)$$

so that we can write (5.1.17) as

$$\frac{d}{d\alpha}\Big[-G(\alpha)e^{-2(\lambda-K_1)\alpha}\Big] \leq K_2\lambda^2 e^{2\lambda\alpha}Q^2. \qquad (5.1.19)$$

Integrating from α to γ leads to

$$G(\alpha) \leq G(\gamma)e^{-2(\lambda-K_1)(\gamma-\alpha)} + \frac{K_2}{2K_1}\lambda^2 e^{2[(\lambda-K_1)\alpha+K_1\gamma]}Q^2. \qquad (5.1.20)$$

If we now assume that u belongs to the set of functions that satisfy for $\lambda > 1$ the constraint

$$G(\gamma) \leq \lambda^2 M^2, \qquad (5.1.21)$$

for a constant M and if we choose

$$\lambda - K_1 = \frac{1}{2\gamma}\ln\Big[\frac{M^2}{Q^2}\Big], \qquad (5.1.22)$$

then we find that

$$G(\alpha) \leq \lambda^2\Big[1 + \frac{K_2}{2K_1}\,e^{2K_1\alpha}\Big]M^{2\alpha/\gamma}Q^{2(1-\alpha/\gamma)}. \qquad (5.1.23)$$

Recalling the definition of $G(\alpha)$, we obtain the desired continuous dependence inequality for solutions u that satisfy (5.1.21), namely

$$\int_{D_\alpha} a\Big(\frac{\partial f}{\partial x}\Big)^4 u^2 dx \leq C_0 M^{2\alpha/\gamma}Q^{2(1-\alpha/\gamma)}, \qquad (5.1.24)$$

for a computable constant C_0.

We point out here that problems similar to (5.1.1), (5.1.2) occur in various physical contexts. For example Payne & Straughan (1990b) consider such a problem in their study of a glacier, which is partly composed of ice at a temperature less than its freezing point, commonly referred to as cold ice, see also Flavin & Rionero (1996). If only the surface of the glacier is accessible for data measurements, then we know the temperature and the heat flux only on a part of the boundary of the glacier. This is reflected in our mathematical model by non-characteristic Cauchy data. More specifically, if we let Ω be a bounded region in the (x,t) plane with $x > 0$, a portion Γ of whose boundary lies along the $t-$axis with $-T \leq t \leq T$, then the temperature $u(x,t)$ can be shown to satisfy the problem (see Payne & Straughan (1990b)),

$$\frac{\partial u}{\partial t} = \frac{\partial}{\partial x}\left[\left(1 + \epsilon_1 u + \epsilon_2 u^2\right)\frac{\partial u}{\partial x}\right], \qquad \text{in } \Omega, \tag{5.1.25}$$

$$u(0,t) = g(t), \qquad \frac{\partial u}{\partial x}(0,t) = h(t), \qquad \text{on } \Gamma, \tag{5.1.26}$$

where g and h are prescribed functions that are measured only on Γ. Equation (5.1.25) is a heat equation with a nonlinear thermal diffusivity appropriate to cold ice, see e.g. Hutter (1984), Payne & Straughan (1990b). Using a weighted energy argument that proceeds in a manner analogous to that described earlier, Payne & Straughan (1990b) establish Hölder continuous dependence estimates on the Cauchy data for solutions of (5.1.25), (5.1.26). Following Payne (1985), they introduce a family of curves $f(x,t) = \alpha$ for $0 < \alpha \leq 1$ that give rise to the regions described in (5.1.4) - (5.1.6) and (5.1.7) ($\gamma = 1$.) In their problem, the restrictions on f consist of two conditions analogous to (5.1.8) and (5.1.9) and a third condition, instead of (5.1.10), that has the effect of reducing the region over which estimates can be made to points that are not too far from Γ. The third condition is necessary due to the nonlinear thermal diffusivity. Consequently, continuous dependence results are again restricted to the regions D_α for $\alpha < 1$.

Let u_1 and u_2 be two solutions of (5.1.25) corresponding to different data $g_i(t)$ and $h_i(t)$, $i = 1,2$, and assume that

$$|u_1|, |u_2|, \left|\frac{\partial u_1}{\partial x}\right|, \left|\frac{\partial u_1}{\partial t}\right|, \left|\frac{\partial u_2}{\partial t}\right| \leq M_1, \tag{5.1.27}$$

are bounded in Ω by the known bound M_1. Suppose also that the Cauchy data are close in the sense that

$$\max\left\{\sup_{\bar{\Gamma}} |g_1 - g_2|, \sup_{\bar{\Gamma}} |h_1 - h_2|\right\} \leq \epsilon, \tag{5.1.28}$$

for some prescribed $\epsilon > 0$. Defining

$$v = (u_1 - u_2)\big[1 + \frac{1}{2}\epsilon_1(u_1 + u_2) + \frac{1}{3}\epsilon_2(u_1^2 + u_1 u_2 + u_2^2)\big], \tag{5.1.29}$$

Payne & Straughan (1990b) use a weighted energy argument to show that the functional

$$G(\alpha) = \int_{D_\alpha} \left(\frac{\partial f}{\partial x}\right)^2 \left(\frac{\partial v}{\partial x}\right)^2 dx + 2\lambda^2 \int_{D_\alpha} \left(\frac{\partial f}{\partial x}\right)^4 v^2\, dx,$$

satisfies an inequality of the form

$$G(\alpha) \le K_1 \lambda^2 \epsilon^{2(1-\alpha)}, \tag{5.1.30}$$

provided the class of allowable solutions is restricted to those such that

$$G(1) \le \lambda^2 M_2^2, \tag{5.1.31}$$

for a constant M_2 and λ is chosen so that it is the positive solution of

$$2\lambda\left(\frac{\lambda - \beta_1}{\lambda + \beta_2}\right) = \ln\left(\frac{M^2}{\epsilon^2}\right). \tag{5.1.32}$$

Here β_1 and β_2 are computable constants that arise from the application of standard inequalities. Recalling the definition of G and v and using the bound (5.1.28), it follows from (5.1.30) that

$$\int_{D_\alpha} (u_1 - u_2)^2 dx \le K_2 \epsilon^{2(1-\alpha)}, \tag{5.1.33}$$

if

$$\left|\frac{\partial f}{\partial x}\right| \ge c > 0,$$

for a constant c. Inequality (5.1.33) is the desired Hölder continuous dependence result in D_α for $0 < \alpha < 1$.

Payne's modified weighted energy argument was also used by Monk (1986) to prove continuous dependence results for discrete solutions of a model non-characteristic Cauchy problem for the heat equation. Using these results, he then obtains error estimates for an optimization method for numerical approximation of solutions of this problem.

The problem that Monk (1986) studies can be described in the following way. Suppose that $\Omega = (0, 1) \times (0, T]$, and denote the boundary segment

$[0,1] \times \{0\}$ by Σ and the segment $\{0\} \times [0,T]$ by Γ. A function $u(x,t)$ is sought to satisfy

$$\frac{\partial^2 u}{\partial x^2} - \frac{\partial u}{\partial t} = \psi(x,t), \qquad \text{in } \Omega, \tag{5.1.34}$$

$$|u - u_0|_{1,\Sigma} \leq \epsilon_0, \tag{5.1.35}$$

$$|u - g|_{1,\Gamma} \leq \epsilon_1, \tag{5.1.36}$$

$$\left| \frac{\partial u}{\partial x} - h \right|_{0,\Gamma} \leq \epsilon_2, \tag{5.1.37}$$

for prescribed functions ψ, g, h and u_0 and positive constants ϵ_i, $(i = 0,1,2)$. Here $|\cdot|_{n,\omega}$ denotes the nth Sobolev norm on the line segment ω.

Monk (1986) begins his study of (5.1.34) - (5.1.37) by proving that solutions to the continuous problem with $\psi = 0$ depend Hölder continuously on the Cauchy data. His analysis follows the argument developed by Payne (1985). As Monk (1986) observes, such results for continuous problems do not directly reflect the behaviour of discrete approximations. A major contribution of his work is a continuous dependence result for an approximate discrete solution of the heat equation. He defines such a discrete solution by introducing a mesh of points (x_n, t_m) in Ω with $x_n = nk$ for $0 \leq n \leq 1/k$, and $t_m = mp$ for $0 \leq m \leq T/p$. Then he denotes by u_i^n the solution satisfying the finite difference scheme

$$u_{i+1}^n - u_i^n = \mu\big[\theta(u_{i+1}^{n+1} - 2u_{i+1}^n + u_{i+1}^{n-1}) + (1-\theta)(u_i^{n+1} - 2u_i^n + u_i^{n-1})\big], \tag{5.1.38}$$

for $0 \leq i \leq (T/p) - 1$ and $1 \leq n \leq (1/k) - 1$. Here $\mu = p/k^2$ and θ is a parameter ranging between 0 and 1. A main result of Monk's (1986) paper is a Hölder continuous dependence inequality for u_i^n which he proves by using a discrete analogoue of the proof for the continuous problem. This inequality can then be employed to establish error estimates for a least squares method that approximates the solution to the problem (5.1.34) - (5.1.37). Monk's (1986) numerical calculations reveal that the discrete solution obtained by this method is accurate away from $x = 1$. Monk (1986) points out that his method is unlikely to be competitive with other methods for one-dimensional problems such as those suggested in the papers of Ginsberg (1963), Cannon (1964), Cannon & Douglas (1967), and Cannon & Ewing (1976). However, what is interesting and appealing about Monk's (1986) work is that his method generalizes naturally to higher dimensions and to more complex parabolic operators.

5.2 The "Sideways" Problem for a Hyperbolic Equation

The ill-posed Cauchy problem for the wave equation with Cauchy data given on a lateral surface has important applications in geophysics and optimal control theory. This problem was studied by John (1960) who found it possible to stabilize the solution of the wave equation in some neighbourhood of a plane data surface if bounds on the absolute value of the solution and its derivatives are imposed. Later, Lavrentiev *et al.* (1983) investigated ill-posed Cauchy problems for a more general class of ultrahyperbolic quasilinear equations. Their work also required bounds on the absolute values of the solution and its first derivatives. Levine & Payne (1985) have established a Hölder stability result for the wave equation when the data are prescribed on the entire lateral boundary. Using the methods of John (1960), they were able to stabilize the problem locally by requiring only L^2 bounds on the solution and its spatial derivatives. These stabilization assumptions are less stringent than the L^∞ hypotheses of previous investigators and thus deserve some discussion.

Consider then the following problem

$$\frac{\partial^2 v}{\partial t^2} - \Delta v = 0, \qquad \text{in } \Omega \times (-T, T), \tag{5.2.1}$$

$$v = f, \qquad \operatorname{grad} v = \mathbf{g}, \qquad \text{on } \Gamma \times [-T, T], \tag{5.2.2}$$

where Ω is the interior of an $n-$ball of radius r centered at the origin, with boundary Γ, and Δ is the Laplace operator in $\mathbf{R}^n$. To stabilize this problem locally, Levine & Payne (1985) introduce a family of surfaces

$$F(x,t) \equiv |x|^2 - t^2 = \epsilon^2, \tag{5.2.3}$$

with $0 < \epsilon_0 \leq \epsilon \leq r$. These surfaces are required to intersect $\Omega \times [-T, T]$ along $\Gamma \times [-T, T]$, the lateral surface, and consequently, we want

$$r^2 - \epsilon_0^2 \leq T^2. \tag{5.2.4}$$

Following Levine & Payne (1985), let us define the set

$$D_\epsilon \equiv \left\{(x,t) \middle| F(x,t) > \epsilon^2, |x| < r\right\},$$

and let S_ϵ denote the portion of the boundary of D_ϵ that lies on $F(x,t) = \epsilon^2$. The point set

$$\left\{(x,t) \middle| \, |x| = r, |t| < \sqrt{r^2 - \epsilon^2}\right\},$$

will be denoted by Σ_ϵ.

Weighted energy arguments are used in Levine & Payne (1985) to derive the inequality

$$\int_{D_\epsilon} v^2 dx \leq CM^{2/3}Q^{1/3}, \tag{5.2.5}$$

for $\frac{1}{2}(\epsilon_0 + r) \leq \epsilon$. Here C is a computable constant and Q consists of the data

$$Q = \int_{-T}^{T} \int_{\Gamma} \Big\{ f^2 + |g|^2 + \Big(\frac{\partial f}{\partial t}\Big)^2 \Big\} d\sigma\, dt. \tag{5.2.6}$$

The solution of (5.2.1), (5.2.2) must satisfy the constraint

$$\int_{D_{\epsilon_0}} \big[v^2 + |\mathrm{grad}\, v|^2\big] dx \leq M, \tag{5.2.7}$$

with M a prescribed constant.

We here present the arguments of Levine & Payne (1985). The proof of (5.2.5) proceeds by setting

$$v = e^{-\lambda F} u, \tag{5.2.8}$$

for λ to be chosen. After some computation, it is found that u will satisfy the differential equation

$$\begin{aligned} \frac{\partial^2 u}{\partial t^2} - \Delta u =& 2\lambda(F_{,t}u_{,t} - F_{,i}u_{,i}) + u\Big[\lambda\Big(\frac{\partial^2 F}{\partial t^2} - \Delta F\Big) \\ &+ \lambda^2\big(F_{,j}F_{,j} - F_{,t}^2\big)\Big], \end{aligned} \tag{5.2.9}$$

or, upon substituting for F,

$$\begin{aligned} 0 =& Lu, \\ =& \frac{\partial^2 u}{\partial t^2} - \Delta u + 4\lambda(tu_{,t} + x_i u_{,i}) \\ &- \big[4\lambda^2(|x|^2 - t^2) - 2\lambda(n+1)\big]u. \end{aligned} \tag{5.2.10}$$

Let us now integrate the expression

$$\int_{D_\epsilon} \{tu_{,t} + x_i u_{,i} + \frac{1}{2}(n-1)u\} Lu\, dx = 0. \tag{5.2.11}$$

Observing that the components of the unit normal on S_ϵ have the form

$$\nu_i = \frac{-x_i}{\sqrt{|x|^2 + t^2}}, \qquad \nu_t = \frac{t}{\sqrt{|x|^2 + t^2}},$$

we can obtain from (5.2.11), after discarding a nonnegative term and employing standard inequalities, the inequality,

$$\left[8\lambda^2 F - 2(n+1)\lambda\right] \int_{D_\epsilon} u^2 dx \leq C_1 \int_{S_\epsilon} \frac{u_{,i}u_{,i} + \lambda^2 u^2}{|\text{grad}\, F|} d\sigma + C_2 \int_{\Sigma_\epsilon} \left(u_{,t}^2 + |\text{grad}\, u|^2 + \lambda^2 u^2\right) ds, \quad (5.2.12)$$

for computable constants C_1 and C_2. Assume that λ is large enough that

$$\lambda \geq \frac{n+1}{\epsilon_0^2(1-\alpha)}, \qquad 0 < \alpha < 1. \quad (5.2.13)$$

Note that we may assume that C_1 and C_2 are independent of ϵ since for $\epsilon_0 < \epsilon < r$ we may replace them by constants that are independent of ϵ. If, moreover, we were to integrate the left side of (5.2.12) over a smaller domain, the sense of the inequality would be preserved so that we can obtain

$$8\lambda^2 \epsilon_0^2 \alpha \int_{D_\beta} u^2 dx \leq C_1 \int_{S_\epsilon} \frac{u_{,i}u_{,i} + \lambda^2 u^2}{|\text{grad}\, F|} d\sigma + C_2 \int_{\Sigma_\epsilon} \left(u_{,t}^2 + |\text{grad}\, u|^2 + \lambda^2 u^2\right) ds, \quad (5.2.14)$$

where $\beta = (2\epsilon + r)/3$. Integrating (5.2.14) with respect to ϵ, returning to the v variable, and assuming that v satisfies the constraint (5.2.7), one is led to

$$\int_{D_\delta} v^2 dx \leq C_3 e^{-\lambda(r-\epsilon_0)/2} M + C_4 e^{3\lambda(r-\epsilon_0)/4} \frac{\hat{Q}}{\lambda^2}, \quad (5.2.15)$$

where $\delta = (r + \epsilon_0)/2$ and

$$\hat{Q} = \int_{-T}^{T} \int_{\Gamma} \left[\left(\frac{\partial f}{\partial t}\right)^2 + |g|^2 + \lambda^2 f^2\right] d\sigma\, dt. \quad (5.2.16)$$

Observe that for $\lambda > 1$, $\hat{Q}/\lambda^2 \leq Q$, where Q is defined in (5.2.6).

Assuming Q is sufficiently small, the choice

$$\lambda = \frac{8}{9(r-\epsilon_0)} \ln \frac{M}{Q}, \quad (5.2.17)$$

leads to

$$\int_{D_\delta} v^2 dx \leq C M^{2/3} Q^{1/3}, \quad (5.2.18)$$

which is the desired inequality, (5.2.5), since for $\epsilon > (r + \epsilon_0)/2$, $D_\epsilon \subset D_\delta$.

The drawback of the preceding argument is that it cannot be applied when the data are prescribed on only a part of the lateral boundary. This difficulty was addressed by Ames & Isakov (1991) who established an explicit stability estimate for the two-dimensional wave equation when the Cauchy data are given on the half - lateral boundary of a cylindrical domain.

If we let Ω be the cylindrical domain

$$\{x_2 > -1, x_1^2 + (x_2 + 1)^2 < 1, -4 < t < 4\},$$

and let Γ be the part of the boundary of Ω for which

$$\{x_1^2 + (x_2 + 1)^2 = 1, -4 < t < 4\},$$

then Ames & Isakov (1991) consider the following Cauchy problem:

$$\frac{\partial^2 u}{\partial t^2} - \frac{\partial^2 u}{\partial x_1^2} - \frac{\partial^2 u}{\partial x_2^2} = f, \qquad \text{in } \Omega, \tag{5.2.19}$$

$$u = \frac{\partial u}{\partial \nu} = 0, \qquad \text{on } \Gamma, \tag{5.2.20}$$

where ν is the unit exterior normal to Γ. Since the details of the argument that Ames & Isakov (1991) used to obtain a stability result are rather lengthy and specific to a particular domain, we shall not describe their analysis here. The basic idea of their argument is to combine Friedrichs - Leray energy integrals with the Carleman type estimates used by Hörmander (1976). The method of Carleman reduces estimates of weighted L^2 norms to estimates of symbols of differential quadratic forms. With these tools, Ames & Isakov (1991) prove:

Theorem 5.2.1 (Ames & Isakov (1991)).
Any solution $u \in H^2(\Omega)$ of problem (5.2.19), (5.2.20) satisfies the estimate

$$\|u\|_{(1)}(\Omega_\epsilon) \leq \max\Big\{ C\|f\|(\Omega_0), \Big[\|f\|(\Omega_0)\Big]^{\epsilon/(20-\epsilon)} \Big[\epsilon^{-2}\|u\|_{(1)}(\Omega_0)\Big]^{1-\epsilon/(20-\epsilon)} \Big\}, \tag{5.2.21}$$

where Ω_ϵ denotes the subsets

$$\Omega \cap \{x_1^2 + (x_2 + 5)^2 - \frac{1}{2}t^2 - 17 > \epsilon\}$$

with $0 < \epsilon < 1$, $\|u\|_{(1)}(\Omega)$ denotes the norm in $H^1(\Omega)$ and $\|u\|$ the usual L^2 norm, and C is a computable constant.

Inequality (5.2.21) represents a stability inequality of the type that can be utilized to construct numerical algorithms which generate approximate solutions to problems such as (5.2.19), (5.2.20). Since there are presently no existence theorems for these problems, solutions obtained via numerical regularization are often the best that can be found. Therefore, it is desirable to have explicit stability estimates with optimal constants. Although the results of Ames & Isakov (1991) provide an explicit estimate, the constants in this estimate are not optimal and, in fact, may be too large to be useful in actual computations.

We close this chapter by briefly describing some results obtained by Payne (1985) for a nonstandard problem for the wave equation. Let us consider the following problem:

$$\frac{\partial^2 u}{\partial t^2} - \Delta u = 0, \qquad \text{in } \Omega, \tag{5.2.22}$$

$$u = f, \qquad \frac{\partial u}{\partial \nu} = g, \qquad \text{on } \Gamma, \tag{5.2.23}$$

$$u(x,0) = 0, \qquad u(x,T) = 0. \tag{5.2.24}$$

In this problem,

$$\Omega = \bigcup_{0<t<T} \Omega_t \times \{t\}, \tag{5.2.25}$$

where each Ω_t is a smoothly varying domain in $\mathbf{R}^n$ and

$$\Gamma = \bigcup_{0<t<T} \Gamma_t \times \{t\}, \tag{5.2.26}$$

where Γ_t denotes an $(n-1)$–dimensional simply connected surface of the boundary of Ω. We further assume that the hyperplanes $x_1 = \alpha$ for $0 < \alpha < \alpha_1$ intersect the boundary of Ω only in points of Γ.

This problem is a special case of a class of problems analysed by Lavrentiev *et al.* (1983) who derive continuous dependence on the data f and g using a modified weighted energy method. Their argument required uniform bounds on the solution u and all of its derivatives. Payne (1985) relaxed these constraints on the solution, obtaining continuous dependence inequalities that required weaker L^2 conditions.

These results were obtained via a modified weighted energy method not dissimilar to that described in section 5.1.

Introducing the convex sets

$$D_\alpha = \{(x,t)\big| x_1 < \alpha\} \cap \Omega, \tag{5.2.27}$$

and the weight

$$u = e^{\lambda x_1} v, \tag{5.2.28}$$

for a constant λ to be chosen, Payne (1985) establishes continuous dependence of the solution of (5.2.22) - (5.2.24) on the data f and g, by deriving the inequality

$$\int_{D_\alpha} u^2 d\tau \leq K(M) Q^{[(3-\sqrt{2})/2]-2x_1/\alpha_1}, \tag{5.2.29}$$

for a computable constant K and data term Q. To deduce (5.2.29), it was necessary for Payne (1985) to constrain solutions u to satisfy the condition

$$\lambda^{-2} \int_{D_{\alpha_1}} \left[\left(\frac{\partial u}{\partial x_1} \right)^2 + \lambda^2 u^2 \right] d\tau \leq M^2. \tag{5.2.30}$$

Inequality (5.2.29) is a Hölder stability result on the region

$$0 \leq x_1 < \left(\frac{3-\sqrt{2}}{2} \right) \alpha_1,$$

for solutions constrained by (5.2.30).

Although Payne (1985) confined his analysis to the wave equation, his argument would extend to the class of equations

$$\frac{\partial^2 u}{\partial t^2} - \Delta u = F\Big(x, t, u, \frac{\partial u}{\partial x_1}\Big),$$

where the function F is assumed to be Lipschitz in its last two arguments. Since the length of the stability interval for these equations as well as for (5.2.22) depends on the choice of the constant λ, a different selection could lead to a larger interval on which stability is guaranteed.

5.3 The Cauchy Problem for the Laplace Equation and Other Elliptic Equations

One of the classical ill-posed problems for partial differential equations is the Cauchy problem for Laplace's equation which was first studied by Hadamard at the turn of the century. This problem was introduced in section 1.4. Not only did Hadamard show that a global solution to this problem will exist only if the Cauchy data satisfy certain compatibility conditions, but he also demonstrated that even if a solution does exist, it will not necessarily depend continuously on the data.

To illustrate this fact, Hadamard used the following example. Consider a sequence of problems for the two-dimensional Laplace equation where the n−th problem ($n = 1, 2, \ldots$) satisfies analytic Cauchy data, i.e.

$$\frac{\partial^2 u}{\partial x^2} + \frac{\partial^2 u}{\partial y^2} = 0, \qquad \text{in } \; 0 < x < a, \; 0 < y < b,$$

$$u(x,0) = \frac{\sin nx}{n}, \qquad \frac{\partial u}{\partial y}(x,0) = 0, \qquad 0 < x < a.$$

We observe that $\sin nx/n$ tends to zero uniformly as n becomes infinite and that the solution of the limit problem is $u \equiv 0$. The method of separation of variables leads to the solution

$$u(x,y) = \frac{\cosh ny \sin nx}{n^2},$$

for the n-th problem. For $y \neq 0$, $u(x,y)$ does not approach the limit solution as $n \to \infty$. In fact, $|u(x,y)|$ grows rapidly as n becomes infinite. This example demonstrates how a small variation in the data might not lead to a corresponding small change in the solution.

The above example may be viewed as an elliptic version of the parabolic case presented in section 1.3.

A method of obtaining stability inequalities for the Cauchy problem for the Laplace equation with general data was introduced by Payne (1960). In fact, Payne (1960) posed the problem in the form

$$\Delta u = 0, \qquad \text{in} \quad \Omega \subset \mathbf{R}^n, \tag{5.3.1}$$

$$\int_\Gamma (u-f)^2 d\sigma \leq \epsilon_1^2, \qquad \int_\Gamma (u_{,i} - g_i)(u_{,i} - g_i) d\sigma \leq \epsilon_2^2. \tag{5.3.2}$$

Here Γ is a portion of the boundary of Ω on which the Cauchy data f and g_i are measured so that ϵ_1 and ϵ_2 are known.

By introducing a set of spherical surfaces of the form $\psi(r) = \alpha$ for a constant α, with the property that the intersection of the boundary of Ω with $\psi(r)$ contains only points of Γ, Payne (1960) uses a logarithmic convexity argument to establish that solutions of (5.3.1), (5.3.2) which satisfy a prescribed constraint depend continuously on the Cauchy data in an appropriate measure. More specifically, defining $S(r)$ as the set consisting of points of the boundary of $\psi(r)$ intersecting Ω, Payne (1960) derives the inequality

$$\int_{S(r)} u^2 d\sigma \leq CM^{2(1-\delta)} \left[\alpha_1 \int_\Gamma u^2 d\sigma + \alpha_2 \int_\Gamma u_{,i} u_{,i} d\sigma \right]^\delta, \tag{5.3.3}$$

for solutions u that satisfy the uniform bound

$$|u| \leq M, \tag{5.3.4}$$

explicit constants C, α_1, α_2, and a function $\delta(r)$ for which $0 < \delta(r) \leq 1$. From inequality (5.3.3), both uniqueness and continuous dependence results can be obtained for harmonic functions satisfying (5.3.4).

Payne's idea to use spherical surfaces to derive stability inequalities was extended by Trytten (1963) who established pointwise bounds for second order elliptic equations that have lower order nonlinearities. Schaefer (1965,1967) generalized the method to higher order quasilinear elliptic equations and to systems of elliptic equations. Because Schaefer used a different measure, the constraint on the solution that he required to stabilize his systems was weaker than the uniform bound Payne needed. Instead of (5.3.4), the condition

$$\int_{\Omega} u^2 dx \leq M^2, \tag{5.3.5}$$

for a constant M was sufficient to guarantee continuous dependence results.

One of the drawbacks of these studies is that the estimates at points in Ω away from the portion of the boundary on which the Cauchy data are prescribed are rather crude because the bounds at distant points depend on bounds at intermediate points. Consequently, the approximation error accumulates quickly. To circumvent this problem, Payne (1970) showed how the introduction of an approximate class of auxiliary surfaces could be used to avoid the accumulation error. He considered the problem

$$Ku \equiv \frac{\partial}{\partial x_j}\left(\kappa_{ij}\frac{\partial u}{\partial x_i}\right) = F, \qquad \text{in } \Omega \subset \mathbf{R}^n, \tag{5.3.6}$$

$$u = f, \qquad \frac{\partial u}{\partial x_i} = g_i, \qquad \text{on } \Gamma, \tag{5.3.7}$$

where the data f and g_i are prescribed on a portion Γ of the boundary of Ω and the matrix κ_{ij} is symmetric and positive definite. To analyse this problem, Payne (1970) replaces the spherical surfaces in his earlier paper by a set of surfaces $h(\mathbf{x}) = \alpha$, for $0 < \alpha \leq 1$, that may not be closed and are chosen so that they intersect Ω to form a closed region Ω_α. The boundary of Ω_α consists only of points of Ω and points on the surface $h = \alpha$. Additional assumptions on the surfaces $h(\mathbf{x}) = \alpha$, $0 < \alpha \leq 1$, include the conditions,

(1) if $\beta \leq \gamma$, then $\Omega_\beta \subset \Omega_\gamma$,

(2) $|\nabla h| > \delta > 0$, in Ω_1,

(3) $Kh \leq 0$ in Ω_1,

(4) Ω_α has nonzero measure; Ω_0 has zero measure,

(5) $h(\mathbf{x})$ has continuous derivatives in $\bar{\Omega}_1$.

Recall that these are the same type of surfaces Bennett used in his modeling investigations which were described in chapter 3. Using these surfaces and assuming the differential equation holds on Γ, Payne (1970) derives bounds on the solution u at points in Ω_α with a logarithmic convexity argument applied to the functional

$$\phi(\alpha) = \int_0^\alpha \int_{\Omega_\eta} (\alpha - \eta)\big[\kappa_{ij}\frac{\partial u}{\partial x_i}\frac{\partial u}{\partial x_j} + u\,Ku\big]dx\,d\eta + Q^2, \tag{5.3.8}$$

where

$$Q^2 = c_1 \int_\Gamma u^2 d\sigma + c_2 \int_\Gamma \frac{\partial u}{\partial x_i} \frac{\partial u}{\partial x_i} d\sigma + c_3 \int_\Gamma (Ku)^2 d\sigma, \tag{5.3.9}$$

for explicit constants c_i to be chosen.

Payne's (1970) argument relies on differentiating the expression for ϕ in (5.3.8) to find that

$$\phi'(\alpha) = \int_0^\alpha \int_{\Omega_\eta} \left[\kappa_{ij} \frac{\partial u}{\partial x_i} \frac{\partial u}{\partial x_j} + u\,Ku\right] dx\, d\eta, \tag{5.3.10}$$

and after a further differentiation,

$$\phi''(\alpha) = \int_{\Omega_\alpha} \left[\kappa_{ij} \frac{\partial u}{\partial x_i} \frac{\partial u}{\partial x_j} + u\,Ku\right] dx. \tag{5.3.11}$$

Rewriting $\phi'(\alpha)$, it can be shown that

$$\begin{aligned} \phi(\alpha) &= \int_0^\alpha \phi'(\eta) d\eta + Q^2, \\ &\geq \frac{1}{2} \int_{\Omega_\alpha} \kappa_{ij} \frac{\partial h}{\partial x_i} \frac{\partial h}{\partial x_j} u^2 dx - \mu_1 \int_\Gamma u^2 d\sigma \\ &\quad - \mu_2 \int_\Gamma \frac{\partial u}{\partial x_i} \frac{\partial u}{\partial x_i} d\sigma + Q^2, \end{aligned} \tag{5.3.12}$$

for computable constants μ_1 and μ_2. Consequently, the constants c_i in (5.3.9) can be chosen so that

$$\begin{aligned} \frac{1}{2} C \left[\int_{\Omega_\alpha} \kappa_{ij} h_{,i} h_{,j} u^2 dx + Q^2 \right] &\geq \phi(\alpha) \\ &\geq \frac{1}{2} \left[\int_{\Omega_\alpha} \kappa_{ij} h_{,i} h_{,j} u^2 dx + Q^2 \right], \end{aligned} \tag{5.3.13}$$

for another computable constant C.

Payne (1970) continues his argument by deriving the two inequalities

$$|\phi'| \leq \phi' + a_1 \phi, \tag{5.3.14}$$

and

$$\int_{\Omega_\alpha} \kappa_{ij} u_{,i} u_{,j} dx - 2 \int_{\Omega_\alpha} \frac{\left[\kappa_{ij} u_{,i} h_{,j}\right]^2}{\kappa_{ij} h_{,i} h_{,j}} dx \geq -a_2 \phi' - a_3 \phi, \tag{5.3.15}$$

where the a_i $(i = 1, 2, 3)$ are computable constants. A judicious use of these inequalities combined with standard ones then help to establish the result

$$\phi\phi'' - (\phi')^2 \geq -\beta_1\phi\phi' - \beta_2\phi^2, \tag{5.3.16}$$

for computable constants β_1 and β_2. As is shown in section 1.5, this inequality can be integrated by introducing the variable

$$\sigma = e^{-\beta_1\alpha}. \tag{5.3.17}$$

Thus, as in section 1.5, one finds

$$\phi(\alpha)\sigma^{-\beta_2/\beta_1^2} \leq \phi(0)^{(\sigma-\sigma_1)/(1-\sigma_1)}\left[\phi(1)\sigma_1^{-\beta_2/\beta_1^2}\right]^{(1-\sigma)/(1-\sigma_1)}, \tag{5.3.18}$$

where $\sigma_1 = e^{-\beta_1}$. Observe that $\phi(0) = Q^2$, an expression involving only data terms. To ensure stability, we now assume that the solution u belongs to that class of sufficiently smooth functions such that

$$\int_{\Omega_1} u^2 dx \leq M^2, \tag{5.3.19}$$

for some prescribed number M. One can then compute an M_1 so that

$$\phi(1)\sigma_1^{-\beta_2/\beta_1^2} \leq M_1^2, \tag{5.3.20}$$

and conclude from (5.3.18) that

$$\phi(\alpha) \leq AM_1^{2\nu(\alpha)}Q^{1(1-\nu(\alpha))}, \tag{5.3.21}$$

for a computable constant A and a function $0 < \nu(\alpha) \leq 1$. Inequality (5.3.21) is the desired continuous dependence estimate which can be used to obtain pointwise bounds on u in the domain Ω_α.

We emphasize that crucial to Payne's (1970) argument is the choice of the surfaces $h = \alpha$. In the earlier work of Payne, Schaefer, and Trytten, the choice was a family of hyperspheres with origin outside of Ω. Although these surfaces were easy to work with, the initial bounds could be computed only at points sufficiently close to the boundary on which the Cauchy data were given. In practice, some problems are better treated with surfaces other than the spherical ones used in this earlier work.

Numerical Studies

Numerical investigations of Cauchy problems for elliptic equations include the early studies of Douglas (1960), Cannon (1964), Cannon & Miller (1965), and Cannon & Douglas (1968). These papers examine the Cauchy problem for Poisson's equation and succeed in deriving stability and error estimates for various numerical schemes. Later, Han (1982) formulated a class of ill-posed problems as variational inequalities and established convergence of finite element discretizations. A different approach was then taken by Falk & Monk (1986) who obtained error estimates for a regularization method that approximates the solution of a Cauchy problem for Poisson's equation on a rectangle by first deriving stability results for discrete harmonic functions that are solutions to a standard piecewise - linear approximation of Laplace's equation.

Falk & Monk (1986) analyse the following problem:

$$-\Delta u = F, \qquad \text{in} \quad \Omega = [0,1] \times [0,1], \tag{5.3.22}$$

$$u(0,y) = u(1,y) = 0, \tag{5.3.23}$$

$$|u - f_1|_{1,\Gamma} \leq \epsilon_1, \qquad |u_n - f_2|_{0,\Gamma} \leq \epsilon_2 \,. \tag{5.3.24}$$

Here Γ is that part of the boundary of Ω that lies on the x-axis and $|\cdot|_{k,\Gamma}$ denotes the $k-$th Sobolev norm on the line segment Γ. The functions F, f_1 and f_2 as well as the constants ϵ_1 and ϵ_2 are all prescribed data. In addition, it is assumed that u satisfies an L^2 bound on the open segment of the boundary of Ω that lies along $y = 1$. This bound serves to stabilize the ill-posed problem (5.3.22) - (5.3.24).

Defining a discrete functional that is the analogue of functionals used in the logarithmic convexity method in the continuous case, Falk & Monk (1986) are able to derive Hölder stability results for discrete harmonic functions v that vanish on the boundary of Ω and belong to a finite element space of continuous piecewise - linear functions. Using these stability results, they then proceed to establish error estimates for a least squares penalty method approximating the problem (5.3.22) - (5.3.24).

Even though stability estimates for problems such as (5.3.22) - (5.3.24) had been obtained by Payne (1960,1970) as described earlier in this section, such estimates are not sufficient for analyzing numerical methods for approximating solutions. Error estimates can be derived from these continuous results, but they are frequently pessimistic and the constraints required to ensure stability are too excessive in actual numerical experiments. It is to circumvent this difficulty that Falk & Monk (1986) devote a part of their paper to proving Hölder continuous dependence inequalities for discrete harmonic functions. Their proof of this result uses logarithmic convexity arguments applied to a functional defined on discrete harmonic functions u_h that satisfy

$$(\nabla u_h, \nabla \phi_h) = 0, \qquad \forall \phi_h \in S_h, \tag{5.3.25}$$

where S_h denotes the finite element space of all continuous piecewise linear functions that vanish on Γ.

Having established stability results, Falk & Monk (1986) then describe a numerical method to approximate the solution of the original problem and prove error estimates for this regularization method which involves minimizing a quadratic form with respect to a penalty parameter. These error estimates are conservative as evidenced by the results of the numerical experiments that are detailed in their work.

A different computational approach to the Cauchy problem for Laplace's equation is taken by Klibanov & Santosa (1991) who use the quasireversibility method proposed by Lattès & Lions (1969) to obtain approximate solutions. A reformulation of this classical problem leads them to consider the problem of finding a solution to

$$-\Delta u = f, \qquad \text{in} \quad \Omega, \tag{5.3.26}$$

$$u = \frac{\partial u}{\partial n} = 0, \qquad \text{on} \quad \Gamma. \tag{5.3.27}$$

Here Ω is a two-dimensional region bounded by a straight line and a parabola and Γ is the straight line portion of the boundary. Introducing the function

$$\phi = \left(x_2 + \frac{x_1^2}{\alpha^2} + \frac{1}{4} \right)^{-\beta} - \frac{1}{2^\beta}, \tag{5.3.28}$$

so that ϕ =constant represents a parabola and defining a region

$$\Omega_\gamma = \{x_2 > 0, \phi > \gamma\},$$

Klibanov & Santosa (1991) solve (5.3.26), (5.3.27) in the region $\Omega = \Omega_{-\sigma}$ for some $\sigma > 0$ with the boundary segment defined by

$$\Gamma = \left\{ x_2 = 0, \frac{x_1^2}{\alpha^2} < \left(2^\beta - \sigma\right)^{-1/\beta} - \frac{1}{4} \right\}. \tag{5.3.29}$$

The quasireversibility method replaces the ill-posed problem (5.3.26), (5.3.27) by a related perturbed problem that is properly posed. The solution of this perturbed problem converges to the solution of the original problem as the perturbation parameter approaches zero and thus constitutes an approximation to the solution of (5.3.26), (5.3.27). Klibanov & Santosa (1991) approximate this solution by a function u_ϵ that satisfies the problem

$$\epsilon(-\nabla\rho\nabla u_\epsilon + \nu u_\epsilon) + \Delta(\rho\Delta u_\epsilon) = \epsilon f - \Delta(\rho f), \qquad \text{in } \Omega, \tag{5.3.30}$$

$$u_\epsilon = 0, \qquad \frac{\partial u_\epsilon}{\partial n} = 0, \qquad \text{on } \Gamma, \tag{5.3.31}$$

where $\epsilon \in (0,1)$ and the function ρ is 1 in Ω_0 and tends to 0 smoothly near the boundary segment $\partial\Omega - \Gamma$. The constant ν is positive and independent of ϵ. The crucial points of this method are that (5.3.30), (5.3.31) constitute a well-posed problem and that u_ϵ converges to u as ϵ and σ tend to zero. Using a Carleman estimate Klibanov & Santosa (1991) prove an estimate for the rate of convergence of u_ϵ to u and thus obtain a measure for the difference between the solution to (5.3.26), (5.3.27) and the quasireversibility solution. The key result needed to establish this error bound is the Carleman estimate,

$$C\Big[\lambda\int_{\Omega_0}|\nabla v|^2 e^{2\lambda\phi}dx + \lambda^3\int_{\Omega_0} v^2 e^{2\lambda\phi}dx + \frac{1}{\lambda}\int_{\Omega_0}\sum_{i,j=1}^{2}\Big(\frac{\partial^2 v}{\partial x_i \partial x_j}\Big)^2 e^{2\lambda\phi}dx\Big] \le \int_{\Omega_0}(\Delta v)^2 e^{2\lambda\phi}dx, \quad (5.3.32)$$

for functions v that are H^2 in Ω_0 and vanish together with their normal derivative on the boundary of Ω_0, and constants λ and β that are sufficiently large. The derivation of this inequality, which can be found in Lavrentiev *et al.* (1983), depends critically on the shape of the domain Ω_0 which is reflected in the constant C. In order to actually compute numerical solutions to this problem, Klibanov & Santosa (1991) must discretize this quasireversibility method. They accomplish this using second order finite differences, obtaining a Carleman - type estimate for this discrete approximation in the process. Convergence of their numerical method and error bounds for their schemes can then be established from the discrete Carleman estimate.

In their numerical experiments involving the discrete quasireversibility scheme, Klibanov & Santosa (1991) pay particular attention to examining the effects of data errors on the approximate solution, since these effects were not studied by Lattès & Lions (1969) in their earlier numerical simulations with this method. By solving the Cauchy problem in a half-space with data given on the axis, they find that their method is rather accurate, robust, and efficient. It seems to work particularly well when additional data are available on a line that is parallel to the boundary in the interior of the domain. This suggests that quasireversibility methods might work effectively for that class of problems in which the Cauchy data are known on only parts of the boundary.

6

Some Further Improperly Posed Problems

6.1 Uniqueness and Continuous Dependence for Singular Equations

This, the final chapter, is devoted to a consideration of some new ill-posed problems for linear differential equations that have application to a number of physical situations. Payne's (1975) monograph provides an excellent source of uniqueness and continuous dependence results which are available for such problems prior to 1975. Because research in the last twenty years has revealed various physical phenomena exhibiting behavior that can be modeled by improperly posed problems containing linear differential equations, we take this opportunity to discuss the relevant analysis that has produced new uniqueness, continuous dependence and related theorems.

We point out here that, despite the fact that the differential equations in the sequel are linear, the spaces in which we can establish continuous dependence for improperly posed problems are nonlinear. Consequently, we are faced with some of the difficulties inherent in nonlinear problems.

We begin this chapter by describing the results of Ames (1984) on singular differential equations. The paper of Dunninger & Levine (1976) is an interesting contribution to the theory for differential equations with a singular term which give rise to improperly posed problems in the field of partial differential equations. This paper also gives several references to other appropriate work on partial differential equations which contain singular terms.

Ames (1984) establishes uniqueness and continuous dependence theorems for solutions to Cauchy problems for operator equations of the form

$$\frac{d^2u}{dt^2} + \frac{\psi(t)}{t}\frac{du}{dt} = Mu, \qquad t \in [0,T), \tag{6.1.1}$$

in a Hilbert space setting and for the singular backward heat equation

$$\frac{\partial u}{\partial t} + \Delta u + \sum_{i=1}^{m} \frac{k_i}{x_i}\frac{\partial u}{\partial x_i} + g(t)u = 0, \qquad t \in [0,T). \tag{6.1.2}$$

If $\psi(t)$ is a real constant, equation (6.1.1) is an abstract Euler - Poisson - Darboux equation. The case where M is the $n-$dimensional Laplacian is used as a model for various physical phenomena, e.g. the motion of a gas when a sound wave of finite amplitude is propagated through it. The equation with $\psi(t) = 1/3$ is known as Tricomi's equation and arises in the study of transonic gas dynamics. If $M = -\Delta$, then equation (6.1.1) is called the generalized axially symmetric potential equation. Since the analysis of Ames (1984) requires no definiteness assumption on the operator M, the Cauchy problems she studies do not have to be well posed. Consequently, logarithmic convexity arguments provide a natural means to analyse such equations.

Following Ames (1984) let D_1 be a dense linear subspace of a real Hilbert space H and let M be a symmetric linear operator that is either bounded or unbounded and maps D_1 into H. (If M is actually bounded then we can take $D_1 = H$.) Consider the problem

$$\frac{d^2u}{dt^2} + \frac{\psi(t)}{t}\frac{du}{dt} = Mu, \qquad 0 < t < T, \tag{6.1.3}$$

$$u(0) = f, \qquad \frac{du}{dt}(0) = 0, \tag{6.1.4}$$

where $f \in H$ and T is finite. The following hypotheses are adopted as well:

(i) M and H are independent of t,

(ii) the solution $u \in C^2([0,T); D_1)$ exists,

(iii) $\psi(t) \in C^1((0,T))$ satisfies for $t \in (0,T)$

$$\psi(t) > 0, \qquad \text{and} \qquad \psi(t) \geq t\psi'(t),$$

i.e. $\psi(t)/t$ is a decreasing function of t.

Suppose v and w are two solutions to (6.1.3), (6.1.4) with different initial data $v(0) = f_1$, $v_t(0) = 0$, and $w(0) = f_2$, $w_t(0) = 0$, then $u = v - w$ satisfies (6.1.3), (6.1.4) with $f = f_1 - f_2$. We shall now prove that the functional

$$\phi(t) = \|u(t)\|^2 + \beta|(f, Mf)|, \tag{6.1.5}$$

where β is a positive constant satisfies the inequality

$$\phi\phi'' - (\phi')^2 \geq -k\phi^2, \tag{6.1.6}$$

for a computable nonnegative constant k. Uniqueness and Hölder continuous dependence on the initial data can then be inferred from (6.1.6), in a manner similar to that outlined in section 1.5.

Differentiating (6.1.5), we obtain

$$\phi'(t) = 2(u, u_t), \tag{6.1.7}$$

and a second differentiation yields

$$\phi''(t) = 2\|u_t\|^2 + 2(u, u_{tt}). \tag{6.1.8}$$

Substitution of the differential equation leads to the expression

$$\phi''(t) = 2\|u_t\|^2 - 2\Big(u, \frac{\psi(t)}{t}\, u_t\Big) + 2(u, Mu). \tag{6.1.9}$$

Upon integrating the identity

$$0 = 2\int_0^t \Big(u_\eta, u_{\eta\eta} + \frac{\psi(\eta)}{\eta} u_\eta - Mu\Big)\, d\eta,$$

and using the result to eliminate the term (u, Mu) in (6.1.9), we find that

$$\phi''(t) = \frac{4t}{\psi(t)}\, F' - \frac{\psi(t)}{t}\, \phi' + 4F + 2(f, Mf), \tag{6.1.10}$$

where

$$F(t) = \int_0^t \|u_\eta\|^2 \frac{\psi(\eta)}{\eta}\, d\eta.$$

Equation (6.1.10) can be rewritten as

$$\frac{d}{dt}\Big[\mu(t)\frac{d\phi}{dt}\Big] = \frac{4t}{\psi(t)}\,\frac{d}{dt}\Big[\mu(t)F\Big] + 2\mu(t)(f, Mf), \tag{6.1.11}$$

with the integrating factor

$$\mu(t) = \exp\Big(\int \frac{\psi(t)}{t}\, dt\Big).$$

Integrating the previous equation from 0 to t, we derive the result

$$\begin{aligned}\frac{\psi(t)}{t}\,\frac{d\phi}{dt} =& 4F - \frac{4\psi(t)}{t\mu(t)}\int_0^t \mu(\eta)F(\eta)\Big[\frac{\psi(\eta) - \eta\psi'(\eta)}{\psi^2(\eta)}\Big]\, d\eta \\ &+ 2(f, Mf) - \frac{2(f, Mf)}{\mu(t)}\int_0^t \mu(\eta)\, d\eta.\end{aligned} \tag{6.1.12}$$

Since $\psi(t) > 0$ for $t \in [0, T)$, it follows that $F(t) \geq 0$ and condition (iii) implies that the second term on the right hand side of (6.1.12) is nonnegative. Hence, we have the lower bound,

$$\phi'' \geq 4\|u_t\|^2 + 2(f, Mf)G(t), \tag{6.1.13}$$

where

$$G(t) = 1 - \frac{1}{\mu(t)} \int_0^t \mu(\eta) d\eta.$$

We now form the expression $\phi\phi'' - (\phi')^2$ to obtain

$$\phi\phi'' - (\phi')^2 \geq 4S^2 + 4\|u_t\|^2 \beta|(f, Mf)| + 2\phi(f, Mf)G(t), \tag{6.1.14}$$

where

$$S^2 = \|u_t\|^2 \|u\|^2 - (u, u_t)^2$$

is nonnegative by the Cauchy - Schwarz inequality. Discarding the first two nonnegative terms on the right side of (6.1.14), we arrive at the inequality

$$\phi\phi'' - (\phi')^2 \geq -2\phi|(f, Mf)| \max_{0 \leq t < T} |G(t)|. \tag{6.1.15}$$

In view of our definition of $\phi(t)$, we have (6.1.6). As we have repeatedly seen, uniqueness of the solution easily follows from this convexity inequality. Hölder continuous dependence on the Cauchy data also follows from (6.1.6) provided we constrain the solution u to belong to the class of functions

$$\Big\{\phi \in C^2([0, T); D_1) \Big| \|\phi(T)\|^2 \leq N^2\Big\}$$

for a prescribed constant N.

Let us point out that the results concerning the uniqueness of solutions to the Cauchy problem for the Euler - Poisson - Darboux (EPD) equation

$$\frac{\partial^2 u}{\partial t^2} + \frac{c}{t}\frac{\partial u}{\partial t} = \Delta u, \qquad \text{in} \quad \Omega \subset \mathbf{R}^n \times [0, T)$$

are consistent with previous work, e.g. Weinstein (1954), where the solution is shown to be unique provided $c > 0$. Uniqueness does not occur when $c < 0$. In fact, if $u_{(\alpha)}(x, t)$ is a solution of the Cauchy problem with $c = \alpha < 0$, then any function of the form $t^{1-c} u_{(2-\alpha)}(x, t)$ which vanishes together with its t derivatives at $t = 0$ may be added to $u_{(\alpha)}(x, t)$ to yield another solution. (Here $u_{(2-\alpha)}(x, t)$ satisfies the EPD equation with $c = 2 - \alpha > 0$.)

Uniqueness and Hölder continuous dependence on the Cauchy data can also be obtained for solutions of (6.1.2), a singular backward heat equation, from the logarithmic convexity method. Since the substitution

$$w(x, t) = u(x, t) \exp\Big(\int_0^t g(\eta) d\eta\Big), \tag{6.1.16}$$

transforms (6.1.2) into the equation

$$\frac{\partial w}{\partial t} + \Delta w + \sum_{i=1}^{m} \frac{k_i}{x_i}\frac{\partial w}{\partial x_i} = 0,$$

we consider instead of (6.1.2), a Cauchy problem for $w(x,t)$, namely

$$\frac{\partial w}{\partial t} + \Delta w + \sum_{i=1}^{m} \frac{k_i}{x_i}\frac{\partial w}{\partial x_i} = 0, \quad \text{in} \quad \Omega \times [0,T), \tag{6.1.17}$$

$$\frac{\partial w}{\partial n} = 0, \quad \text{on} \quad \partial\Omega \times [0,T], \tag{6.1.18}$$

$$w(x,0) = f(x), \quad \text{in} \quad \Omega. \tag{6.1.19}$$

Here Ω is the domain

$$\left\{ x \in \mathbf{R}^m \middle| 0 < x_i < a_i, i = 1, \ldots, m \right\},$$

and $\partial/\partial n$ denotes the outward directed normal derivative to the boundary of Ω. The k_i are real constants, T is again finite, and $f(x) \in C(\Omega)$. Classical solutions to (6.1.17) - (6.1.19) are assumed to exist.

Using the Cauchy-Schwarz inequality, it is easy to show that the functional

$$\phi(t) = \int_\Omega w^2 \left(\prod_{i=1}^{m} x_i^{k_i} \right) dx, \tag{6.1.20}$$

satisfies the standard logarithmic convexity inequality

$$\phi\phi'' - (\phi')^2 \geq 0, \tag{6.1.21}$$

so that if w is the difference of two solutions with identical initial data, uniqueness follows from (6.1.21). In addition, if w is the difference of two solutions with different initial data, then we can obtain from (6.1.21) the bound

$$\phi(t) \leq \left[\phi(0)\right]^{1-t/T} \left[\phi(T)\right]^{t/T}, \tag{6.1.22}$$

for $0 \leq t < T$, using arguments outlined in section 1.5. This is a stability inequality if w is restricted to lie in the correct class of functions, i.e. the set of all functions

$$\psi(x,t) \in C(\bar{\Omega} \times [0,T]) \cap C^2(\Omega \times (0,T)),$$

such that

$$\int_\Omega \psi^2(x,T) \left(\prod_{i=1}^{m} x_i^{k_i} \right) dx \leq N^2, \tag{6.1.23}$$

for some prescribed constant N. We then conclude that solutions of (6.1.17) - (6.1.19) depend Hölder continuously on the Cauchy data on compact subsets of $[0,T)$. In view of the transformation (6.1.16), the same conclusion can be reached for solutions $u(x,t)$ of the related Cauchy problem for (6.1.2).

We point out that similar results can be obtained for several generalizations of the problems for (6.1.1) and (6.1.2). For example, the operator M in (6.1.1) may depend on t, in which case we must define $M'(t)$ appropriately and assume that for each $t \in [0,T)$ and $\psi \in D_1$,

$$(\psi, M'(t)\psi) \leq 0.$$

For example $M(t) = p(t)\Delta$ meets such a criterion if the function $p(t)$ is increasing on $[0,T)$. Also, if we include a second derivative u_{tt} in equation (6.1.2), then logarithmic convexity arguments can be used to demonstrate that a solution to a suitably defined boundary - initial value problem for the resulting equation is unique whenever it exists and depends Hölder continuously on the initial data in an appropriate measure.

6.2 Spatial Decay in Improperly Posed Parabolic Problems

The analysis of spatial decay of solutions to partial differential equations in various shaped cylinders has attracted much recent attention, see e.g. the account and the references in the book by Flavin & Rionero (1995), and we especially mention the novel work of Lin & Payne (1993) on improperly posed problems and that of Lin & Payne (1994) who investigated a variety of novel effects. It is worthy of note that apart from the paper of Lin & Payne (1993), these studies have typically been for well posed problems. In the field of improperly posed problems the study of spatial decay estimates is very recent having been initiated by Lin & Payne (1993) and continued by Franchi & Straughan (1994). Since this book is primarily devoted to improperly posed problems and also because the literature on decay estimates in well posed problems is so vast, we simply restrict attention here to an account of decay estimates as applied to improperly posed problems.

To describe this work we use the notation of Lin & Payne (1993). Let R be the semi - infinite cylindrical domain given by

$$R = \{(x_1, \ldots, x_N)\big|(x_2, \ldots, x_N) \in D, x_1 > 0\},$$

where D is a bounded region in $\mathbf{R}^{N-1}$, where N typically is 2 or 3. The boundary of D is denoted by ∂D and is supposed smooth enough that Poincaré like inequalities may be used.

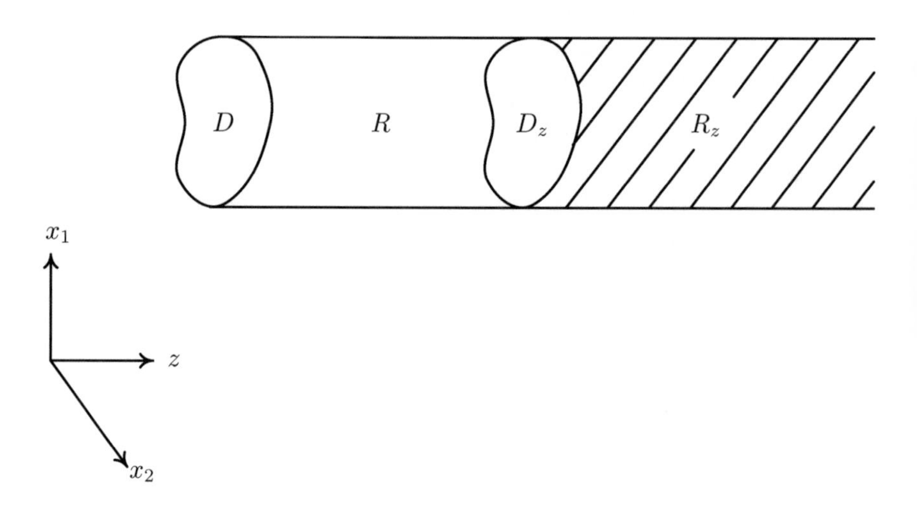

Figure 4. Notation for the Cylinder R.

The notation used is as indicated in figure 4, where $N = 3$, so R_z is the region

$$R_z = \{(x_1, x_2, x_3)\big|(x_2, x_3) \in D,\ x_1 > z\},$$

where $z \geq 0$, and D_z denotes the region D but for $x_1 = z$.

Lin & Payne (1993) study two problems. The first is one for the backward heat equation in an $\mathbf{R}^N$ dimensional cylinder, so they consider

$$\begin{aligned}
&\frac{\partial u}{\partial t} + \Delta u = 0, \qquad \text{in } R \times (0, T),\\
&u(\mathbf{x}, t) = 0, \quad \text{on } \partial D,\ x_1 \geq 0, t \in [0, T],\\
&u(\mathbf{x}, 0) = g(\mathbf{x}), \quad \mathbf{x} \in R,\\
&u(0, x_2, \ldots, x_N, t) = f(x_2, \ldots, x_N, t), \quad (x_2, \ldots, x_N) \in D.
\end{aligned} \tag{6.2.1}$$

Since (6.2.1) is an improperly posed problem, to establish spatial decay Lin

& Payne (1993) need to define a constraint set and they require

$$\int_0^T \int_R |\nabla u|^2 dx\, dt \leq M, \tag{6.2.2}$$

for some constant M.

The second problem considered by Lin & Payne (1993) concerns the stream function for Stokes flow in a domain in $\mathbf{R}^2$. They investigate

$$\begin{aligned}
&\frac{\partial \Delta v}{\partial t} + \Delta^2 v = 0, \qquad \text{in } R \times (0,T),\\
&v = 0, \frac{\partial v}{\partial x_2} = 0, \quad \text{when} \quad x_2 = 0, h;\ x_1 \geq 0, t \in [0,T],\\
&v(x_1,x_2,0) = \tilde{g}(x_1,x_2), \quad (x_1,x_2) \in D,\\
&v(0,x_2,t) = \tilde{f}(x_2,t), \quad 0 \leq x_2 \leq h,\ t \in [0,T].
\end{aligned} \tag{6.2.3}$$

The constraint set for this problem consists of those functions satisfying

$$\int_0^T \int_R (\xi - z)^3 (v_{,ij}v_{,ij} + h^2 v_{,it}v_{,it}) dx\, dt \leq M_1, \tag{6.2.4}$$

for another constant M_1. Here z is fixed, ξ is the integration variable in the z direction and dx denotes integration over x and ξ, i.e. x_1 and x_2.

The result of Lin & Payne (1993) for the backward heat equation problem (6.2.1) commences with the function

$$E(z,t) = \int_0^t \int_{R_z} u_{,i}u_{,i} dx\, d\eta, \qquad 0 < t \leq T.$$

By various devices they establish that $\exists$ a constant k such that an inequality holds of form

$$\frac{\partial E}{\partial z} - \frac{1}{k}\frac{\partial E}{\partial t} + 2k\left(E + \frac{1}{2}\int_{R_z} g^2 dx\right) \leq 0. \tag{6.2.5}$$

They then introduce the distance s along a line

$$z + kt = \text{constant},$$

and convert (6.2.5) to an ordinary differential inequality in the variable s, i.e.

$$\frac{dE}{ds} + \frac{2k^2}{\sqrt{1+k^2}}\left(E + \frac{1}{2}\int_{R_z} g^2 dx\right) \leq 0.$$

From this inequality they are able to show that

$$E(z,t) \le E(0,t+z/k)e^{-kz} - \frac{1}{2}(1-e^{-2kz})\int_{R_z} g^2 dx. \tag{6.2.6}$$

At this point they employ the constraint (6.2.2) to bound the first term on the right hand side of (6.2.6). Thus they establish a spatial decay estimate which shows that for $0 < z < kT$,

$$E\Big(z, T - \frac{z}{k}\Big) \le Me^{-2kz}. \tag{6.2.7}$$

The spatial decay estimate for the backward Stokes flow problem is technically much more involved. It uses the function

$$\Phi(z,t) = \int_0^t \int_{R_z} (\xi - z)^3 (v_{,ij}v_{,ij} + \tilde{k}h^2 v_{,it}v_{,it}) dx\, dt,$$

where $\tilde{k}$ is a constant to be selected. After much calculation Lin & Payne (1993) show that Φ satisfies the inequality

$$\frac{\partial \Phi}{\partial z} - \frac{3h}{20\pi}\frac{\partial \Phi}{\partial t} + \frac{\pi}{5h}[\Phi + Q(z)] \le 0, \tag{6.2.8}$$

where $Q(z)$ is a data term. If s now denotes distance along a line

$$z + \frac{20\pi}{3h}t = \text{constant},$$

then (6.2.8) is rearranged as an ordinary differential inequality

$$\frac{d\Phi}{ds} + \frac{4\pi}{h\sqrt{400 + 9h^2/\pi^2}}[\Phi + Q(z)] \le 0.$$

This inequality is integrated to yield the result

$$\Phi(z,t) + \Big[1 - \exp\Big(-\frac{\pi z}{5h}\Big)\Big]Q(z) \le \Phi\Big(0, \frac{3hz}{20\pi} + t\Big)\exp\Big(-\frac{\pi z}{5h}\Big),$$

and then with the constraint this gives

$$\Phi(z,t) + \Big[1 - \exp\Big(-\frac{\pi z}{5h}\Big)\Big]Q(z) \le M_1 \exp\Big(-\frac{\pi z}{5h}\Big). \tag{6.2.9}$$

This is the spatial decay estimate for the problem (6.2.3).

Franchi & Straughan (1994) consider an equation which is motivated by a problem for the toroidal part of a magnetic field in a dynamo. Further details of the dynamo aspect may be found in Lortz & Meyer Spasche (1982). The problem studied by Franchi & Straughan (1994) is

$$\begin{aligned}
&\frac{\partial h}{\partial t} = -\frac{\partial}{\partial x_i}\Big(\eta(\mathbf{x})\frac{\partial h}{\partial x_i}\Big) + \frac{\partial}{\partial x_i}\Big(v_i(\mathbf{x})h\Big), \quad \text{in } R\times(0,T),\\
&h = 0, \quad \text{on } \partial D,\ x_1\ge 0, t\in[0,T],\\
&h(\mathbf{x},0) = h_0(\mathbf{x}), \quad \mathbf{x}\in R,\\
&h(0,x_2,x_3,t) = f(x_2,x_3,t), \quad (x_2,x_3)\in D.
\end{aligned} \tag{6.2.10}$$

The diffusivity η and velocity $\mathbf{v}$ are here prescribed.

Franchi & Straughan (1994) assume that h belongs to the class of solutions which satisfy the constraint

$$\int_0^T\int_R |\nabla h|^2 dx\, dt \le M, \tag{6.2.11}$$

for a positive constant M. Additionally the diffusivity is uniformly bounded below, hence

$$\eta(\mathbf{x}) \ge \eta_0 > 0, \qquad \forall\mathbf{x}\in R. \tag{6.2.12}$$

Define the quantities m and Q by

$$Q = \sup_R |v_{i,i}|, \qquad m = \sup_R |v_1|\,. \tag{6.2.13}$$

The decay rate of Franchi & Straughan (1994) depends on the numbers m and Q, and v_i is such that

$$Q < 2\eta_0\lambda_1\,, \tag{6.2.14}$$

where λ_1 is the first eigenvalue in the membrane problem for D.

Franchi & Straughan (1994) also need the conditions

$$\begin{aligned}
&\lim_{x_1\to\infty}\int_D h^2 dA = 0,\\
&\lim_{x_1\to\infty}\int_D h_{,1}^2 dA = 0,
\end{aligned} \tag{6.2.15}$$

which are to be expected from (6.2.11).

The analysis employs the function $E(z,t)$, defined by

$$E(z,t) = \int_0^t\int_{R_z} \eta h_{,i}h_{,i} dx\, ds, \qquad 0 < t \le T. \tag{6.2.16}$$

By integrating this expression by parts and using the differential equation we find

$$\begin{aligned}
E(z,t) =& \frac{1}{2}\int_{R_z} h^2 dx - \frac{1}{2}\int_{R_z} h_0^2 dx - \int_0^t\int_{D_z} \eta h h_{,1} dA\, ds\\
&- \frac{1}{2}\int_0^t\int_{R_z} v_{i,i}h^2 dx\, ds + \frac{1}{2}\int_0^t\int_{D_z} v_1 h^2 dA\, ds.
\end{aligned} \tag{6.2.17}$$

Estimates for the last three terms in (6.2.17) are obtained using Poincaré's inequality. The results are

$$
\begin{aligned}
-\frac{1}{2}\int_0^t\int_{R_z} v_{i,i}h^2 dx\, ds \leq & \frac{Q}{2\lambda_1}\int_0^t\int_z^\infty\int_{D_s} h_{,\alpha}h_{,\alpha} dA\, ds\, d\tau, \\
\leq & \frac{Q}{2\lambda_1\eta_0}E; \qquad (6.2.18)
\end{aligned}
$$

$$
\begin{aligned}
\frac{1}{2}\int_0^t\int_{D_z} v_1 h^2 dA\, ds \leq & \frac{m}{2\lambda_1}\int_0^t\int_{D_z} h_{,\alpha}h_{,\alpha} dA\, ds, \\
\leq & \frac{m}{2\lambda_1\eta_0}\int_0^t\int_{D_z} \eta h_{,i}h_{,i} dA\, ds, \\
= & -\frac{m}{2\lambda_1\eta_0}\frac{\partial E}{\partial z}; \qquad (6.2.19)
\end{aligned}
$$

and

$$
\begin{aligned}
-\int_0^t\int_{D_z} \eta h h_{,1} dA\, ds \leq & \left[\int_0^t\int_{D_z} \eta h^2 dA\, ds \int_0^t\int_{D_z} \eta h_{,1}^2 dA\, ds\right]^{1/2} \\
\leq & \frac{1}{\sqrt{\mu_1}}\int_0^t\int_{D_z} \eta h_{,i}h_{,i} dA\, ds \\
= & -\frac{1}{\sqrt{\mu_1}}\frac{\partial E}{\partial z}, \qquad (6.2.20)
\end{aligned}
$$

where μ_1 denotes the value

$$
\mu_1 = \min_{\phi\in H_0^1(D)} \frac{\int_D \eta|\nabla\phi|^2 dA}{\int_D \eta\phi^2 dA}.
$$

Upon using (6.2.18) - (6.2.20) in (6.2.17) one finds

$$
\begin{aligned}
\left(1-\frac{Q}{2\lambda_1\eta_0}\right)E(z,t) \leq & \frac{1}{2}\int_{R_z} h^2 dx - \frac{1}{2}\int_{R_z} h_0^2 dx \\
& -\left(\frac{m}{2\lambda_1\eta_0}+\frac{1}{\sqrt{\mu_1}}\right)\frac{\partial E}{\partial z}. \qquad (6.2.21)
\end{aligned}
$$

Another application of Poincaré's inequality shows that

$$
\begin{aligned}
\int_{R_z} h^2 dx \leq & \frac{1}{\lambda_1}\int_{R_z} h_{,\alpha}h_{,\alpha} dx \\
\leq & \frac{1}{\lambda_1\eta_0}\int_{R_z} \eta h_{,i}h_{,i} dx \\
= & \frac{1}{\lambda_1\eta_0}\frac{\partial E}{\partial t},
\end{aligned}
$$

and then using this in (6.2.21) we obtain

$$\Big(1-\frac{Q}{2\lambda_1\eta_0}\Big)E(z,t)\le-\Big(\frac{m}{2\lambda_1\eta_0}+\frac{1}{\sqrt{\mu_1}}\Big)\frac{\partial E}{\partial z}+\frac{1}{2\lambda_1\eta_0}\frac{\partial E}{\partial t}-\frac{1}{2}\int_{R_z}h_0^2dx. \tag{6.2.22}$$

Franchi & Straughan (1994) define $A, B, C, k(>0)$ by

$$A=\frac{m}{2\lambda_1\eta_0}+\mu^{-1/2},\quad B=\frac{1}{2\lambda_1\eta_0},\quad C=1-\frac{Q}{2\lambda_1\eta_0},\quad k=\frac{A}{B},$$

and rewrite (6.2.22) as

$$\frac{\partial E}{\partial z}-\frac{1}{k}\frac{\partial E}{\partial t}+\frac{C}{A}\Big(E+\frac{1}{2C}\int_{R_z}h_0^2dx\Big)\le 0. \tag{6.2.23}$$

Let now s measure the distance along the line

$$z+kt=\text{constant},$$

in the direction of increasing z. The analysis is facilitated by introducing the function $J(z,t)$ as

$$J(z,t)=E(z,t)+\frac{1}{2C}\int_{R_z}h_0^2dx,$$

and from (6.2.23) observing that J satisfies the inequality:

$$\frac{\partial J}{\partial z}+\frac{1}{2C}\int_{D_z}h_0^2dA-\frac{1}{k}\frac{\partial J}{\partial t}+\frac{C}{A}J\le 0, \tag{6.2.24}$$

or

$$\frac{dJ}{ds}+\frac{Ck}{A\sqrt{1+k^2}}J+\frac{k}{2C\sqrt{1+k^2}}\int_{D_z}h_0^2dA\le 0. \tag{6.2.25}$$

If the points P_0 and P are $P_0=(0,t+z/k)$ and $P=(z,t)$, then since

$$z=\frac{kd}{\sqrt{1+k^2}},$$

where d is the distance from P_0 to P, inequality (6.2.25) may be integrated and we find

$$J(z,t)-J(0,t+z/k)e^{-Cz/A}+\frac{1}{2C}e^{-Cz/A}\Big(\int_0^z\int_{D_\xi}h_0^2dA\,d\xi\Big)\le 0.$$

After rearrangement this gives

$$E(z,t) \leq E(0,t+z/k)e^{-Cz/A} - \frac{1}{2C}\left(1-e^{-Cz/A}\right)\left(\int_{R_z} h_0^2 dx\right). \quad (6.2.26)$$

The first term on the right of (6.2.26) is bounded using (6.2.11) so that for $0 < z < kT$,

$$E(z, T-z/k) \leq Me^{-Cz/A}. \quad (6.2.27)$$

The above (exponential) spatial decay estimate was obtained in the work of Franchi & Straughan (1994). The constant C/A is explicitly

$$\frac{C}{A} = \frac{2\lambda_1\eta_0 - Q}{m + (2\lambda_1\eta_0/\sqrt{\mu_1})},$$

and this shows how the spatial decay rate depends on the size of the maximum values of $|v_1|$ and $|v_{i,i}|$ in the region R.

Continuous Dependence on Spatial Geometry and on Modeling in Cylindrical Geometry Spatial Decay Problems

Lin & Payne (1994) is an influential paper which introduces the idea of studying continuous dependence on the spatial geometry itself into problems involving spatial decay in a cylinder like that in figure 4. They use the equation of heat conduction in a bar to illustrate their ideas. Additionally, they study continuous dependence on the model itself by studying continuous dependence on the thermal diffusivity, or on the specific heat. While these analyses are not specifically for improperly posed problems we believe they are worth reporting here due to the connection with the work in chapters 2 and 3. Of course, we could have included an account of their work in chapter 2 and in chapter 3. However, we believe it fits in well here with the account of other spatial decay problems.

Let D be a simply - connected domain in $\mathbf{R}^2$ with boundary ∂D, and let R, R_z, be as defined earlier in this section. Lin & Payne (1994) study the partial differential equation,

$$\nu\frac{\partial u}{\partial t} = \Delta u, \qquad \text{in } R\times(0,\infty). \quad (6.2.28)$$

The constant coefficient $\nu(>0)$ may be regarded as the specific heat, or alternatively, ν^{-1} can be thought of as the thermal diffusivity. The function u is required to satisfy the initial conditions,

$$u(x_1,x_2,x_3,0) = 0, \qquad \mathbf{x}\in\mathbf{R}^3, \quad (6.2.29)$$

and the boundary conditions,

$$\begin{aligned} &u(x_1,x_2,x_3,t)=0, \qquad (x_1,x_2)\in\partial D, x_3\geq 0, t\geq 0,\\ &u(x_1,x_2,0,t)=g(x_1,x_2,t), \qquad (x_1,x_2)\in D, t\geq 0. \end{aligned} \tag{6.2.30}$$

Additionally, Lin & Payne (1994) assume that

$$\sup_{t\geq 0}\left[\int_0^t\int_R \frac{\partial u}{\partial x_i}\frac{\partial u}{\partial x_i}\,dx\,d\tau+\frac{1}{2}\int_R u^2dx\right] \tag{6.2.31}$$

is bounded.

Lin & Payne (1994) begin by showing the solution u satisfies a Phragmén - Lindelöf type theorem, so that u either must grow exponentially or decay exponentially, in a suitable measure.

To establish continuous dependence on the coefficient ν, Lin & Payne (1994) let u solve (6.2.28) - (6.2.31) for the coefficient ν, while v solves the same problem for the coefficient $\tilde{\nu}$. They set

$$w=u-v,$$

and show w satisfies the boundary - initial value problem

$$\begin{aligned} &\nu\frac{\partial w}{\partial t}+(\nu-\tilde{\nu})\frac{\partial v}{\partial t}=\Delta w, \qquad \text{in } R\times(0,\infty),\\ &w(x_1,x_2,x_3,0)=0, \qquad \text{in } R,\\ &w(x_1,x_2,x_3,t)=0, \qquad \text{on } \partial D\times[0,\infty),\\ &w(x_1,x_2,0,t)=0, \qquad (x_1,x_2)\in D, t\geq 0. \end{aligned}$$

By deriving a differential inequality for the quantity Φ defined by

$$\Phi(z,t)=\int_0^t\int_{D_z} w^2dA\,d\tau,$$

Lin & Payne (1994) establish the following theorem which shows explicit continuous dependence of the solution on the variation in ν.

Theorem 6.2.1. (Lin & Payne (1994).)
Let u be the solution to problem (6.2.28) - (6.2.31) and let v be the solution to the same problem with ν replaced by $\tilde{\nu}$. Then for arbitrary $z\geq 0$, $t\geq 0$, the following inequalities hold,

$$\int_0^t\int_{D_z}(u-v)^2ds\,d\tau\leq\frac{(\nu-\tilde{\nu})^2}{4k^3}Q_0(t)(2kz+1)e^{-2kz}, \tag{A}$$

and

$$\int_0^t \int_{R_z} (u-v)^2 dx\, d\tau \le \frac{(\nu-\tilde{\nu})^2}{16k^5} Q_0(t)[4k^2z^2+6kz+3]e^{-2kz}. \qquad (B)$$

In equations (A) and (B), k is the square root of the first eigenvalue for the fixed membrane problem for D, and the function Q_0 is given by

$$Q_0(t) = \int_0^t \int_D \left(\frac{\partial g}{\partial \tau}\right)^2 dA\, d\tau.$$

To study continuous dependence on the spatial geometry Lin & Payne (1994) define regions R_1, R_2, by

$$R_1 = \{(x_1,x_2,x_3)\big|(x_1,x_2) \in D_1, x_3 \ge 0\},$$
$$R_2 = \{(x_1,x_2,x_3)\big|(x_1,x_2) \in D_2, x_3 \ge 0\},$$

where D_1 and D_2 are different geometries as in figure 2; D_2 may be thought of as a perturbation to D_1. The function u satisfies (6.2.28) - (6.2.31) with $R = R_1$ and $g = g_1$, while v satisfies the same problem with $R = R_2$ and $g = g_2$.

Define

$$D = D_1 \cap D_2 \ne 0,$$
$$D' = D_1 \cup D_2.$$

Lin & Payne (1994) define the difference w by

$$w = u - v$$

and then show w satisfies the boundary - initial value problem

$$\begin{aligned} &\nu \frac{\partial w}{\partial t} = \Delta w, && \text{in } R \times (0,\infty),\\ &w(x_1,x_2,x_3,0) = 0, && (x_1,x_2,x_3) \in R,\\ &w(x_1,x_2,x_3,t) = u - v, && \text{on } \partial D \times [0,\infty),\\ &w(x_1,x_2,0,t) = g_1 - g_2, && (x_1,x_2) \in D, t \ge 0. \end{aligned}$$

They define a quantity δ which is a suitable measure of the "distance" between D_1 and D_2 and set

$$c = \max_{\partial D} \left(\frac{1}{n_\rho}\right),$$

where n_ρ is the radial component of the unit normal vector on ∂D.

By developing a somewhat involved analysis they establish the following theorem.

Theorem 6.2.2. (Lin & Payne (1994).)
Let u be a solution to (6.2.28) - (6.2.31) in $R_1 \times (0,\infty)$ with data g_1, and let v be a solution to the same problem in $R_2 \times (0,\infty)$ with data g_2. Then, if $g_1 - g_2$ is defined in $D_1 \cap D_2$ it follows that for arbitrary $z \geq 0$, $t \geq 0$,

$$\sqrt{\int_0^t \int_{R_z} (u-v)^2 dx\, d\tau} \leq \delta^{1/2} \Big\{ A_1 \sqrt{(2kz+1)z} \sqrt{B_1(t) + B_2(t)} + A_2 \sqrt{Q_1(z)} \sqrt{B_3(t) + B_4(t)} \Big\} e^{-pz}. \quad (6.2.32)$$

In this theorem, A_1 and A_2 are constants computed in Lin & Payne (1994), and the decay rate p is given by

$$p = \min\{k_1, k_2\},$$

where k_1 and k_2 are the square roots of the first eigenvalues in the membrane problems for D_1 and D_2, Q_1 is a quadratic function of z and B_i are data terms given by

$$B_1(t) = \int_0^t \int_{D_1} \left[4k_1 g_1^2 + \frac{1}{2k_1} \frac{\partial g_1}{\partial x_\alpha} \frac{\partial g_1}{\partial x_\alpha} + \frac{1}{16k_1^3} \left(\frac{\partial g_1}{\partial \tau} \right)^2 \right] dA\, d\tau,$$

$$B_2(t) = \int_0^t \int_{D_2} \left[4k_2 g_2^2 + \frac{1}{2k_2} \frac{\partial g_2}{\partial x_\alpha} \frac{\partial g_2}{\partial x_\alpha} + \frac{1}{16k_2^3} \left(\frac{\partial g_2}{\partial \tau} \right)^2 \right] dA\, d\tau,$$

$$B_3(t) = \int_0^t \int_{D_1} \left[4k_1 \left(\frac{\partial g_1}{\partial \tau} \right)^2 + \frac{1}{2k_1} \frac{\partial^2 g_1}{\partial x_\alpha \partial \tau} \frac{\partial^2 g_1}{\partial x_\alpha \partial \tau} + \frac{1}{16k_1^3} \left(\frac{\partial^2 g_1}{\partial \tau^2} \right)^2 \right] dA\, d\tau,$$

$$B_4(t) = \int_0^t \int_{D_2} \left[4k_2 \left(\frac{\partial g_2}{\partial \tau} \right)^2 + \frac{1}{2k_2} \frac{\partial^2 g_2}{\partial x_\alpha \partial \tau} \frac{\partial^2 g_2}{\partial x_\alpha \partial \tau} + \frac{1}{16k_2^3} \left(\frac{\partial^2 g_2}{\partial \tau^2} \right)^2 \right] dA\, d\tau,$$

where a repeated α denotes summation over 1 and 2.

Note that inequality (6.2.32) establishes continuous dependence on the spatial geometry by virtue of the $\delta^{1/2}$ term which measures the "difference" in geometry of D_1 and D_2.

6.3 Uniqueness for the Backward in Time Navier - Stokes Equations on an Unbounded Spatial Domain

The backward in time Navier -Stokes equations present a famous improperly posed problem for a fully nonlinear system of equations. The fundamental result on uniqueness backward in time when the spatial domain Ω is bounded was demonstrated by Serrin (1963) who used a weighted energy method following Lees & Protter (1961). Continuous dependence for the Navier - Stokes equations backward in time followed by Knops & Payne (1968) who used a logarithmic convexity argument. Other results are due to Payne (1971), Bardos & Tartar (1973), Straughan (1983), and Ghidaglia (1986).

A continuous dependence result for the backward in time Navier - Stokes equations which occupy an unbounded spatial region Ω, which is exterior to a bounded domain $\Omega^*(\subset \mathbf{R}^3)$, without requiring extraneous decay assumptions at infinity is due to Galdi & Straughan (1988). By using a weighted logarithmic convexity method Galdi & Straughan (1988) are able to demonstrate continuous dependence in the class of solutions for which the velocity perturbation and acceleration satisfy

$$u_i \in L^{6-\epsilon}(\Omega), \qquad \frac{\partial u_i}{\partial t} \in L^{6-\epsilon}(\Omega),$$

for any $\epsilon(> 0)$, while for the pressure there holds

$$p = O(r^{1-\epsilon}), \qquad \frac{\partial p}{\partial x_i} = O(r^{1-\epsilon}).$$

Of course, the above work also predicts uniqueness. It is important to realize that this is non-trivial in such a backward in time improperly posed problem for a nonlinear system of equations. For example, Franchi & Straughan (1993c) show that another nonlinear fluid dynamical theory, namely that for a generalized second grade material, the solution may even fail to exist, let alone be unique.

The interior boundary of Ω we denote by $\partial\Omega$. For the purpose of this chapter the boundary - initial value problem for the Navier - Stokes equations backward in time may be written:

$$\begin{aligned}
&\frac{\partial v_i}{\partial t} = v_j \frac{\partial v_i}{\partial x_j} - \Delta v_i + \frac{\partial p}{\partial x_i}, \qquad \text{in} \quad \Omega \times (0,T), \\
&\frac{\partial v_i}{\partial x_i} = 0, \qquad \text{in} \quad \Omega \times (0,T), \\
&v_i(\mathbf{x},t) = v_i^B(\mathbf{x},t), \qquad \text{on} \quad \partial\Omega \times [0,T], \\
&v_i(\mathbf{x},0) = v_i^0(\mathbf{x}), \qquad \mathbf{x} \in \Omega.
\end{aligned} \tag{6.3.1}$$

The functions v_i^B and v_i^0 are prescribed and the upper limit of time, T, is a constant.

To study continuous dependence on the initial data Galdi & Straughan (1988) let (v_i, p) and (v_i^*, p^*) be two solutions which satisfy (6.3.1) for the same v_i^B, but for different v_i^0 and v_i^{0*}. The idea as usual is to produce an inequality for $u_i = v_i^* - v_i$ which bounds a measure of this difference in terms of $u_i(\mathbf{x}, 0) = v_i^{0*}(\mathbf{x}) - v_i^0(\mathbf{x})$. The difference solution (u_i, π) satisfies

$$\begin{aligned}
&\frac{\partial u_i}{\partial t} = v_j^* \frac{\partial u_i}{\partial x_j} + u_j \frac{\partial v_i}{\partial x_j} - \Delta u_i + \frac{\partial \pi}{\partial x_i}, \qquad \text{in} \quad \Omega \times (0, T), \\
&\frac{\partial u_i}{\partial x_i} = 0, \qquad \text{in} \quad \Omega \times (0, T), \\
&u_i(\mathbf{x}, t) = 0, \qquad \text{on} \quad \partial\Omega \times [0, T], \\
&u_i(\mathbf{x}, 0) = u_i^0(\mathbf{x}), \qquad \mathbf{x} \in \Omega.
\end{aligned} \tag{6.3.2}$$

The constraint set employed by Galdi & Straughan (1988) is that for which $\exists$ a constant $M(< \infty)$ such that

$$|u_i|, |v_i|, |\frac{\partial u_i}{\partial t}|, |\frac{\partial v_i}{\partial t}|, |\frac{\partial u_i}{\partial x_j}|, |\frac{\partial v_i}{\partial x_j}| \leq M, \tag{6.3.3}$$

$\forall (\mathbf{x}, t) \in \Omega \times [0, T]$.

The proof in Galdi & Straughan (1988) begins with the requirement that $u_i \in L^{6-\epsilon}(\Omega)$ and $\pi = O(r^{1-\epsilon})$. By differentiating $(6.3.2)_1$ one shows

$$\Delta \pi = \frac{\partial \psi_i}{\partial x_i}, \tag{6.3.4}$$

where the divergence function ψ_i is

$$\psi_i = -u_k \Big(\frac{\partial u_i}{\partial x_k} + 2 \frac{\partial v_i}{\partial x_k} \Big).$$

Equation (6.3.4) is in standard form for application of the theory of singular integrals and then one uses the Calderon - Zygmund theorem together with work of Galdi & Maremonti (1988) to deduce that $\pi \in L^{6-\epsilon}(\Omega)$. Likewise, when one starts with $u_i, u_{i,t} \in L^{6-\epsilon}(\Omega)$ and $\partial\pi/\partial t = O(r^{1/2-\epsilon})$, the function $\partial\pi/\partial t$ satisfies

$$\Delta \frac{\partial \pi}{\partial t} = \frac{\partial}{\partial x_i} \Big(\frac{\partial \psi_i}{\partial t} \Big),$$

and the same procedure leads to the conclusion that $\partial\pi/\partial t \in L^{6-\epsilon}(\Omega)$.

What has been achieved is important. By starting with the pressure perturbation and its derivative being allowed to grow at infintity we have shown that π and $\partial\pi/\partial t$ must belong to an integrability class required in application of the weighted logarithmic convexity method, namely $L^{6-\epsilon}(\Omega)$.

The weighted logarithmic convexity argument commences with the function

$$F(t) = \int_\Omega e^{-\alpha r} u_i u_i \, dx + A\alpha + K_1\alpha^\mu + K_2\alpha^{2\epsilon}, \tag{6.3.5}$$

where K_1, K_2 are constants calculated in Galdi & Straughan (1988), α is a constant (related to the initial data measure), μ is given by

$$\mu = \frac{\epsilon}{6-\epsilon},$$

and A is a data term

$$A = \left| \oint_{\partial\Omega} n_i x_i e^{-\alpha r} \pi^2 dA + 2\oint_{\partial\Omega} x_i n_j e^{-\alpha r} \pi \frac{\partial u_i}{\partial x_j} \, dS \right|.$$

The proof of continuous dependence calculates F' and F'' in the usual way and then computes $FF'' - (F')^2$ to find

$$\begin{aligned}
FF'' - (F')^2 &= 4S^2 + 2F\int_\Omega \frac{\partial e^{-\alpha r}}{\partial x_i} \frac{\partial u_i}{\partial t} \pi \, dx \\
&- 2F\int_\Omega \frac{\partial e^{-\alpha r}}{\partial x_i} u_i \frac{\partial \pi}{\partial t} \, dx - 2F\int_\Omega \Delta(e^{-\alpha r})\, u_i \frac{\partial u_i}{\partial t} \, dx \\
&- 4F\int_\Omega \frac{\partial e^{-\alpha r}}{\partial x_k} \frac{\partial u_i}{\partial x_k} \frac{\partial u_i}{\partial t} \, dx - F\int_\Omega \frac{\partial e^{-\alpha r}}{\partial x_k} \frac{\partial v_k^*}{\partial t} u_i u_i \, dx \\
&- 2F\int_\Omega \frac{\partial e^{-\alpha r}}{\partial x_m} v_m^* u_i \frac{\partial u_i}{\partial t} \, dx - 2F\int_\Omega \frac{\partial e^{-\alpha r}}{\partial x_m} u_m u_i \frac{\partial v_i}{\partial t} \, dx \\
&- \frac{1}{4}\left(\int_\Omega \frac{\partial e^{-\alpha r}}{\partial x_k} v_k^* u_i u_i \, dx \right)^2 \\
&- 2\int_\Omega \frac{\partial e^{-\alpha r}}{\partial x_k} v_k^* u_i u_i \, dx \int_\Omega e^{-\alpha r} u_i \left(\frac{\partial u_i}{\partial t} - \frac{1}{4} u_k \omega_{ik} - \frac{1}{2} v_k^* \frac{\partial u_i}{\partial x_k} \right) dx \\
&+ 4(A\alpha + K_1\alpha^\mu + K_2\alpha^{2\epsilon}) \Bigg\{ \int_\Omega e^{-\alpha r} \frac{\partial u_i}{\partial t} \frac{\partial u_i}{\partial t} \, dx \\
&\qquad - \int_\Omega e^{-\alpha r} v_k^* \frac{\partial u_i}{\partial x_k} \frac{\partial u_i}{\partial t} \, dx - \frac{1}{2}\int_\Omega e^{-\alpha r} u_k \frac{\partial u_i}{\partial t} \omega_{ik} \, dx \Bigg\} \\
&- \int_\Omega e^{-\alpha r} u_i u_i \, dx \int_\Omega e^{-\alpha r} \left(v_r^* \frac{\partial u_i}{\partial x_r} + \frac{1}{2} u_r \omega_{ir} \right) \left(v_q^* \frac{\partial u_i}{\partial x_q} + \frac{1}{2} u_q \omega_{iq} \right) dx \\
&- 2F\int_\Omega e^{-\alpha r} u_k \frac{\partial u_i}{\partial x_k} \frac{\partial v_i}{\partial t} \, dx,
\end{aligned} \tag{6.3.6}$$

where ω_{is} is the skew part of $v_{i,s}$,

$$\omega_{is} = v_{i,s} - v_{s,i},$$

and S^2 which is positive is

$$
\begin{aligned}
S^2 = \int_\Omega e^{-\alpha r} u_i u_i\, dx \int_\Omega e^{-\alpha r} \Big(\frac{\partial u_i}{\partial t} - \frac{1}{2} v_k^* \frac{\partial u_i}{\partial x_k} - \frac{1}{4} u_k \omega_{ik} \Big) \\
\times \Big(\frac{\partial u_i}{\partial t} - \frac{1}{2} v_p^* \frac{\partial u_i}{\partial x_p} - \frac{1}{4} u_p \omega_{ip} \Big)\, dx \\
- \Big[\int_\Omega e^{-\alpha r} u_i \Big(\frac{\partial u_i}{\partial t} - \frac{1}{2} v_k^* \frac{\partial u_i}{\partial x_k} - \frac{1}{4} u_k \omega_{ik} \Big)\, dx \Big]^2 .
\end{aligned}
$$

The calculation from (6.3.6) involves estimating each term, but one can arrive at

$$
\begin{aligned}
FF'' - (F')^2 \geq & - c_1 F^2 - c_2 FF' - \alpha A F \\
& - c_3 \alpha^2 F \int_\Omega e^{-\alpha r} \pi^2 dx - \alpha^2 F \int_\Omega e^{-\alpha r} \Big(\frac{\partial \pi}{\partial t} \Big)^2 dx \\
& - 4\alpha^4 F \int_\Omega e^{-\alpha r} \frac{\partial \pi}{\partial x_i} \frac{\partial \pi}{\partial x_i}\, dx \\
& - 16 \alpha^2 F \int_\Omega \frac{e^{-\alpha r}}{r^2} \frac{\partial \pi}{\partial x_i} \frac{\partial \pi}{\partial x_i}\, dx. \qquad (6.3.7)
\end{aligned}
$$

The first two terms in (6.3.7) are standard in studies of logarithmic convexity as we saw in section 1.5. However, the last five are due to the pressure or boundary terms and need handling differently. In particular, the pressure terms are estimated using the fact that $\pi, \partial\pi/\partial t \in L^{6-\epsilon}(\Omega)$ and $\partial\pi/\partial x_i = O(r^{1/2-\epsilon})$. The last four (pressure) terms in (6.3.7) are bounded below by a term of form

$$
-K_1 \alpha^\mu F - K_2 \alpha^{2\epsilon} F. \qquad (6.3.8)
$$

This analysis allows us to deduce F satisfies the standard logarithmic convexity inequality

$$
FF'' - (F')^2 \geq -m_1 F^2 - m_2 FF', \qquad (6.3.9)
$$

for constants m_1, m_2 independent of α. As shown in section 1.5 this integrates to

$$
F(t) \leq K(\delta) \big[F(T) \big]^{1-\delta} \big[F(0) \big]^{\delta}, \qquad t \in [0, T), \qquad (6.3.10)
$$

for

$$
K(\delta) = \exp \Big\{ \frac{m_1}{m_2} \big[T(1-\delta) - t \big] \Big\}
$$

and

$$
\delta = \frac{e^{-m_2 t} - e^{-m_2 T}}{1 - e^{-m_2 T}} .
$$

To derive an unweighted bound for u_i Galdi & Straughan (1988) bound $F(t)$ below by

$$e^{-\alpha R}\int_{\Omega\cap B} u_i u_i\,dx$$

for R a fixed number and where B is the ball center 0, radius R. Then

$$\int_{\Omega\cap B} u_i u_i\,dx \le K(\delta)e^{\alpha R}\big[F(T)\big]^{1-\delta}\big[F(0)\big]^{\delta}. \tag{6.3.11}$$

When the initial data are such that

$$|u_i(\mathbf{x},0)| < \eta,$$

for a constant η, then it can be shown $\exists$ constants K_3, K_4 such that

$$F(0) \le K_3\frac{\eta^2}{\alpha^3} + K_4\alpha^a, \tag{6.3.12}$$

where $a = \min\{\mu, 2\epsilon\}$. Upon using the constraint set,

$$F(T) \le \frac{K_5}{\alpha^{3(4-\epsilon)/(6-\epsilon)}}\,. \tag{6.3.13}$$

If η and α are related so that

$$\eta \le \alpha^{(3+a)/2},$$

then from (6.3.11) - (6.3.13) one may show, provided $\alpha \le 1/R$,

$$\int_{\Omega\cap B} u_i u_i\,dx \le K\alpha^{a\delta-3(1-\delta)(4-\epsilon)(6-\epsilon)}. \tag{6.3.14}$$

Then provided $t\in[0,t^*]$, with $t^* < T$ given by

$$t^* = -\frac{1}{m_2}\log\left[\frac{3(1-e^{-m_2T})}{3+a(6-\epsilon)/(4-\epsilon)} + e^{-m_2T}\right],$$

Galdi & Straughan (1988) show (6.3.14) leads to the Hölder continuous dependence on the initial data inequality

$$\int_{\Omega\cap B} u_i(\mathbf{x},t)u_i(\mathbf{x},t)\,dx \le K\alpha^b, \tag{6.3.15}$$

for $b>0$, $t\in[0,t^*]$ where t^* is independent of α.

6.4 Improperly Posed Problems for Dusty Gases

In the last section we studied an improperly posed problem for the nonlinear Navier - Stokes system of equations. We continue in this vein and report work for another nonlinear system arising in fluid dynamics which generalizes the Navier - Stokes equations to the situation where dust particles may be present in the fluid. This system is the Saffman dusty gas model. The work we report on is due to Crooke (1972,1973) and Calmelet - Eluhu & Crooke (1990).

The Saffman dusty gas model incorporates the Navier - Stokes equations for the solenoidal velocity field u_i, but allows for added dust particles by an interaction force term in the momentum equation. A separate equation is postulated for the particle velocity v_i, and a continuity equation is included for the number density of dust particles $N(\mathbf{x},t)(\geq 0)$. The model is then

$$\begin{aligned}
&\rho\Big(\frac{\partial u_i}{\partial t}+u_j\frac{\partial u_i}{\partial x_j}\Big)=-\frac{\partial p}{\partial x_i}+\mu\Delta u_i+kN(v_i-u_i)\,,\\
&\frac{\partial u_i}{\partial x_i}=0,\\
&mN\Big(\frac{\partial v_i}{\partial t}+v_j\frac{\partial v_i}{\partial x_j}\Big)=kN(u_i-v_i)\,,\\
&\frac{\partial N}{\partial t}+\frac{\partial}{\partial x_i}(Nv_i)=0,
\end{aligned}\tag{6.4.1}$$

where k is a positive constant, the $kN(v_i-u_i)$ piece represents the interaction force common in mixture theories, ρ is the constant fluid density, and m is the uniform mass of a dust particle. The work we report all studies the dusty gas equations on a bounded spatial domain $\Omega(\subset\mathbf{R}^3)$.

Crooke (1972) establishes a variety of results for (6.4.1) when time is positive. We do not include an exposition of these and refer only to his work on the improperly posed problem when $t<0$. In this case he was interested in a uniqueness result for the zero solution to (6.4.1). He adopted the constraint set

$$\sup_{\Omega\times[0,T]}\big(|u_i|+|v_i|+N\big)\leq M.\tag{6.4.2}$$

If we define the function $E(t)$ by

$$E(t)=\frac{1}{2}\int_\Omega(\rho u_iu_i+mNv_iv_i)dx,$$

then Crooke (1972) proves the theorem:

Theorem 6.4.1. ((Crooke (1972).)
Suppose at some time - level T, $E(T) = 0$. If $\exists$ a positive real number M such that (6.4.2) holds, then $E(t) \equiv 0 \ \forall t \in [0, T]$.

This is essentially a backward uniqueness result for the zero solution to (6.4.1). Crooke (1972) first showed a similar result is true for the forward in time problem and then interpreted the above result by arguing that $E(t)$ can never be zero at any time unless it is zero for all time. Crooke's (1972) proof is based on a logarithmic convexity argument for the function

$$J(t) = \int_t^T \int_\Omega \rho u_i u_i \, dx \, ds.$$

Crooke (1973) extended his uniqueness result to one of continuous dependence type. He supposes on the boundary of Ω, Γ,

$$u_i = 0, \quad N v_i n_i = f(\mathbf{x}), \qquad \mathbf{x} \in \Gamma. \tag{6.4.3}$$

He works with a function of form

$$F(t) = \frac{1}{2} \int_t^T \int_\Omega \rho u_i u_i \, dx \, ds + \frac{1}{2} t \int_{\Omega(T)} \rho u_i u_i \, dx + aQ, \tag{6.4.4}$$

for a constant $a > 1$ and Q being the data term

$$Q = \frac{1}{2} \int_{\Omega(T)} (\rho u_i u_i + m N v_i v_i) dx + \oint_\Gamma f \, dS.$$

The argument used is one of logarithmic convexity type and shows F satisfies the inequality

$$FF'' - (F')^2 \geq c_1 F F' - c_2 F^2,$$

for $c_1, c_2 > 0$ and this may be integrated to show

$$F(t) \leq \sigma^\delta \big[F(T) \sigma_1^{-\delta}\big]^{(1-\sigma)/(1-\sigma_1)} \big[F(0)\big]^{(\sigma - \sigma_1)/(1-\sigma_1)},$$

from which continuous dependence of the zero solution follows provided $F(T)$ is suitably restricted by a constraint set. The continuous dependence demonstrated is on both the initial and boundary data. In fact the following theorem is proved.

Theorem 6.4.2. (Crooke (1973).)
Suppose u_i, v_i and N are bounded solutions to (6.4.1). Then $\exists$ positive constants r_1, r_2, r_3, ζ and Λ and a function $\epsilon(t)$ such that $\forall t \in [0,T]$, $\epsilon(t) \geq 1$,

$$\begin{aligned}\frac{1}{2}\int_t^T\int_\Omega \rho u_i u_i\,dx\,ds \leq &\Lambda\Big[\Big(\frac{T+\zeta}{2}\Big)\int_{\Omega(T)}\rho u_i u_i\,dx + \zeta\oint_\Gamma f\,dS\\ &+\frac{1}{2}\int_{\Omega(T)} mNv_iv_i\,dx\Big]^\epsilon,\end{aligned}$$

and

$$\begin{aligned}\frac{1}{2}\int_t^T\int_\Omega mNv_iv_i\,dx\,ds \leq &r_1\Lambda\Big[\Big(\frac{T+\zeta}{2}\Big)\int_{\Omega(T)}\rho u_i u_i\,dx + \zeta\oint_\Gamma f\,dS\\ &+\frac{1}{2}\zeta\int_{\Omega(T)} mNv_iv_i\,dx\Big]^\epsilon\\ &+\frac{1}{2}r_2\int_{\Omega(T)} mNv_iv_i\,dx + r_3\oint_\Gamma f\,dS.\end{aligned}$$

Calmelet - Eluhu & Crooke (1990) extended this work to the continuous dependence of an arbitrary solution by considering the difference of two solutions to (6.4.1) but they also extended the work to a dusty gas model where the fluid (gas) is acted upon by an applied magnetic field. The interaction with a suitable form of Maxwell's equations is not considered. It is assumed the magnetic field is prescribed and only enters in the Lorentz force. The system considered by Calmelet - Eluhu & Crooke (1990) is

$$\begin{aligned}\frac{\partial U_i}{\partial t} + U_j\frac{\partial U_i}{\partial x_j} =& -\frac{1}{\rho}\frac{\partial P}{\partial x_i} + \nu\Delta U_i + \frac{kN}{\rho}(V_i - U_i)\\ &+\frac{\sigma}{\rho}(U_jB_jB_i - B_jB_jU_i),\\ \frac{\partial U_i}{\partial x_i} =&0,\\ \frac{\partial V_i}{\partial t} + V_j\frac{\partial V_i}{\partial x_j} =&\frac{k}{m}(U_i - V_i),\\ \frac{\partial V_i}{\partial x_i} =&0,\\ \frac{\partial N}{\partial t} + \frac{\partial}{\partial x_i}(Nv_i) =&0,\end{aligned}\qquad(6.4.5)$$

where B_i denotes the magnetic field. These equations are studied on $\Omega \times (0,T)$ with Ω bounded, and the boundary conditions are

$$U_i = U_i^s, \qquad n_iV_i = n_iV_i^s, \qquad N = N_s, \qquad \mathbf{x}\in\Gamma,$$

while the initial conditions are

$$U_i = U_i^0, \qquad V_i = V_i^0, \qquad N = N_0, \qquad \mathbf{x} \in \Omega.$$

The functions (U_i, V_i, N, P) are thought of as base variables and are perturbed by (u_i, v_i, n, p) and the boundary - initial value problem satisfied by the perturbation variables is

$$\begin{aligned}
&\frac{\partial u_i}{\partial t} + U_j \frac{\partial u_i}{\partial x_j} + u_j \frac{\partial U_i}{\partial x_j} + u_j \frac{\partial u_i}{\partial x_j} = -\frac{1}{\rho}\frac{\partial p}{\partial x_i} + \nu \Delta u_i \\
&\qquad\qquad + \frac{k}{\rho}(N+n)(v_i - u_i) + \frac{k}{\rho} n(V_i - U_i), \\
&\frac{\partial u_i}{\partial x_i} = 0, \\
&\frac{\partial v_i}{\partial t} + V_j \frac{\partial v_i}{\partial x_j} + v_j \frac{\partial V_i}{\partial x_j} + v_j \frac{\partial v_i}{\partial x_j} = \frac{k}{m}(u_i - v_i), \\
&\frac{\partial v_i}{\partial x_i} = 0, \\
&\frac{\partial n}{\partial t} + \frac{\partial}{\partial x_i}(n v_i + N v_i + n V_i) = 0,
\end{aligned} \tag{6.4.6}$$

with on Γ

$$u_i = 0, \qquad n_i v_i = 0, \qquad n = 0, \tag{6.4.7}$$

while at $t = 0$,

$$u_i = u_i^0, \qquad v_i = v_i^0, \qquad n = n_0. \tag{6.4.8}$$

The proof of the theorem of continuous dependence involves a logarithmic convexity method and as might be imagined from the system (6.4.6), it is quite involved. In fact, Calmelet - Eluhu & Crooke (1990) define a function $F(t)$ by

$$\begin{aligned}
F(t) =& \frac{1}{2}\int_t^T \int_\Omega u_i u_i dx\, ds + \frac{1}{2}\int_t^T \int_\Omega v_i v_i dx\, ds \\
&+ \frac{1}{2}\int_t^T \int_\Omega n^2 dx\, ds + \frac{1}{2} t \int_{\Omega(T)} u_i u_i dx \\
&+ \frac{1}{2}\zeta \left[\int_{\Omega(T)} (u_i u_i + v_i v_i + n^2) dx \right].
\end{aligned} \tag{6.4.9}$$

They show that for suitably constrained solutions this function satisfies the inequality, for c_1, c_2 computable constants

$$F F'' - (F')^2 \geq c_1 F F' - c_2 F^2.$$

From this they derive the theorem:

Theorem 6.4.3. (Calmelet - Eluhu & Crooke (1990).)
Consider a solution (U_i, V_i, N, P) with perturbations (u_i, v_i, n, p) with

$$U_i, V_i, N + n, V_i + v_i, B_i \qquad \text{bounded on} \qquad \Omega \times [0, T]$$

and N, v_i with bounded gradients on $\Omega \times [0, T]$. Then the function $F(t)$ defined by (6.4.9) depends continuously on the data $F(0)$ and $F(T)$ and satisfies

$$F(t) \leq \exp\left\{\frac{c_2}{c_1}\left(t - \frac{T}{q}\right)\right\}\left[F(T)\right]^{1/q}\left[F(0)\right]^{1-(1/q)},$$

where

$$\frac{1}{q} = \frac{e^{at} - 1}{e^{aT} - 1}.$$

As a corollary they establish a backward uniqueness result.

We close this chapter by observing that since B_i is given in (6.4.5) an interesting problem would be to investigate continuous dependence of the solution (U_i, V_i, N, P) on the function B_i. If the model is to be useful then the solution ought to depend continuously on the applied field B_i, unless there is some *a priori* reason to expect discontinuous bifurcation behavior. This would be a result of continuous dependence on modeling type.

6.5 Linear Thermoelasticity

Various ill-posed problems of thermoelasticity have been investigated in the literature. Among the earliest studies were those of Brun (1965b,1969) which also brought to the fore the importance of the Lagrange identity method in the field of improperly posed problems for partial differential equations. These investigations dealt with questions of uniqueness of solutions of the forward equations, assuming that the strain energy function is indefinite. The technique of Brun was extended by Rionero & Chirita (1987) who studied uniqueness and continuous dependence questions in linear thermoelasticity. Rionero & Chirita (1987) obtain a weighted Lagrange identity from which they establish uniqueness and Hölder continuous dependence on the body force and heat supply for solutions to various boundary initial value problems forward in time. They treat problems that are defined on both bounded and unbounded domains, including a half space region. As a consequence of these continuous dependence inequalities, they are also able to deduce theorems about continuous dependence on the initial data and the thermoelastic coefficients. The novel contribution of Rionero & Chirita (1987) is that their results are obtained without recourse to an energy conservation law or to any boundedness or definiteness assumptions on the thermoelastic coefficients.

The system of equations adopted by Rionero & Chirita (1987) is the usual set of partial differential equations derived by a linearisation process as described in e.g. Green (1962), Iesan (1989), and can be written as

$$\rho\frac{\partial^2 u_i}{\partial t^2} = \frac{\partial}{\partial x_j}\Big(c_{ijk\ell}\frac{\partial u_k}{\partial x_\ell}\Big) - \frac{\partial}{\partial x_j}\Big(\beta_{ij}\theta\Big) + \rho F_i, \tag{6.5.1}$$

$$a\frac{\partial \theta}{\partial t} + \beta_{ij}\frac{\partial^2 u_i}{\partial x_j \partial t} = \frac{\partial}{\partial x_i}\Big(\kappa_{ij}\frac{\partial \theta}{\partial x_j}\Big) + \rho S, \tag{6.5.2}$$

for the displacement u_i and temperature θ in $\Omega\times(0,T)$ where Ω is a bounded domain in $\mathbf{R}^3$ with a smooth boundary $\partial\Omega$. As everywhere else in the book, the notation above conforms with standard indicial notation employed in the continuum mechanics literature. The coefficients $\rho, F_i, a, \kappa_{ij}$ and S denote the density, body force, specific heat, thermal conductivity, and heat supply, respectively, while β_{ij} and $c_{ijk\ell}$ represent the thermoelastic coupling coefficients and the elastic coefficients which are assumed to satisfy the so-called major symmetry condition,

$$c_{ijk\ell} = c_{k\ell ij}\,. \tag{6.5.3}$$

Moreover, the thermal conductivity is a symmetric form and the Clausius-Duhem inequality implies that the thermal conductivity tensor is nonnegative in the sense that

$$\kappa_{ij}\xi_i\xi_j \geq 0, \qquad \forall \xi_i\,.$$

Associated with these equations, Rionero & Chirita (1987) assume various standard boundary and initial conditions.

Lagrange identity arguments

The argument that Rionero & Chirita (1987) use to establish their uniqueness and continuous dependence results is based on the application of the following Lagrange identity,

$$\begin{aligned}\int_\Omega \rho\Big[w_i(\mathbf{x},t)\frac{\partial v_i}{\partial t}(\mathbf{x},t) - v_i(\mathbf{x},t)\frac{\partial w_i}{\partial t}(\mathbf{x},t)\Big]\,dx \\ = \int_0^t\int_\Omega \rho\Big[w_i(\mathbf{x},\eta)\frac{\partial^2 v_i}{\partial \eta^2}(\mathbf{x},\eta) - v_i(\mathbf{x},\eta)\frac{\partial^2 w_i}{\partial \eta^2}(\mathbf{x},\eta)\Big]\,dx\,d\eta \\ + \int_\Omega \rho\Big[w_i(\mathbf{x},0)\frac{\partial v_i}{\partial t}(\mathbf{x},0) - v_i(\mathbf{x},0)\frac{\partial w_i}{\partial t}(\mathbf{x},0)\Big]\,dx\,.\end{aligned} \tag{6.5.4}$$

If u_i and θ are assumed to be the differences between two solutions to (6.5.1), (6.5.2) that satisfy the same initial and boundary conditions but have different body forces and heat supplies, then we can set

$$w_i(\mathbf{x},\eta) = u_i(\mathbf{x},\eta), \qquad v_i(\mathbf{x},\eta) = u_i(\mathbf{x},2t-\eta),$$

for $0 \le t \le T/2$. After suitable manipulations Rionero & Chirita (1987) then deduce (6.5.4) leads to the equation

$$2\int_\Omega \rho u_i(\mathbf{x},t)\frac{\partial u_i}{\partial t}(\mathbf{x},t)dx = \int_0^t\int_\Omega \rho\big[u_i(\mathbf{x},2t-\eta)\ddot{u}_i(\mathbf{x},2t-\eta)\big]dx\,d\eta, \quad (6.5.5)$$

where a superposed dot denotes differentiation with respect to the underlying time argument. Rionero & Chirita (1987) then show that use of the partial differential equations (6.5.1) and (6.5.2) in (6.5.5) yields the identity,

$$\begin{aligned}
2\int_\Omega \rho u_i(\mathbf{x},t)\frac{\partial u_i}{\partial t}(\mathbf{x},t)\,dx &+ \int_\Omega \kappa_{ij}\Big(\int_0^t \theta_{,i}(\eta)d\eta\Big)\Big(\int_0^t \theta_{,j}(\eta)d\eta\Big)\,dx\\
&= \int_0^t\int_\Omega \rho\Big[F_i(\eta)u_i(2t-\eta) - F_i(2t-\eta)u_i(\eta)\Big]dx\,d\eta\\
&\quad + \int_0^t\int_\Omega \rho\left[\theta(\eta)\int_0^{2t-\eta} S(\xi)d\xi - \theta(2t-\eta)\int_0^\eta S(\xi)d\xi\right]dx\,d\eta, \qquad (6.5.6)
\end{aligned}$$

where F_i and S now denote the differences in the two body forces and heat supplies, respectively, and $0 \le t < T/2$. Rionero & Chirita (1987) deduce uniqueness and continuous dependence results to various boundary inital value problems for (6.5.1), (6.5.2). For example, the following theorem can be obtained from (6.5.6).

Theorem 6.5.1. (Rionero & Chirita (1987).)
Assume that the thermal conductivity tensor κ_{ij} is positive definite so that

$$\kappa_{ij}\xi_i\xi_j \ge \kappa_0\xi_i\xi_i, \qquad \forall \xi_i, \qquad (6.5.7)$$

for a positive constant κ_0. The solution u_i, θ to the boundary - initial value problem consisting of (6.5.1) and (6.5.2) and the conditions

$$\begin{aligned}
&u_i(\mathbf{x},t) = u_i^*(\mathbf{x},t), \qquad \theta(\mathbf{x},t) = \theta^*(\mathbf{x},t), \qquad \text{on } \partial\Omega\times[0,T),\\
&\theta(\mathbf{x},0) = \theta_0(\mathbf{x}), \qquad \mathbf{x}\in\Omega, \qquad\qquad (6.5.8)\\
&u_i(\mathbf{x},0) = f_i(\mathbf{x}), \qquad \frac{\partial u_i}{\partial t}(\mathbf{x},0) = g_i(\mathbf{x}), \qquad \mathbf{x}\in\Omega,
\end{aligned}$$

is unique in $\Omega\times[0,T)$. Moreover, if we assume that u_i, θ represent the difference between two solutions satisfying the same boundary and initial conditions but correspond to different body forces and heat supplies, and that there exists a number $\mathcal{T} \in (0, T)$, such that

$$\int_0^{\mathcal{T}} \int_\Omega \rho u_i u_i dx\, d\eta \le M_1^2, \qquad \int_0^{\mathcal{T}} \int_\Omega \rho \theta^2 dx\, d\eta \le M_2^2, \tag{6.5.9}$$

and

$$\int_0^{\mathcal{T}} \int_\Omega \rho F_i F_i dx\, d\eta \le M_3^2, \qquad \int_0^{\mathcal{T}} \int_\Omega \rho \Big(\int_0^\eta S\, d\xi \Big)^2 dx\, d\eta \le M_4^2, \tag{6.5.10}$$

for prescribed constants M_i $(i = 1, \dots, 4)$, then

$$\begin{aligned} \int_\Omega \rho u_i u_i dx + \int_0^t \int_\Omega \kappa_{ij} \Big(\int_0^\eta \theta_{,i} d\xi \Big) \Big(\int_0^\eta \theta_{,j} d\xi \Big) dx\, d\eta \\ \le M_1 \mathcal{T} \Big(\int_0^{\mathcal{T}} \int_\Omega \rho F_i F_i dx\, d\eta \Big)^{1/2} \\ + M_2 \mathcal{T} \Big(\int_0^{\mathcal{T}} \int_\Omega \rho \Big[\int_0^\eta S\, d\xi \Big]^2 dx\, d\eta \Big)^{1/2}, \end{aligned} \tag{6.5.11}$$

for $0 \le t \le \mathcal{T}/2$.

As a consequence of the above theorem, Rionero & Chirita (1987) are additionally able to establish continuous dependence on the initial data and on the thermoelastic coefficients.

Another contribution of Rionero & Chirita (1987) is the derivation of uniqueness and continuous dependence theorems for a thermoelastic solid that occupies either the exterior of a bounded region in three-space or a half-space region. To obtain such results, Rionero & Chirita (1987) have modified the Lagrange identity (6.5.4) by introducing a weight function of the form

$$h(\mathbf{x}) = \exp(-\alpha |\mathbf{x}|),$$

for a constant α, in each of the three integrals and imposing growth conditions on solutions to (6.5.1), (6.5.2) at large spatial distances. In fact, for the exterior domain case, the restrictions they need take the form

$$|u_i|, |\theta|, |c_{ijk\ell} u_{k,\ell}|, |\kappa_{ij} \int_0^t \theta_{,j}(\mathbf{x}, s) ds| = O(e^{\alpha |\mathbf{x}|}), \qquad \text{as } |\mathbf{x}| \to \infty,$$

and additionally

$$\lim_{\alpha \to 0} \alpha^2 Q_1(\mathcal{T}) = 0, \qquad \text{and} \qquad \lim_{\alpha \to 0} \alpha^2 Q_2(\mathcal{T}) = 0,$$

where the functions Q_1 and Q_2 take the form

$$Q_1(T) = \int_0^T \int_\Omega \rho^{-1} e^{\alpha|\mathbf{x}|} \Big(c_{ijk\ell} \frac{\partial u_k}{\partial x_\ell} + \beta_{ij}\theta \Big) \Big(c_{ijrs} \frac{\partial u_r}{\partial x_s} + \beta_{ij}\theta \Big) dx \, d\eta,$$

$$Q_2(T) = \int_0^T \int_\Omega (\rho\theta_0)^{-1} e^{\alpha|\mathbf{x}|} \Big(\kappa_{ij} \int_0^\eta \theta_{,j} d\xi \Big) \Big(\kappa_{ir} \int_0^\eta \theta_{,r} d\xi \Big) dx \, d\eta.$$

A weighted Lagrange identity analysis then forms the basis of the argument needed to establish continuous dependence of the solution on the body force and the heat supply as well as the initial data and thermoelastic coefficients. As in the case of a bounded domain, no definiteness assumptions on the thermoelastic coefficients other than (6.5.7) are required in their analysis. In addition, the methods used by Rionero & Chirita (1987) are not restricted to the conditions (6.5.8) but can be applied to a large class of boundary initial value problems of linear thermoelasticity.

Logarithmic convexity arguments

Continuous dependence of solutions to (6.5.1), (6.5.2) on the body force, heat supply and thermoelastic coupling coefficients was also obtained by Ames & Straughan (1992) who employ a different method of analysis, namely the logarithmic convexity method, to prove their inequalities.

To establish continuous dependence in this case, we let u_i, θ and u_i^*, θ^* be two solutions to (6.5.1), (6.5.2) on $\Omega \times (0,T)$ corresponding to body force, heat supply, and thermoelastic coupling coefficients F_i, S, β_{ij} and F_i^*, S^*, β_{ij}^*, respectively. We assume that the displacements $u_i(\mathbf{x},t)$ and $u_i^*(\mathbf{x},t)$ satisfy the same data on the boundary of Ω and the same prescribed initial data as do the temperatures $\theta(\mathbf{x},t)$ and $\theta^*(\mathbf{x},t)$.

Defining the difference variables

$$\begin{aligned} &v_i = u_i^* - u_i, \qquad \phi = \theta^* - \theta, \qquad \epsilon_{ij} = \beta_{ij}^* - \beta_{ij}, \\ &f_i = \rho(F_i^* - F_i), \qquad H = \rho(S^* - S), \end{aligned} \tag{6.5.12}$$

the following boundary - initial value problem results

$$\rho \frac{\partial^2 v_i}{\partial t^2} = f_i + \frac{\partial}{\partial x_j} \Big(c_{ijk\ell} \frac{\partial v_k}{\partial x_\ell} \Big) - \frac{\partial}{\partial x_j} \Big(\epsilon_{ij} \theta^* \Big) - \frac{\partial}{\partial x_j} \Big(\beta_{ij} \phi \Big), \tag{6.5.13}$$

$$a \frac{\partial \phi}{\partial t} + \epsilon_{ij} \frac{\partial^2 u_i^*}{\partial x_j \partial t} + \beta_{ij} \frac{\partial^2 v_i}{\partial x_j \partial t} = \frac{\partial}{\partial x_i} \Big(\kappa_{ij} \frac{\partial \phi}{\partial x_j} \Big) + H, \tag{6.5.14}$$

holding on $\Omega \times (0,T)$, with

$$v_i = \phi = 0, \qquad \text{on} \quad \partial\Omega \times [0,T], \tag{6.5.15}$$

and

$$v_i = \frac{\partial v_i}{\partial t} = \phi = 0, \qquad \text{at} \quad t = 0. \tag{6.5.16}$$

We make the physically correct assumptions that the density and specific heat are strictly positive, i.e.

$$\rho(\mathbf{x}) \geq \rho_0 > 0, \qquad a(\mathbf{x}) \geq a_0 > 0, \tag{6.5.17}$$

as well as assuming condition (6.5.7) holds.

Continuous dependence results can be obtained by showing that the logarithm of the functional

$$G(t) = \int_0^t \int_\Omega \rho v_i v_i dx\, d\eta + \int_0^t \int_\Omega (t - \eta)\kappa_{im}\psi_{,i}\psi_{,m} dx\, d\eta + Q, \tag{6.5.18}$$

is convex. Here Q is a constant data term to be selected and

$$\psi(\mathbf{x}, t) = \int_0^t \phi(\mathbf{x}, \eta) d\eta. \tag{6.5.19}$$

The key idea in this logarithmic convexity analysis is the integration of (6.5.14) with respect to t and the introduction of the function $\psi(\mathbf{x}, t)$ in the definition of $G(t)$. Ames & Straughan (1992) show that if the displacements u_i^*, the temperature θ^*, and the coupling coefficients β_{ij} are suitably constrained and the term Q is chosen appropriately, the functional $G(t)$ satisfies the inequality,

$$GG'' - (G')^2 \geq -cG^2, \tag{6.5.20}$$

for a computable constant c. More specifically, it is required that the quantities

$$\sup|\theta^*|^2, \quad \sup|\nabla\theta^*|^2, \quad \sup|u_{i,t}^*|^2, \quad \sup|\nabla u_i^* - \nabla u_i^*(\mathbf{x}, 0)|^2, \tag{6.5.21}$$

where the supremum is taken over $\Omega \times [0, T]$, as well as

$$\sup_\Omega|\beta_{ij}\beta_{ij}|, \qquad \sup_\Omega|\beta_{ij,j}\beta_{im,m}|, \tag{6.5.22}$$

be bounded by a constant M_0 and Q is selected as

$$Q = \int_0^T \int_\Omega \left[f_i f_i + 2\epsilon_{ij}\epsilon_{ij} + \epsilon_{ij,j}\epsilon_{im,m} + H^2 + \left(\int_0^t H d\eta \right)^2 \right] dt. \tag{6.5.23}$$

The constant c depends on $T, M_0, \rho_0, \kappa_0, a_0$, and λ_1 where λ_1 is the constant in Poincaré's inequality for Ω.

As has been shown in section 1.5, the inequality (6.5.20) may be integrated to yield the basic estimate

$$G(t) \leq \exp\Big[\frac{1}{2}ct(T-t)\Big]\big[G(0)\big]^{1-t/T}\big[G(T)\big]^{t/T}, \tag{6.5.24}$$

from which continuous dependence results can be deduced, provided the class of admissible solutions is suitably restricted. Namely, if a bound for $G(T)$ is known, say $G(T) \leq M$, for some constant M, then inequality (6.5.24) becomes

$$G(t) \leq KQ^{1-t/T}M^{t/T}. \tag{6.5.25}$$

In view of the definitions for G and Q, this inequality may be explicitly written as

$$\begin{aligned}\int_0^t \int_\Omega \rho v_i v_i dx\, d\eta &+ \int_0^t \int_\Omega (t-\eta)\kappa_{im}\psi_{,i}\psi_{,m} dx\, d\eta \\ \leq & KM^{t/T}\bigg\{\int_0^T \int_\Omega \Big[f_i f_i + 2\epsilon_{ij}\epsilon_{ij} + \epsilon_{ij,j}\epsilon_{im,m} \\ &+ H^2 + \Big(\int_0^t H d\eta\Big)^2\Big] dx\, dt\bigg\}^{1-t/T}.\end{aligned} \tag{6.5.26}$$

We thus conclude that solutions v_i, ϕ depend Hölder continuously on the data, in the form Q, on compact subintervals of $[0,T)$. Since the term Q involves changes in the body force, f_i, the heat supply, H, and the coupling coefficients ϵ_{ij}, inequality (6.5.26) yields the desired continuous dependence results.

Extension to a theory capable of admitting second sound

It is worth remarking here that we may extend the logarithmic convexity analysis just outlined to the system studied by Green (1972). Green's (1972) system contains a (time) second derivative of the temperature field in the governing equation for that field and is thus capable of allowing for a heat wave travelling with a finite speed of propagation. In fact, Green (1972) uses a Lagrange identity method to show that his system possesses a unique solution even when the strain energy function is not sign-definite.

The linearised system for a thermoelastic body derived by Green (1972) is, when the body has a centre of symmetry at each point but is otherwise non-isotropic,

$$\begin{aligned}\rho\frac{\partial^2 u_i}{\partial t^2} &= \frac{\partial}{\partial x_k}\Big(k_{ikrs}\frac{\partial u_r}{\partial x_s}\Big) + \frac{\partial}{\partial x_k}\Big(a_{ik}(\theta + \alpha\frac{\partial\theta}{\partial t})\Big) + \rho F_i, \\ h\frac{\partial^2\theta}{\partial t^2} + d\frac{\partial\theta}{\partial t} - a_{ij}\frac{\partial^2 u_i}{\partial x_j \partial t} &= \frac{\partial}{\partial x_i}\Big(k_{ij}\frac{\partial\theta}{\partial x_j}\Big) + \rho r.\end{aligned}$$

In these equations u_i, θ are the displacement and temperature fields. The coefficients $k_{ikrs}(\mathbf{x}), a_{ij}(\mathbf{x}), k_{ik}(\mathbf{x})$, are the elasticities, coupling coefficients, and thermal conductivity, respectively. They are subject to symmetries as outlined by Green (1972). The terms F_i, r represent body force and heat supply.

The point of including this system here is to draw attention to the fact that it is not difficult to modify the above logarithmic convexity analysis to be applicable to the second sound system, and thereby establish stability, structural stability, and uniqueness results, even with sign indefinite elastic coefficients, although they need to be symmetric in that

$$k_{ikrs} = k_{rsik},$$

as required by Green (1972).

Linear thermoelasticity backward in time

A somewhat different class of problems was studied by Ames & Payne (1991) who derived stabilizing criteria for solutions of the Cauchy problem for the standard equations of dynamical linear thermoelasticity (i.e. without second sound effects) backward in time. These writers also obtained inequalities establishing continuous dependence on the coupling parameter that links the elastic displacement field and the temperature.

The formulation of the equations that govern linear thermoelasticity that Ames & Payne (1992) used can be found in Green (1962) and can be written as,

$$\rho \frac{\partial^2 u_i}{\partial t^2} - \frac{\partial}{\partial x_j}\Big(c_{ijk\ell}\frac{\partial u_k}{\partial x_\ell}\Big) + \frac{\partial}{\partial x_j}\Big(F_{ij}\theta\Big) = 0, \qquad \text{in } \Omega \times (0,T), \quad (6.5.27)$$

$$\frac{\partial \theta}{\partial t} + bF_{ij}\frac{\partial^2 u_i}{\partial x_j \partial t} + \frac{\partial}{\partial x_i}\Big(a_{ij}\frac{\partial \theta}{\partial x_j}\Big) = 0, \qquad \text{in } \Omega \times (0,T). \qquad (6.5.28)$$

This formulation is, of course, the same as that given earlier in system (6.5.1), (6.5.2), although the body force and heat supply terms are omitted. In addition, the notation adhered to is more in keeping with Ames & Payne (1992). However, the major difference with (6.5.27), (6.5.28) and (6.5.1), (6.5.2), is that effectively time has been reversed. For equation (6.5.27), since this is an equation with only second order time derivatives, it remains unchanged. Equation (6.5.28), on the other hand, does change due to the presence of first order time derivatives. The components of displacement are denoted by $u_i(\mathbf{x},t)$ and $\theta(\mathbf{x},t)$ is the temperature. The functions $\rho, c_{ijk\ell}, F_{ij}$, and a_{ij} depend on $\mathbf{x}$ while b is a prescribed constant. The following hypotheses are again made:

$$\rho(\mathbf{x}) \geq \rho_0 > 0, \qquad (6.5.29)$$

$$c_{ijk\ell} = c_{k\ell ij}, \tag{6.5.30}$$
$$a_{ij} = a_{ji}, \tag{6.5.31}$$
$$a_{ij}\xi_i\xi_j \geq a_0\xi_i\xi_i, \qquad \forall \xi_i, \quad \text{some } a_0 > 0, \tag{6.5.32}$$
$$F_{ij}F_{ij} \leq K_1^2, \qquad F_{ij,j}F_{im,m} \leq K_2^2. \tag{6.5.33}$$

In addition, it is assumed that the strain energy form is positive-definite so that

$$\int_\Omega c_{ijk\ell}\phi_{ij}(\mathbf{x})\phi_{k\ell}(\mathbf{x})dx \geq 0, \tag{6.5.34}$$

for all tensors $\phi_{ij}(\mathbf{x})$.

We now describe the work of Ames & Payne (1992). Following Ames & Payne (1992), we now establish that an appropriately constrained solution of (6.5.27) and (6.5.28) with the boundary conditions

$$u_i = 0, \qquad \theta = 0, \qquad \text{on } \partial\Omega \times [0, T], \tag{6.5.35}$$

and the initial conditions

$$u_i(\mathbf{x}, 0) = f_i(\mathbf{x}), \qquad \frac{\partial u_i}{\partial t}(\mathbf{x}, 0) = g_i(\mathbf{x}), \qquad \theta(\mathbf{x}, 0) = h(\mathbf{x}), \tag{6.5.36}$$

depends continuously on the initial data if $b > 0$; (note that $b > 0$ holds when the equations of linearised thermoelasticity are followed *backward* in time.) We point out that since the system (6.5.27), (6.5.28) is linear, continuous dependence on arbitrary initial data is equivalent to the stability of the zero solution. The method we employ is again the Lagrange identity method but combined with an energy equality. Without loss, let us assume that $b = 1$ since (6.5.27), (6.5.28) can be reduced to an equivalent system with $b = 1$.

Let u_i^* and θ^* represent solutions to the adjoint equations

$$\rho\frac{\partial^2 u_i^*}{\partial t^2} - \frac{\partial}{\partial x_j}\left(c_{ijk\ell}\frac{\partial u_k^*}{\partial x_\ell}\right) + \frac{\partial}{\partial x_j}\left(F_{ij}\theta^*\right) = 0, \tag{6.5.37}$$

$$\frac{\partial \theta^*}{\partial t} + bF_{ij}\frac{\partial^2 u_i^*}{\partial x_j \partial t} - \frac{\partial}{\partial x_i}\left(a_{ij}\frac{\partial \theta^*}{\partial x_j}\right) = 0, \tag{6.5.38}$$

Form the identities

$$\int_0^t \int_\Omega \Big\{ u_{i,\eta}^*\Big[\rho u_{i,\eta\eta} - (c_{ijk\ell}u_{k,\ell})_{,j} + (F_{ij}\theta)_{,j}\Big] + u_{i,\eta}\Big[\rho u_{i,\eta\eta}^* - (c_{ijk\ell}u_{k,\ell}^*)_{,j} + (F_{ij}\theta^*)_{,j}\Big]\Big\} dx\, d\eta = 0, \tag{6.5.39}$$

and

$$\int_0^t \int_\Omega \Big\{ \theta^* \Big[\theta_{,\eta} + F_{ij} u_{i,j\eta} + (a_{ij}\theta_{,j})_{,i} \Big] + \theta \Big[\theta^*_{,\eta} + F_{ij} u^*_{i,j\eta} - (a_{ij}\theta^*_{,j})_{,i} \Big] \Big\} dx\, d\eta = 0. \quad (6.5.40)$$

Integrating (6.5.39) and (6.5.40) by parts and combining the resulting expressions, we obtain

$$\int_\Omega \big[\rho u^*_{i,\eta} u_{i,\eta} + c_{ijk\ell} u^*_{k,\ell} u_{i,j} + \theta^* \theta \big] dx \Big|_0^t = 0. \quad (6.5.41)$$

If we choose

$$\theta^*(\mathbf{x}, \eta) = \theta(\mathbf{x}, 2t - \eta), \qquad u^*_i(\mathbf{x}, \eta) = u_i(\mathbf{x}, 2t - \eta),$$

and define

$$G(t_1, t_2) = \frac{1}{2} \int_\Omega \Big[-\rho \frac{\partial u_i}{\partial t}(t_1) \frac{\partial u_i}{\partial t}(t_2) + c_{ijk\ell} \frac{\partial u_k}{\partial x_\ell}(t_1) \frac{\partial u_i}{\partial x_j}(t_2) + \theta(t_1)\theta(t_2) \Big] dx, \quad (6.5.42)$$

then (6.5.41) can be written as

$$G(t, t) = G(0, 2t). \quad (6.5.43)$$

We next derive an energy identity. This is done by multiplying (6.5.27) by $u_{i,t}$ and (6.5.28) by θ, integrating over Ω and over the time interval $(0, t)$, and combining the results. We thus find that

$$E_1(t) + \frac{1}{2} \int_\Omega \theta^2 dx = E_1(0) + \int_0^t \int_\Omega a_{ij} \theta_{,i} \theta_{,j} dx\, d\eta + \frac{1}{2} \int_\Omega h^2 dx, \quad (6.5.44)$$

where the function E_1 has been defined by

$$E_1(t) = \frac{1}{2} \int_\Omega \big[\rho u_{i,t} u_{i,t} + c_{ijk\ell} u_{i,j} u_{k,\ell} \big] dx. \quad (6.5.45)$$

The next step is to set

$$E(t) = E_1(t) + \frac{1}{2} \int_\Omega \theta^2 dx. \quad (6.5.46)$$

Then (6.5.44) becomes

$$E(t) = E(0) + \int_0^t \int_\Omega a_{ij}\theta_{,i}\theta_{,j} dx\, d\eta. \tag{6.5.47}$$

Combining (6.5.43) and (6.5.47) leads to the identity

$$-\int_\Omega \rho u_{i,t}(t)u_{i,t}(t)dx + \int_0^t \int_\Omega a_{ij}\theta_{,i}\theta_{,j} dx\, d\eta = G(0,2t) - E(0). \tag{6.5.48}$$

Observing that since

$$E_1(t) = E_1(0) - \int_0^t \int_\Omega (F_{ij}\theta)_{,j} u_{i,\eta} dx\, d\eta,$$

it follows from the Cauchy - Schwarz inequality and the arithmetic - geometric mean inequality that

$$\begin{aligned} E_1(t) \leq E_1(0) + \frac{K}{\sqrt{a_0}}\Big(\frac{1}{2}\sigma_1 \int_0^t \int_\Omega \rho u_{i,\eta}u_{i,\eta} dx\, d\eta \\ + \frac{1}{2\sigma_1}\int_0^t \int_\Omega a_{ij}\theta_{,i}\theta_{,j} dx\, d\eta\Big), \end{aligned} \tag{6.5.49}$$

where the constant K is defined by

$$K = \max_\Omega \left\{ \sqrt{\frac{F_{ij}F_{ij}}{\rho}}, \sqrt{\frac{F_{ij,j}F_{im,m}}{\rho\lambda_1}} \right\}, \tag{6.5.50}$$

and σ_1 is a positive constant which arises through use of the arithmetic - geometric mean inequality and may be selected judiciously. The quantity λ_1 is the smallest eigenvalue for the fixed membrane problem for Ω..

Using (6.5.48) in (6.5.49), we are led to the inequality

$$\begin{aligned} \frac{1}{2}(1 - K a_0^{-1/2}\sigma_1^{-1}) \int_\Omega \rho u_{i,t}u_{i,t} dx + \frac{1}{2}\int_\Omega c_{ijk\ell}u_{i,j}u_{k,\ell} dx \\ \leq E_1(0) + \frac{1}{2}K a_0^{-1/2}\sigma_1^{-1}\big[G(0,2t) - E(0)\big] \\ + \frac{1}{2}K\sigma_1 a_0^{-1/2} \int_0^t \int_\Omega \rho u_{i,\eta}u_{i,\eta} dx\, d\eta. \end{aligned} \tag{6.5.51}$$

An appropriate choice for σ_1 in inequality (6.5.51) is

$$\sigma_1 = \frac{2K}{\sqrt{a_0}},$$

and then an integration with respect to t results in the bound

$$\int_0^t \int_\Omega \rho u_{i,\eta} u_{i,\eta} dx\, d\eta \leq \frac{3a_0}{4K^2} E_1(0) \left[e^{4K^2 t/a_0} - 1 \right] + \int_0^t G(0, 2\eta) e^{4K^2(t-\eta)/a_0} d\eta, \qquad (6.5.52)$$

which is then substituted back into (6.5.51) to find

$$\int_\Omega \rho u_{i,t} u_{i,t} dx \leq 3E_1(0) e^{4K^2 t/a_0} + \frac{4K^2}{a_0} \int_0^t G(0, 2\eta) e^{4K^2(t-\eta)/a_0} d\eta + G(0, 2t). \quad (6.5.53)$$

For the current system of partial differential equations one may also derive the expression

$$\int_\Omega (c_{ijk\ell} u_{i,j} u_{k,\ell} + \theta^2) dx = 2G(0, 2t) + \int_\Omega \rho u_{i,t} u_{i,t} dx. \qquad (6.5.54)$$

Hence, (6.5.53) and (6.5.54) give bounds for $E(t)$ in terms of $G(0, 2t)$.

It is now necessary to derive a bound for $G(0, 2t)$ in terms of data. In order to do this, Ames & Payne (1992) introduce the class of functions ψ defined on $\Omega \times [0, T]$ that satisfy

$$\max_{t \in [0,T]} \int_\Omega \psi^2(\mathbf{x}, t) dx \leq M^2, \qquad (6.5.55)$$

for a prescribed constant M and they assume that the temperature θ belongs to this class. Now,

$$G(0, 2t) \leq \big[E(0) E(2t) \big]^{1/2}, \qquad (6.5.56)$$

and from (6.5.48) we have

$$E(2t) = E(0) + \int_0^{2t} \int_\Omega a_{ij} \theta_{,i} \theta_{,j} dx\, d\eta. \qquad (6.5.57)$$

From the energy equation one may deduce the chain of inequalities,

$$\begin{aligned} \int_0^{2t} \int_\Omega a_{ij} \theta_{,i} \theta_{,j} dx\, d\eta &\leq \int_\Omega \theta^2(\mathbf{x}, 2t) dx - \int_\Omega h^2 dx \\ &\quad + \frac{K^2}{a_0} \int_0^{2t} \int_\Omega \rho u_{i,\eta} u_{i,\eta} dx\, d\eta, \\ &\leq \int_\Omega \theta^2(\mathbf{x}, 2t) dx + \frac{2K^2}{a_0} \int_0^{2t} E(\eta) d\eta. \qquad (6.5.58) \end{aligned}$$

Thus, we obtain the inequality

$$E(2t) \leq E(0) + \int_\Omega \theta^2(\mathbf{x}, 2t)dx + \frac{2K^2}{a_0} \int_0^{2t} E(\eta)d\eta.$$

Since,

$$\int_\Omega \theta^2(\mathbf{x}, 2t)dx \leq M^2,$$

for $2t \leq T$, by assumption, the previous inequality can be rewritten as

$$\frac{d}{dt}\left[e^{-2K^2 t a_0^{-1}} \int_0^{2t} E(\eta)d\eta\right] \leq \left[M^2 + E(0)\right]e^{-2K^2 t a_0^{-1}}, \tag{6.5.59}$$

so that we can conclude, after an integration, that

$$E(2t) \leq \left[M^2 + E(0)\right]e^{4K^2 t a_0^{-1}}. \tag{6.5.60}$$

Consequently, (6.5.56) yields

$$G(0, 2t) \leq \left[M^2 + E(0)\right]^{1/2} E(0)^{1/2} e^{2K^2 t a_0^{-1}}, \tag{6.5.61}$$

as well as

$$\int_0^t G(0, 2\eta)\exp\frac{4K^2}{a_0}(t - \eta)\, d\eta \leq \frac{a_0}{2K^2}\left[M^2 + E(0)\right]^{1/2} E(0)^{1/2} e^{4K^2 t a_0^{-1}}, \tag{6.5.62}$$

for $t \leq T/2$. In view of (6.5.53), (6.5.54) and (6.5.62), we obtain the continuous dependence inequality, valid for $0 \leq t \leq T/2$,

$$E(t) \leq \left\{4\left[M^2 + E(0)\right]^{1/2} E(0)^{1/2} + 3E_1(0)\right\}e^{4K^2 t/a_0}. \tag{6.5.63}$$

Ames & Payne (1992) also establish a continuous dependence result when $b < 0$. In this case, no definiteness condition need be imposed on the strain energy nor a constraint on the temperature that is any stronger then (6.5.55). They complete their investigation of equations (6.5.27), (6.5.28) by comparing the solution θ of the temperature equation with the solution of the heat equation which is obtained by setting $b = 0$. This is, in a sense, a continuous dependence on modeling result. Inequalities that imply continuous dependence on the coupling parameter b are then obtained under slightly different constraints which do not require the strain energy to be definite. These results are also derived from a Lagrange identity argument similar to the one described for the case $b > 0$.

The initial-time geometry problem in linear thermoelasticity

The contribution of Ames & Payne (1995a) treats the initial - time geometry problem for the equations of linear thermoelasticity, forward in time, although no definiteness need be imposed on the elastic coefficients. Their work could rightly appear in chapter 2, although we include it at this point to keep the section on thermoelasticity self contained.

The system treated by Ames & Payne (1995a) is (6.5.1), (6.5.2) with $F_i \equiv 0, S \equiv 0$, and $\rho = a = \tau = 1$. They further suppose,

$$
\begin{aligned}
&(i) \qquad \kappa_{ij} = \kappa_{ji}, \\
&(ii) \qquad \kappa_{ij}\xi_i\xi_j \geq \kappa_0\xi_i\xi_i, \qquad \forall \xi_i, \quad \text{some } \kappa_0 > 0, \\
&(iii) \qquad \sup_\Omega |\beta_{ij}\beta_{ij}|, \qquad \sup_\Omega |\beta_{ij,j}\beta_{im,m}| \leq M^2,
\end{aligned}
$$

for some known constant M, where Ω is a bounded domain of $\mathbf{R}^2$ or $\mathbf{R}^3$. The major symmetry condition is required of the elasticities, i.e.

$$
c_{ijk\ell} = c_{k\ell ij}.
$$

They study continuous dependence on the initial - time geometry as in section 2.1. Thus, they suppose (w_i^1, ϕ^1), (w_i^2, ϕ^2), are solutions to (6.5.1), (6.5.2), which both satisfy homogeneous boundary data on Γ, the boundary of Ω. The functions (w_i^1, ϕ^1) satisfy the initial data,

$$
\begin{aligned}
w_i^1(\mathbf{x}, t) &= f_i(\mathbf{x}), \qquad \frac{\partial w_i^1}{\partial t}(\mathbf{x}, t) = g_i(\mathbf{x}), \\
\phi^1(\mathbf{x}, t) &= h(\mathbf{x}), \qquad \text{on } \Sigma,
\end{aligned}
\tag{6.5.64}
$$

where Σ is the initial data surface $t = \epsilon f(\mathbf{x})$, $|f| < 1$. Whereas, (w_i^2, ϕ^2) satisfy the same initial data on the line $t = 0$, i.e.

$$
\begin{aligned}
w_i^2(\mathbf{x}, t) &= f_i(\mathbf{x}), \qquad \frac{\partial w_i^2}{\partial t}(\mathbf{x}, t) = g_i(\mathbf{x}), \\
\phi^2(\mathbf{x}, t) &= h(\mathbf{x}), \qquad \text{on } t = 0.
\end{aligned}
\tag{6.5.65}
$$

To study continuous dependence on the initial - time geometry they need to derive a stability inequality bounding the difference of (w_i^1, ϕ^1) and (w_i^2, ϕ^2) in terms of the initial - time geometry perturbation ϵ.

In fact, Ames & Payne (1995a) choose to work with an integrated version of (6.5.2), because of the particular variant of the Lagrange identity method they develop. They set

$$v_i = w_i^2 - w_i^1, \qquad \theta = \phi^2 - \phi^1,$$

and define

$$\psi = \int_\epsilon^t \theta(\mathbf{x}, \eta) d\eta.$$

Then from (6.5.2), ψ satisfies the partial differential equation

$$\frac{\partial \psi}{\partial t} + \beta_{ij} \frac{\partial v_i}{\partial x_j} = \frac{\partial}{\partial x_i}\Big(\kappa_{ij} \frac{\partial \psi}{\partial x_i}\Big) + H(\mathbf{x}, \epsilon), \tag{6.5.66}$$

in $\Omega \times (\epsilon, T)$, where the function H is given by,

$$H(\mathbf{x}, \epsilon) = \theta(\mathbf{x}, \epsilon) + \beta_{ij} \frac{\partial v_i}{\partial x_j}(\mathbf{x}, \epsilon).$$

The constraint set needed by Ames & Payne (1995a) requires $(w_i^{(\alpha)}, \phi^{(\alpha)})$ to satisfy the following inequalities, for known constants M_1, M_2, M_3, with $M_2 > 0, M_3 > 0$,

$$\sup_{\epsilon f \le t \le T} \int_\Omega w_i^1 w_i^1 dx + \sup_{0 \le t \le T} \int_\Omega w_i^2 w_i^2 dx \le M_1^2,$$

$$\int_\epsilon^T \left\{ \sqrt{\int_\Omega (\phi^1)^2 dx} + \sqrt{\int_\Omega (\phi^2)^2 dx} \right\} dt \le M_2,$$

$$\int_\epsilon^T \left\{ \sqrt{\int_\Omega \phi_{,j}^1 \phi_{,j}^1 \, dx} + \sqrt{\int_\Omega \phi_{,j}^2 \phi_{,j}^2 \, dx} \right\} dt \le M_3.$$

In this section of the book we do not include details of the involved calculations of Ames & Payne (1995a) and refer to the original article for such technicalities. However, Ames & Payne (1995a) establish the following theorem, where the constant C may be computed using the arguments given by those writers.

Theorem 6.5.2. (Ames & Payne (1995a).)
The solutions to the partial differential equations (6.5.1), (6.5.2), (6.5.66), which satisfy the conditions outlined in this section and belong to the constraint set outlined above, satisfy the continuous dependence on the initial - time geometry estimate

$$\int_\epsilon^t \int_\Omega \kappa_{ij} \psi_{,i} \psi_{,j} dx \, d\eta + \int_\Omega v_i v_i dx \le C\sqrt{\epsilon}, \tag{6.5.67}$$

where t lies in the time interval,

$$0 < \epsilon \leq t \leq \frac{1}{2}(T + \epsilon).$$

6.6 Overcoming Hölder Continuity, and Image Restoration

It has been seen in much of this book that the best continuous dependence result that we can expect for many ill - posed problems is Hölder continuity. Recently, Carasso (1994) has done some very interesting work indicating that, provided a further restriction on the class of admissible solutions is imposed, one can actually establish an improved stability result in a number of ill - posed continuation problems. A typical example of such a problem is the backward - in - time continuation problem for the heat equation, with the spatial domain, Ω, being a bounded one in $\mathbf{R}^n$. The problem is, given $f(x) \in L^2(\Omega)$, find u so that

$$\frac{\partial u}{\partial t} = \kappa \Delta u, \qquad x \in \Omega, \quad 0 < t \leq 1, \tag{6.6.1}$$

$$u = 0, \qquad x \in \partial\Omega, \quad 0 \leq t \leq 1, \tag{6.6.2}$$

$$\|u(1) - f\| \leq \epsilon, \qquad \|u(0)\| \leq M, \tag{6.6.3}$$

where $\|\cdot\|$ as usual denotes the $L^2(\Omega)$ norm. Here $f(x)$ represents noisy data at $t = 1$ and ϵ is an upper bound for the norm of the error in this data at $t = 1$. A logarithmic convexity argument can be used to establish that the norm of the difference of any two solutions $u_1(x,t)$ and $u_2(x,t)$ satisfies the bound,

$$\|u_1(t) - u_2(t)\| \leq 2M^{1-t}\epsilon^t, \tag{6.6.4}$$

for $t \in [0,1]$. It is important to note that at $t = 0$, the dependence on ϵ is lost so that the inequality yields no new information at this boundary. We would like to find useful error estimates near $t = 0$. Carasso (1994) observes that it can be shown that if $\|\Delta u(0)\| \leq M$, a logarithmic continuity result can be obtained at $t = 0$, producing an error of the order of $KM/\log(M/\epsilon)$. As pointed out by Carasso (1994), a bound on the derivatives of u requires knowledge about the structure of u that is often not available in applications. Moreover, a logarithmic continuity result is considerably weaker than the Hölder continuity obtained for $t > 0$.

Carasso (1994) introduces and analyses a supplementary *a priori* constraint on the solution that leads to stronger error bounds near the continuation boundary at $t = 0$. This constraint, called the "slow evolution from the boundary" (SECB) constraint, can be described as follows. Let the continuation variable be t, $0 \leq t \leq 1$, with $t = 0$ being the continuation boundary, and let X be an appropriate Banach space so that the continuation $u(t)$ is $X-$ valued and has norm $\|u(t)\|$ for fixed t. If $u_1(t)$ and $u_2(t)$ are any two continuations from the data at $t = 1$ that satisfy $\|u(0)\| \leq M$, $\|u(1) - f\| \leq \epsilon$, and the estimate

$$\|v(t)\| \leq \|v(0)\|^{1-t}\|v(1)\|^{t}, \tag{6.6.5}$$

where $v(t) = u_1(t) - u_2(t)$, then Carasso (1994) shows that

$$\|u_1(t) - u_2(t)\| \leq 2C^{1-t}\epsilon, \tag{6.6.6}$$

for $0 \leq t \leq 1$, provided there exists a known small constant β with $0 < \beta << M/\epsilon$, and a fixed σ, such that whenever $\sigma > \sigma^*$,

$$\|u(\sigma) - u(0)\| \leq \beta\epsilon. \tag{6.6.7}$$

The threshold parameter σ^* is shown by Carasso (1994) to have value,

$$\sigma^*(\epsilon, M, \beta) = \frac{\log\left(M/(M - \beta\epsilon)\right)}{\log\left(M/\epsilon\right)}. \tag{6.6.8}$$

As Carasso (1994) points out, given any $\epsilon > 0$ the inequality (6.6.7) always holds with a small β by continuity since a sufficiently small σ always exists. What further limits the class of admissible solutions is the requirement that σ is known and $\sigma > \sigma^*$. One would like to have β as small as possible while making $\sigma < 1$ as large as possible, so that $\sigma/\sigma^* >> 1$. In the case of the backward in time heat equation, problem (6.6.1) - (6.6.3), the SECB constraint is an a *priori* condition requiring that the values of the solution at time σ approximate the initial values closely. This is distinctly different from prescribing bounds on spatial derivatives of $u(x, 0)$; neither condition implies the other. In fact, the SECB constraint requires less knowledge about the structure of $u(x, 0)$ than is needed to impose derivative bounds. The SECB condition serves to suppress noise contamination as $t \to 0$, leading to the improved stability result (6.6.6) for $0 \leq t \leq 1$. The term C is a function of the constants β and σ and $C << M/\epsilon$ provided σ/σ^* is large.

A major result of Carasso's (1994) work can be summarized in the following theorem:

Theorem 6.6.1. (Carasso (1994).)
Let the Banach space X have norm $\|\cdot\|$ and let $f \in X$. Suppose there exist positive constants ϵ, M and β with $\epsilon << M$ and $\beta << M/\epsilon$ so that the continuation $u(t)$ from the data f satisfies

$$\|u(0)\| \leq M, \qquad \text{and} \qquad \|u(1) - f\| \leq \epsilon.$$

Moreover, the difference $v(t)$ of any two continuations satisfies the convexity inequality

$$\|v(t)\| \leq \|v(0)\|^{1-t}\|v(1)\|^{t}, \tag{6.6.9}$$

for $0 \leq t \leq 1$. If $\|u(\sigma) - u(0)\| \leq \beta\epsilon$ for some fixed $\sigma > \sigma^*$ with σ^* defined by (6.6.8), then

$$\|v(t)\| \leq 2C^{1-t}\epsilon, \tag{6.6.10}$$

where $C(\beta, \sigma)$ is the unique root of the transcendental equation

$$\theta = \beta + \theta^{1-\sigma}.$$

We emphasize that while prescribing the bound $\|u(0)\| \leq M$ serves to prevent explosive growth in the error in continued solutions, such bounds may not eliminate contamination of the desired solution by spurious high frequency oscillations that data noise induces. The convexity estimates reflect the potential for such spurious noise with their loss of accuracy as $t \to 0$. Carasso (1994) quite clearly demonstrates the existence of such noise as $t \to 0$ and then continues to show how the SECB constraint effectively reduces this contamination.

Carasso's (1994) proof of theorem 6.6.1 proceeds by noting that the inequality (6.6.9) can be written as

$$\|v(t)\| \leq K^{1-t}\delta^{t}, \tag{6.6.11}$$

for $0 \leq t \leq 1$, where $K = 2M$ and $\delta = 2\epsilon$. One may also replace the SECB constraint with the equivalent inequality,

$$\|v(\sigma) - v(0)\| \leq \beta\delta, \qquad \sigma > \sigma^*. \tag{6.6.12}$$

Now, from the convexity inequality (6.6.9), it follows

$$\begin{aligned} \|v(t)\| \leq& \|v(0)\|^{1-t}\|v(1)\|^{t}, \\ \leq& \big[\|v(\sigma) - v(0)\| + \|v(\sigma)\|\big]^{1-t}\|v(1)\|^{t}, \end{aligned} \tag{6.6.13}$$

so that, using (6.6.11) and (6.6.12) one may conclude that

$$\|v(\sigma)\| \leq \big[\beta\delta + K^{1-\sigma}\delta^{\sigma}\big]^{1-\sigma}\delta^{\sigma}. \tag{6.6.14}$$

The initial estimate (6.6.11) has been used to find a new estimate given by (6.6.14). Let us call each estimate $v_n(\sigma)$ starting with $n = 1$. Continual resubstitution leaves us at the n–th iteration with

$$\|v_n(\sigma)\| \leq \alpha_n^{1-\sigma}\delta^{\sigma}, \tag{6.6.15}$$

where

$$\alpha_1 = K, \qquad \frac{\alpha_j}{\delta} = \beta + \left(\frac{\alpha_{j-1}}{\delta}\right)^{1-\sigma}, \qquad j > 1. \tag{6.6.16}$$

A standard method called fixed point iteration, namely if $\beta+1 \le x_0 \le M/\epsilon$, with the iteration

$$x_{n+1} = \beta + x_n^{1-\sigma}, \qquad n = 0, 1, \dots,$$

converges. This leads to the conclusion that α_n will converge to $C\delta$ as $n \to \infty$, where $C(\beta, \sigma)$ is the unique root of $\theta = \beta + \theta^{1-\sigma}$ provided $\sigma^* < \sigma < 1$, for

$$\sigma^* = \frac{\log\left(M/(M - \beta\epsilon)\right)}{\log\left(M/\epsilon\right)}.$$

Hence,

$$\|v(\sigma)\| \le C^{1-\sigma}\delta.$$

Resubstitution of this into (6.6.13) together with (6.6.11) and (6.6.12) results in the estimate

$$\|v(t)\| \le 2C^{1-t}\epsilon, \tag{6.6.17}$$

for $0 \le t \le 1$.

A beautiful illustration of the theory of improperly posed problems is made by Carasso (1994) because of the precise application he has in mind. To illustrate the utility of his results, Carasso (1994) applies them in the latter half of the paper to a class of image restoration problems in which an original image is reconstructed from a noisy blurred version. This deblurring problem can be formulated as an integral equation. If $h(x, y)$ represents the blurring kernel, $g(x, y)$ is the unblurred image, $f(x, y)$ denotes the blurred image that would have been recorded in the absence of noise, $\hat{f}(x, y)$ is the actual recorded image, and $n(x, y)$ represents the cumulative effects of all noise processes, then Carasso (1994) shows that mathematically these quantities are related by the equation

$$\begin{aligned} \hat{f}(x, y) &= f(x, y) + n(x, y), \\ &= \int_{\mathbf{R}^2} h(x - u, y - v) g(u, v) du\, dv. \end{aligned} \tag{6.6.18}$$

Both $f(x, y)$ and $n(x, y)$ are unknown, and n may be a function of f so that the case of multiplicative noise is included in the analysis. Carasso (1994) limits his investigation to shift - invariant blurring kernels $h(x, y)$ that have the Fourier transform

$$\mathcal{F}\{h(\xi, \eta)\} = \exp\left[-a(\xi^2 + \eta^2)^b\right], \tag{6.6.19}$$

where a is positive and $0 < b \leq 1$. This structure includes several special cases that occur in a number of imaging applications. For example, $b = 1$ corresponds to the Gaussian distribution while the case $b = 1/2$, corresponding to the Cauchy distribution, appears in models of x-ray scattering. If the kernel $h(x, y)$ satisfies (6.6.19), then the blurred noiseless image $f(x, y) = u(x, y, 1)$, where $u(x, y, t)$ is the unique bounded solution of the well posed problem

$$\frac{\partial u}{\partial t} = -\mu(-\Delta)^b u, \qquad (x, y) \in \mathbf{R}^2, t > 0, \tag{6.6.20}$$

$$u(x, y, 0) = g(x, y), \tag{6.6.21}$$

where $\mu = a(4\pi^2)^{-b}$. Observe that if $b = 1$, equation (6.6.20) is the classical heat conduction equation while for $0 < b < 1$, it models a generalised diffusion process. Consequently, the image restoration problem is equivalent to backward - in - time continuation of the solution of (6.6.20), (6.6.21) from noisy data $\hat{f}(x, y)$ at $t = 1$. For a given fixed t with $0 < t < 1$, $u(x, y, t)$ is a partial restoration which becomes sharper but noisier as $t \to 0$.

In oder to solve the deblurring problem by the SECB restoration method, Carasso (1994) writes the integral equation (6.6.18) in the form $Hg = \hat{f}$, where H is the integral operator in $L^2(\mathbf{R}^2)$ with kernel $h(x - u, y - v)$. Now define H^t for $0 \leq t \leq 1$ to be the convolution integral operator in $L^2(\mathbf{R}^2)$ with kernel $h(x - u, y - v; t)$ where $h(x, y; t)$ is the inverse Fourier transform of $\exp\left[-\lambda t(\xi^2 + \eta^2)^b\right]$. Noting that H^0 is the identity operator and $H^t H^\tau = H^{t+\tau}$, $t, \tau \geq 0$, one can write the solution of (6.6.20), (6.6.21) in the form $u = H^t g$, for $0 \leq t \leq 1$. In this context, Carasso's (1994) restoration approach assumes the a *priori* constraints,

$$1. \quad \|Hg - \hat{f}\| \leq \epsilon, \tag{6.6.22}$$

$$2. \quad \|g\| \leq M, \tag{6.6.23}$$

$$3. \quad \|g - H^\sigma g\| \leq \beta\epsilon, \qquad \sigma > \sigma^* \text{ fixed}, \tag{6.6.24}$$

for given $\epsilon, M > 0$, $\epsilon/M << 1$, $0 < \beta << M/\epsilon$, and σ^* defined by (6.6.8). Ideally, we would like (6.6.24) to be satisfied with a small β and a relatively large σ so that σ/σ^* is large. The SECB restoration is then defined as the unique function $k(x, y)$ that minimizes the functional

$$\mathcal{H}q \equiv \|Hq - \hat{f}\|^2 + \left(\frac{\epsilon}{M}\right)^2 \|q\|^2 + \frac{1}{\beta^2}\|q - H^\sigma q\|^2, \tag{6.6.25}$$

over all $q \in L^2(\mathbf{R}^2)$. If $\sigma = 0$ in (6.6.25), the SECB restoration reduces to the classical Tikhonov - Miller restoration (see e.g. Miller (1970)), one of the most widely used deblurring methods.

The SECB solution can also be obtained explicitly in Fourier spaces. Moreover, the Fourier transform technique forms the basis of an efficient, noniterative computational procedure which is based on two-dimensional fast Fourier transform (FFT) algorithms. Carasso (1994) concludes his paper with a numerical deblurring experiment that is designed to determine how effective the SECB constraint is in suppressing noise. He accomplishes this by comparing SECB with three other deblurring procedures: Tikhonov - Miller restoration (see e.g. Miller (1970)), the backward beam method (see Carasso *et al.* (1978)), and Hansen's L-curve method (see Hansen (1992)). The experiment consists of artificially blurring an undegraded sharp image of the inside of a human head, expressed mathematically by $g(x, y)$, by convolution with a point spread function $h(x, y)$ whose Fourier transform is given by (6.6.19) with $a = 0.075$ and $b = 0.5$. What Carasso (1994) quite clearly demonstrates both quantitatively and qualitatively, is that not only does the SECB constraint sharply reduce noise contamination but that it is the most effective of the four methods in preventing the occurrence of noise as $t \to 0$. In particular, the relative L^2 error reported by Carasso (1994) is 7.7% for the SECB method as against $13.8, 25.2$, and 27.4% for each of the Tikhonov - Miller, Hansen's L-curve, and backward beam methods, respectively.

The innovative method of Carasso (1994) would appear to have the potential to play a useful role in image restoration problems, and doubtless other practical improperly posed problems. In this section we have only been able to give a brief flavour of all the details of Carasso's (1994) contribution. However, we believe a detailed study of Carasso's (1994) paper is certainly worthwhile.

Bibliography

Adelson, L. 1973. "Singular perturbations of improperly posed problems." *SIAM J. Math. Anal.* **4**, 344–366.

Adelson, L. 1974. "Singular perturbation of an improperly posed Cauchy problem." *SIAM J. Math. Anal.* **5**, 417–424.

Agmon, S. 1966. *Unicité et Convexité dans les Problèmes Différentiels.* Sem. Math. Sup., University of Montreal Press.

Agmon, S. & Nirenberg, L. 1967. "Lower bounds and uniqueness theorems for solutions of differential equations in a Hilbert space." *Comm. Pure Appl. Math.* **20**, 207–229.

Ames, K.A. 1982a. "Comparison results for related properly and improperly posed problems for second order operator equations." *J. Differential Equations* **44**, 383–399.

Ames, K.A. 1982b. "On the comparison of solutions of properly and improperly posed Cauchy problems for first order sytems." *SIAM J. Math. Anal.* **13**, 594–606.

Ames, K.A. 1984. "Uniqueness and continuous dependence results for solutions of singular differential equations." *J. Math. Anal. Appl.* **103**, 172–183.

Ames, K.A. 1992. "Improperly posed problems for nonlinear partial differential equations." In *Nonlinear Equations in the Applied Sciences*, ed. W.F.Ames & C.Rogers. Academic Press, Series in Mathematics in Science and Engineering, vol. **185**.

Ames, K.A. & Cobb, S. 1994. Penetrative convection in a porous medium with internal heat sources. *Int. J. Engng. Sci.* **32**, 95–105.

Ames, K.A. & Isakov, V. 1991. "An explicit stability estimate for an ill-posed Cauchy problem for the wave equation." *J. Math. Anal. Appl.* **156**, 597–610.

Ames, K.A., Levine, H.A. & Payne, L.E. 1987. "Improved continuous dependence results for a class of evolutionary equations." In *Inverse and Ill-Posed Problems*, Engl, H.W. & Groetsch, C.W. eds., pp. 443–450, Academic Press, New York.

Ames, K.A. & Payne, L.E. 1992. "Stabilizing solutions of the equations of dynamical linear thermoelasticity backward in time." *Stab. Appl. Anal. Cont. Media* **1**, 243–260.

Ames, K.A. & Payne, L.E. 1994a. "Continuous dependence results for solutions of the Navier-Stokes equations backward in time." *Nonlinear Anal., Theory, Meths., Applicns.* **23**, 103–113.

Ames, K.A. & Payne, L.E. 1994b. "Stabilizing the backward heat equation against errors in the initial time geometry." From *Inequalities and Applications,* **3**, 47–52, World Scientific.

Ames, K.A. & Payne, L.E. 1994c. "On stabilizing against modeling errors in a penetrative convection problem for a porous medium." *Math. Models Meth. Appl. Sci.* **4**, 733–740.

Ames, K.A. & Payne, L.E. 1995a. "Continuous dependence on initial - time geometry for a thermoelastic system with sign - indefinite elasticities." *J. Math. Anal. Appl.* **189**, 693–714.

Ames, K.A. & Payne, L.E. 1995b. "Continuous dependence results for a problem in penetrative convection." *Quart. Appl. Math.*, to appear.

Ames, K.A. & Straughan, B. 1992. "Continuous dependence results for initially prestressed thermoelastic bodies." *Int. J. Engng. Sci.* **30**, 7–13.

Ames, K.A. & Straughan, B. 1995. "Estimates of the error in the initial - time geometry for a parabolic equation from dynamo theory." *J. Differential Equations* **123**, 153–170.

Bampi, F., Morro, A. & Jou, D. 1981. "Two continuum approaches to a wavelength - dependent description of heat conduction." *Physica* **107A**, 393–403.

Bardos, C. & Tartar, L. 1973. "Sur l'unicité retrograde des equations parabol -iques et quelques questions voisines." *Arch. Rational Mech. Anal.* **50**, 10–25.

Bell, J.B. 1981a. "The noncharacteristic Cauchy problem for a class of equations with time dependence. I. Problems in one space dimension." *SIAM J. Math. Anal.* **12**, 759–777.

Bell, J.B. 1981b. "The noncharacteristic Cauchy problem for a class of equations with time dependence. II. Multidimensional problems." *SIAM J. Math. Anal.* **12**, 778–797.

Bellomo, N. & Preziosi, L. 1995. *Modelling Mathematical Methods and Scientific Computation.* CRC Press, Boca Raton.

Bennett, A.D. 1986. *Continuous Dependence on Modeling in the Cauchy Problem for Second Order Nonlinear Partial Differential Equations.* Ph.D. thesis, Cornell University.

Bennett, A.D. 1991. "Continuous dependence on modeling in the Cauchy problem for nonlinear elliptic equations." *Differential and Integral Equations*, **4**, 1311–1324.

Bloom, F. 1981. *Ill-posed Problems for Integrodifferential Equations in Mechanics and Electromagnetic Theory.* SIAM, Philadelphia.

Brun, L. 1965a. "Sur deux expressions, analogues à la formule de Clapeyron, donnant l'énergie libre et la puissance dissipée pour un corps visco - élastique." *Comptes Rend. Acad. Sci. Paris* **261**, 41–44.

Brun, L. 1965b. "Sur l'unicité en thermoélasticité dynamique et diverses expressions analogues la formule de Clapeyron." *Comptes Rend. Acad. Sci. Paris* **261**, 2584–2587.

Brun, L. 1967. "Sur l'unicité en viscoélasticité dynamique." *Comptes Rend. Acad. Sci. Paris* **264**, 135–137.

Brun, L. 1969. "Méthodes énergtiques dans les systèmes évolutifs linéaires, Deuxième Partie: Théorèmes d'unicité." *J. Mécanique* **8**, 167–192.

Calmelet, C. 1987. *Stability and Special Solutions to the Conducting Dusty Gas Model.* Ph.D. Thesis, Vanderbilt Univ.

Calmelet-Eluhu, C. & Crooke, P.S. 1990. "Continuous dependence on data for solutions of a conducting dusty gas model backwards in time." *Math. Meth. Appl. Sci.* **12**, 183–197.

Cannon, J.R. 1964a. "A priori estimates for continuation of the solution of the heat equation in the space variable." *Ann. Matem. Pura Appl.* **65**, 377–388.

Cannon, J.R. 1964b. "A Cauchy problem for the heat equation." *Ann. Matem. Pura Appl.* **66**, 155–166.

Cannon, J.R. 1964c. "Error estimates for some unstable continuation problems." *J. Soc. Indust. Appl. Math.* **12**, 270–284.

Cannon, J.R. 1984. *The One-Dimensional Heat Equation.* Encyclopedia of Mathematics and its Applications. Vol. **23**. Addison-Wesley. Menlo Park, California.

Cannon, J.R. & Douglas, J. 1967. "The Cauchy problem for the heat equation." *SIAM J. Numer. Anal.* **4**, 317–336.

Cannon, J.R. & Douglas, J. 1968. "The approximation of harmonic and parabolic functions on half-spaces from interior data." Numerical Analysis of Partial Differential Equations. C.I.M.E. 2° Ciclo, Ispra, 1967. Edizioni Cremonese, Roma, pp. 193–230.

Cannon, J.R. & Ewing, R.E. 1976. "A direct numerical procedure for the Cauchy problem for the heat equation." *J. Math. Anal. Appl.* **56**, 7–17.

Cannon, J.R. & Knightly, G.H. 1969. "The approximation of the solution to the heat equation in a half-strip from data specified on the bounding characteristics." *SIAM J. Numer. Anal.* **6**, 149–159.

Cannon, J.R. & Miller, K. 1965. "Some problems in numerical analytic continuation." *SIAM J. Numer. Anal.* **2**, 87–98.

Carasso, A. 1976. "Error bounds in the final value problem for the heat equation." *SIAM J. Appl. Math.* **7**, 195–199.

Carasso, A. 1977. "Computing small solutions of Burgers' equation backwards in time." *J. Math. Anal. Appl.* **59**, 169–209.

Carasso, A. 1982. "Determining surface temperatures from interior observations." *SIAM J. Appl. Math.* **42**, 558–574.

Carasso, A. 1987. "Infinitely divisible pulses, continuous deconvolution, and the characterization of linear time invariant systems." *SIAM J. Appl. Math.* **47**, 892–927.

Carasso, A. 1993. "Slowly divergent space marching schemes in the inverse heat conduction problem." *Num. Heat Transfer B* **23**, 111–126.

Carasso, A. 1994. "Overcoming Hölder continuity in ill-posed continuation problems." *SIAM J. Numer. Anal.* **31**, 1535–1557.

Carasso, A., Sanderson, J.G. & Hyman, J.M. 1978. "Digital removal of random media image degradations by solving the diffusion equation backwards in time." *SIAM J. Numer. Anal.* **15**, 344–367.

Carasso, A. & Stone, A.P. 1975. *Improperly Posed Boundary Value Problems.* Pitman Press, London.

Cattaneo, C. 1948. "Sulla conduzione del calore." *Atti Sem. Mat. Fis. Univ. Modena* **3**, 83–101.

Caviglia, G., Morro, A. & Straughan, B. 1992. "Thermoelasticity at cryogenic temperatures." *Int. J. Nonlinear Mech.* **27**, 251–263.

Chandrasekhariah, D.S. 1986. "Thermoelasticity with second sound: a review." *Appl. Mech. Review* **39**, 355–376.

Chirita, S. & Rionero, S. 1991. "Lagrange identity in linear viscoelasticity." *Int. J. Engng. Sci.* **29**, 1181–1200.

Chorin, A.J. 1985. "Curvature and solidification." *J. of Computational Physics* **57**, 472–490.

Courant, R. & Hilbert, D. 1962. *Methods of Mathematical Physics*, vol. 2, New York, Interscience.

Cowling, T.G. 1976. *Magnetohydrodynamics*, Adam Hilger, The Institute of Physics, Bristol.

Crooke, P.S. 1972. "On growth properties of solutions of the Saffman dusty gas model." *Z.A.M.P.* **23**, 182–200.

Crooke, P.S. 1973. "On the dependence of solutions to a dusty gas model on initial and boundary data backward in time." *J. Math. Anal. Appl.* **42**, 536–544.

Crooke, P.S. & Payne, L.E. 1984. "Continuous dependence on geometry for the backward heat equation." *Math. Meth. Appl. Sci.* **6**, 433–448.

Demmel, J. & Kagstrom, B. 1988. "Accurate solutions of ill-posed problems in control theory." *SIAM J. Math. Anal. Appl.* **9**, 126–145.

Diaz, J.I. & Lions, J.L. 1993. *Mathematics, Climate and Environment.* Masson, Paris.

Di Benedetto, E. & Friedman, A. 1984. "The ill - posed Hele - Shaw model and the Stefan problem for supercooled water." *Trans. Amer. Math. Soc.* **282**, 183–204.

Doering, C.R. & Gibbon, J.D. 1995. *Applied Analysis of the Navier - Stokes Equations.* Cambridge University Press.

Douglas, J. 1960. "A numerical method for analytic continuation, Boundary Value Problems in Differential Equations," University of Wisconsin Press, pp. 179–189.

Drazin, P.G. & Reid, W.H. 1981. *Hydrodynamic Stability.* Cambridge University Press.

Dreyer, W. & Struchtrup, H. 1993. "Heat pulse experiments revisited." *Continuum Mech. Thermodyn.* **5**, 3–50.

Dunninger, D.R. & Levine, H.A. 1976. "Uniqueness criteria for solutions of singular boundary value problems." *Trans. Amer. Math. Soc.* **221**, 289–301.

Elden, L. 1987. "Approximations for a Cauchy problem for the heat equation." *Inverse Problems* **3**, 263–273.

Elden, L. 1988. "Hyperbolic approximations for a Cauchy problem for the heat equation." *Inverse Problems* **4**, 59–70.

Elden, L. 1990. "Algorithms for the computation of functionals defined on the solution of a discrete ill-posed problem." *BIT* **30**, 466-483.

Engl, H.W. & Groetsch, C.W. 1987. *Inverse and Ill-Posed Problems.* Notes and Reports in Mathematics in Science and Engineering. Vol. **4**. Academic Press.

Eringen, A.C. 1966. "Theory of micropolar fluids." *J. Math. Mech.* **16**, 1–18.

Ewing, R.E. 1975. "The approximation of certain parabolic equations backward in time by Sobolev equations." *SIAM J. Math. Anal.* **6**, 283–294.

Ewing, R.E. 1984. "Problems arising in the modeling of processes for hydrocarbon recovery." Vol. I, *Mathematics of Reservoir Simulation.* R.E.Ewing (ed.). Research Frontiers in Applied Math. SIAM, Philadelphia, pp. 3–34.

Ewing, R.E. & Falk, R.S. 1979. "Numerical approximation of a Cauchy problem for a parabolic partial differential equation." *Math. Computation* **33**, 1125–1144.

Ewing, R.E., Lin, T. & Falk, R. 1987. "Inverse and ill-posed problems in reservoir simulation." In *Inverse and Ill-posed Problems. Notes and Reports on Mathematics in Science and Engineering.* Academic Press, pp. 483–497.

Falk, R.S. & Monk, P.B. 1986. "Logarithmic convexity for discrete harmonic functions and the approximation of the Cauchy problem for Poisson's equation." *Math. Computation* **47**, 135–149.

Fichera, G. 1992. "Is the Fourier theory of heat propagation paradoxical?" *Rend. Circolo Matem. Palermo* **41**, 5–28.

Flavin, J.N. 1989. "Upper estimates for a class of non-linear partial differential equations." *J. Math. Anal. Appl.* **144**, 128–140.

Flavin, J.N. 1991. "Some connections between velocity - related and pressure - related quantities in axisymmetric fluid flow." *Int. J. Engng. Sci.* **29**, 969–972.

Flavin, J.N. 1992. "Convexity considerations for the biharmonic equation in plane polars with applications to elasticity." *Q. Jl. Mech. Appl. Math.* **45**, 555–566.

Flavin, J.N. 1993. "Almost uniqueness for a non-linear Dirichlet problem." *Proc. Roy. Irish Acad.* **93A**, 203–207.

Flavin, J.N. 1995. "Qualitative estimates for laminate - like elastic materials." *IUTAM Symp. Anisotropy, Inhomogeneity and Nonlinearity in Solid Mechanics*, ed. D.F. Parker & A.H. England. Kluwer Academic, 339–344.

Flavin, J.N. 1996a. "Integral inequality estimates for pde's in unbounded domains." *Bull. Irish Math. Soc.* **36**, 46–53.

Flavin, J.N. 1996b. "The method of cross sectional integrals for pde's." *Rend. Circolo Matem. Palermo*, in the press.

Flavin, J.N. & Rionero, S. 1993. "Decay and other estimates for an elastic cylinder." *Q. Jl. Mech. Appl. Math.* **46**, 299-309.

Flavin, J.N. & Rionero, S. 1995. *Qualitative Estimates for Partial Differential Equations.* CRC Press, Boca Raton.

Flavin, J.N. & Rionero, S. 1996. "On the temperature distribution in cold ice." To appear.

Flavin, J.N. & Straughan, B. 1997. "Continuous dependence and uniqueness in a theory for simultaneous diffusion of heat and moisture." To appear.

Fox, D. & Pucci, C. 1958. "The Dirichlet problem for the wave equation." *Ann. Mat. Pura Appl.* **46**, 155–182.

Franchi, F. 1984. "Growth estimates in linear elasticity with a sublinear body force without definiteness conditions on the elasticities." *Proc. Edinburgh Math. Soc.* **27**, 223–228.

Franchi, F. 1985. "Wave propagation in heat conducting dielectric solids with thermal relaxation and temperature dependent permittivity." *Riv. Matem. Univ. Parma* **11**, 443–461.

Franchi, F. 1995. "Stabilization estimates for penetrative motions in porous media." *Math. Meth. Appl. Sci.* **17**, 11–20.

Franchi, F. & Straughan, B. 1993a. "Continuous dependence on the body force for solutions to the Navier-Stokes equations and on the heat supply in a model for double diffusive porous convection." *J. Math. Anal. Appl.* **172**, 117–129.

Franchi, F. & Straughan, B. 1993b. "Continuous dependence on modelling in penetrative convection with a nonlinear equation of state." *Riv. Matem. Univ. Parma*, (Ser. 5) **2**, 57–65.

Franchi, F. & Straughan, B. 1993c. "Stability and nonexistence results in the generalized theory of a fluid of second grade." *J. Math. Anal. Appl.* **180**, 122–137.

Franchi, F. & Straughan, B. 1994a. "Continuous dependence on the relaxation time and modelling, and unbounded growth, in theories of heat conduction with finite propagation speeds." *J. Math. Anal. Appl.* **185**, 726–746.

Franchi, F. & Straughan, B. 1994b. "Spatial decay estimates and continuous dependence on modelling for an equation from dynamo theory." *Proc. Roy. Soc. London A* **445**, 437–451.

Franchi, F. & Straughan, B. 1995. "Effects of errors in the initial-time geometry on the solution of an equation from dynamo theory in an exterior domain." *Proc. Roy. Soc. London A* **450**, 109–121.

Franchi, F. & Straughan, B. 1996. "Structural stability for the Brinkman equations of porous media." *Math. Meth. Appl. Sci.* , in the press.

Franklin, J.N. 1974. "On Tikhonov's method for ill-posed problems." *Math. Comp.* **28**, 889–907.

Galdi, G.P., Knops, R.J. & Rionero, S. 1986. "Uniqueness and continuous dependence in the linear elastodynamic exterior and half-space problems." *Math. Proc. Camb. Phil. Soc.* **99**, 357–366.

Galdi, G.P. & Maremonti, P. 1986. "A uniqueness theorem for viscous fluid motions in exterior domains." *Arch. Rational Mech. Anal.* **91**, 375–384.

Galdi, G.P. & Straughan, B. 1988. "Stability of solutions to the Navier-Stokes equations backward in time." *Arch. Rational Mech. Anal.* **101**, 107–114.

Ghidaglia, J.M. 1986. "Some backward uniqueness results." *Nonlin. Anal., Theory, Meths., Applicns.* **10**, 777–790.

Ginsberg, F. 1963. "On the Cauchy problem for the one-dimensional heat equation." *Math. Comp.* **17**, 257–269.

Green, A.E. 1962. "Thermoelastic stresses in initially stressed bodies." *Proc. Roy. Soc. London A* **266**, 1–19.

Green, A.E. 1972. "A note on linear thermoelasticity." *Mathematika* **19**, 69–75.

Han, H. 1982. "The finite element method in a family of improperly posed problems." *Math. Comp.* **38**, 55–65.

Hansen, P.C. 1992. "Analysis of discrete ill-posed problems by means of the L-curve." *SIAM Review* **34**, 561–580.

Hawking, S. 1993. *Black Holes and Baby Universes and Other Essays.* Bantam Press, New York.

Hörmander, L. 1976. *Linear Partial Differential Operators.* Springer-Verlag, New York - Berlin.

Hutter, K. 1984. *Theoretical Glaciology.* D. Reidel publishing Co.

Iesan, D. 1989. *Prestressed Bodies.* Pitman Res. Notes Math. vol. **195**. Longman; Harlow.

Ingham, D.B., Yuan, Y. & Han, H. 1991. "The boundary element method for an improperly posed problem." *IMA J. Appl. Math.* **47**, 61–79.

John, F. 1955. "A note on improper problems in partial differential equations." *Comm. Pure Appl. Math.* **8**, 494–495.

John, F. 1960. "Continuous dependence on data for solutions of partial differential equations with a prescribed bound." *Comm. Pure Appl. Math.* **13**, 551–585.

Joseph, D.D., Renardy, M. & Saut, J.C. 1985. "Hyperbolicity and change of type in the flow of viscoelastic fluids." *Arch. Rational Mech. Anal.* **87**, 213–251.

Joseph, D.D. & Saut, J.C. 1990. "Short-wave instabilities and ill-posed initial-value problems." *Theoret. Comput. Fluid Dyn.* **1**, 191–227.

Jou, D., Casas - Vazquez, J. & Lebon, G. 1988. "Extended irreversible thermodynamics." *Rep. Prog. Phys.* **51**, 1105–1179.

Kazemi, A. & Klibanov, M.V. 1993. "Stability estimates for ill - posed Cauchy problems involving hyperbolic equations and inequalities." *Applicable Analysis* **50**, 93–102.

Klibanov, M.V. & Rakesh. 1992. "Numerical solution of a time-like Cauchy problem for the wave equation." *Math. Meth. Appl. Sci.* **15**, 559–570.

Klibanov, M.V. & Santosa, F. 1991. "A computational quasi-reversibility method for Cauchy problems for Laplace's equation." *SIAM J. Appl. Math.* **51**, 1653–1675.

Knabner, P. & Vessella, S. 1987a. "Stability estimates for ill-posed Cauchy problems for parabolic equations." In *Inverse and Ill-Posed Problems* Engl, H.W. & Groetsch, C.W. eds., pp. 351–368, Academic Press, New York.

Knabner, P. & Vessella, S. 1987b. "Stabilization of ill-posed Cauchy problems for parabolic equations." *Ann. Mat. Pura Appl.* **149**, 383–409.

Knabner, P. & Vessella, S. 1988. "The optimal stability estimate for some ill-posed Cauchy problems for a parabolic equation." *Math. Meth. Appl. Sci.* **10**, 575–583.

Knops, R.J. & Payne, L.E. 1968. "On the stability of the Navier-Stokes equations backward in time." *Arch. Rational Mech. Anal.* **29**, 331–335.

Knops, R.J. & Payne, L.E. 1969. "Continuous data dependence for the equations of classical elastodynamics." *Proc. Camb. Phil. Soc.* **66**, 481–491.

Knops, R.J. & Payne, L.E. 1988. "Improved estimates for continuous data dependence in linear elastodynamics." *Math. Proc. Camb. Phil. Soc.* **103**, 535–559.

Kutev, N. & Tomi, F. 1995. "Nonexistence and instability in the exterior Dirichlet problem for the minimal surface equation in the plane." *Pacific J. Math.* **170**, 535–542.

Landis, E.M. 1984. "On the dependence on the geometry of the domain of the uniqueness classes of the solution of the second initial - boundary value problem for the heat equation in an unbounded domain." *Soviet Math. Dokl.* **29**, 292–295.

Lattès, R. & Lions, J.L. 1969. *The Method of Quasireversibility, Applications to Partial Differential Equations.* Elsevier, New York.

Lavrentiev, M.M. 1956. "On the Cauchy problem for the Laplace equation" (In Russian). *Izvest. Akad. Nauk SSSR (Ser. Matem.)* **20**, 819–842.

Lavrentiev, M.M. 1967. *Some Improperly Posed Problems in Mathematical Physics.* Springer - Verlag, New York.

Lavrentiev, M.M., Romanov, V.G. & Sisatskii, S.P. 1983. *Problemi non ben Posti in Fisica, Matematica e Analisi.* Pubbl. dell'Ist. Analisi Globale e Appl., Vol. 12, Firenze.

Lees, M. & Protter, M.H. 1961. "Unique continuation for parabolic differential equations and differential inequalities." *Duke Math. J.* **28**, 369–382.

Levine, H.A. 1970. "Logarithmic convexity and the Cauchy problem for some abstract second order differential inequalities." *J. Differential Equations* **8**, 34–55.

Levine, H.A. 1972. "Some uniqueness and growth theorems in the Cauchy problem for $Pu_{tt}+Mu_t+Nu=0$ in Hilbert space." *Math. Z.* **126**, 345–360.

Levine, H.A. 1977. "An equipartition of energy theorem for weak solutions of evolutionary equations in Hilbert space: the Lagrange identity method." *J. Differential Equations* **24**, 197–210.

Levine, H.A. 1983. "Continuous data dependence, regularization, and a three lines theorem for the heat equation with data in a space like direction." *Ann. Mat. Pura Appl.* **134**, 267–286.

Levine, H.A. & Payne, L.E. 1985. "On an ill posed Cauchy problem for the wave equation." *Office of Naval Research Report.*

Levine, H.A. & Vessella, S. 1985a. "Stabilization and regularization for solutions of an ill-posed problem for the wave equation." *Math. Meth. Appl. Sci.* **7**, 202–209.

Levine, H.A. & Vessella, S. 1985b. "Estimates and regularization for solutions of some ill-posed problems of elliptic and parabolic type." *Rend. Circolo Matem. Palermo* **34**, 141–160.

Lin, C. & Payne, L.E. 1993. "On the spatial decay of ill-posed parabolic problems." *Math. Models Meth. Appl. Sci.* **3**, 563–575.

Lin, C. & Payne, L.E. 1994. "The influence of domain and diffusivity perturbations on the decay of end effects in heat conduction." *SIAM J. Math. Anal.* **25**, 1242–1258.

Lin, T. & Ewing, R.E. 1992. "Direct numerical method for an inverse problem of hyperbolic equations." *Numer. Meth. Part. Diff. Equns.* **8**, 551–574.

Lortz, D. & Meyer-Spasche, R. 1982. "On the decay of symmetric toroidal dynamo fields." *Z. Naturforschung* **37a**, 736–740.

McKay, G. 1992. *Nonlinear Stability Analyses of Problems in Patterned Ground Formation and Penetrative Convection.* Ph.D. Thesis, Univ. Glasgow.

Maiellaro, M., Palese, L. & Labianca, A. 1989. "Instabilizing - stabilizing effects of MHD anisotropic currents." *Int. J. Engng. Sci.* **27**, 1353–1359.

Manselli, P. & Miller, K. 1980. "Calculation of the surface temperature and heat flux on one side of a wall from measurements on the opposite side." *Ann. Mat. Pura Appl.* **123**, 161–183.

Maxwell, J.C. 1867. "On the dynamical theory of gases." *Philos. Trans. Roy. Soc. London A* **157**, 49–88.

Miller, K. 1970. "Least squares methods for ill-posed problems with a prescribed bound." *SIAM J. Math. Anal.* **1**, 82–89.

Miller, K. 1973a. "Non-unique continuation for certain ODE's in Hilbert space and for uniformly parabolic elliptic equations in self-adjoint divergence form." *Symposium on Non-Well Posed problems and Logarithmic Convexity.* Springer Lecture Notes in Math. **316**, 85–101.

Miller, K. 1973b. "Stabilized quasi-reversibility and other nearly best possible methods for non well posed problems." *Symposium on Non-Well Posed problems and Logarithmic Convexity.* Springer Lecture Notes in Math. **316**, 161–176.

Miller, K. 1974. "Nonunique continuation for uniformly parabolic and elliptic equations in self-adjoint divergence form with Hölder continuous coefficients." *Arch. Rational Mech. Anal.* **54**, 105–117.

Miller, K. 1975. "Logarithmic convexity results for holomorphic semigroups." *Pacific J. Math.* **58**, 549–551.

Monk, P. 1986. "Error estimates for a numerical method for an ill-posed Cauchy problem for the heat equation." *SIAM J. Numer. Anal.* **23**, 1155–1172.

Morozov, V.A. 1984. *Methods for Solving Incorrectly Posed Problems.* Springer-Verlag, Berlin-Heidelberg-New York.

Morro, A., Payne, L.E. & Straughan, B. 1990. "Decay, growth, continuous dependence and uniqueness results in generalized heat conduction theories." *Applicable Analysis* **38**, 231–243.

Morro, A., Payne, L.E. & Straughan, B. 1993. "Growth and pulse propagation in a fluid mixture." *J. Non-Equilm. Thermodyn.* **18**, 135–146.

Morro, A., & Ruggeri, T. 1984. *Propagazione del Calore ed Equazioni Costitutive.* Quaderni del C.N.R., G.N.F.M. Pitagora, Bologna.

Morro, A., & Ruggeri, T. 1988. "Non-equilibrium properties of solids obtained from second-sound measurements." *J. Phys. C: Solid State Phys.* **21**, 1743–1752.

Morro, A. & Straughan, B. 1990a. "Time-invariant uniqueness for reacting viscous fluids near chemical equilibrium." *Math. Methods Appl. Sci.* **13**, 137–141.

Morro, A. & Straughan, B. 1990b. "Equations for reacting viscous fluids near chemical equilibrium." *Riv. Mat. Univ. Parma* **16**, 173–182.

Morro, A. & Straughan, B. 1991. "A uniqueness theorem in the dynamical theory of piezoelectricity." *Math. Methods Appl. Sci.* **14**, 295–299.

Morro, A. & Straughan, B. 1992. "Continuous dependence on the source parameters for convective motion in porous media." *Nonlinear Analysis, Theory, Meths., Applicns.* **18**, 307–315.

Müller, I. 1966. *Zur Ausbreitungsgeschwindigkeit von Störungen in kontinuierlichen Medien.* Dissertation, Technische Hochschule, Aachen.

Müller, I. 1967. "Zum Paradox der Wärmeleitungstheorie." *Zeit. Phys.* **198**, 329–344.

Müller, I. 1985. *Thermodynamics.* Pitman: Boston - London - Melbourne.

Müller, I. & Villaggio, P. 1976. "Conditions of stability and wave speeds for fluid mixtures." *Meccanica* **11**, 191–195.

Mulone, G., Rionero, S. & Straughan, B. 1992. "Continuous dependence on modeling for an improperly posed problem for the equations of magnetohydrodynamics." *Ricerche Matem.* **41**, 197–207.

Murray, A.C. & Protter, M.H. 1973. "The asymptotic behaviour of solutions of second order systems of partial differential equations." *J.Differential Equations* **13**, 57–80.

Nield, D.A. & Bejan, A. 1992. *Convection in Porous Media.* Springer - Verlag, New York.

Papi Frosali, G. 1979. "On the stability of the Dirichlet problem for the vibrating string equation." *Ann. Scuola Norm. Sup. Pisa* (Ser VI), **6**, 719–728.

Payne, L.E. 1960. "Bounds in the Cauchy problem for Laplace's equation." *Arch. Rational Mech. Anal.* **5**, 35–45.

Payne, L.E. 1964. "Uniqueness criteria for steady state solutions of the Navier-Stokes equations." In *Atti del Simposio Internazionale sulle Applicazioni dell'Analaisi alla Fisica Matematica,* Cagliari - Sassari, 130–153.

Payne, L.E. 1966. "On some non well posed problems for partial differential equations." In *Numerical Solutions of Nonlinear Differential Equations,* ed. D. Greenspan. Wiley.

Payne, L.E. 1967. "On the stability of solutions to the Navier-Stokes equations and convergence to steady state." *SIAM J. Appl. Math.* **15**, 392–405.

Payne, L.E. 1970. "On a priori bounds in the Cauchy problem for elliptic equations." *SIAM J. Math. Anal.* **1**, 82–89.

Payne, L.E. 1971. "Uniqueness and continuous dependence criteria for the

Navier-Stokes equations." *Rocky Mtn. J. Math.* **2**, 641–660.

Payne, L.E. 1975. *Improperly Posed Problems in Partial Differential Equations.* Regional Conf. Ser. Appl. Math., SIAM.

Payne, L.E. 1985. "Improved stability estimates for classes of ill-posed Cauchy problems." *Applicable Analysis* **19**, 63–74.

Payne, L.E. 1987a. "On stabilizing ill-posed problems against errors in geometry and modeling." *Proc. Conf. on Inverse and Ill-posed Problems: Strobhl.* H.Engel and C.W.Groetsch (eds.) pp. 399–416. Academic Press, New York.

Payne, L.E. 1987b. "On geometric and modeling perturbations in partial differential equations." In *Proc. L.M.S. Symp. on Non-Classical Continuum Mechanics*, Knops, R.J. & Lacey, A.A. (eds.) pp. 108–128. Cambridge University Press.

Payne, L.E. 1989. "Continuous dependence on geometry with applications in continuum mechanics." In *Continuum Mechanics and its Applications*, Graham, G.A.C. & Malik, S.K., eds., pp. 877–890. Hemisphere Publ. Co.

Payne, L.E. 1992. "Some remarks on ill-posed problems for viscous fluids." *Int. J. Engng. Sci.* **30**, 1341–1347.

Payne, L.E. 1993a. "Continuous dependence on spatial geometry for solutions of the Navier-Stokes equations backward in time." *Nonlinear Anal., Theory, Meths., Applicns.* **21**, 651–664.

Payne, L.E. 1993b. "On stabilizing ill-posed Cauchy problems for the Navier-Stokes equations." In, *Differential Equations with Applications to Mathematical Physics.* Academic Press, 261–271.

Payne, L.E. & Sather, D. 1967a. "On singular perturbations of non-well posed problems." *Ann. Mat. Pura Appl.* **75**, 219–230.

Payne, L.E. & Sather, D. 1967b. "On some improperly posed problems for the Chaplygin equation." *J. Math. Anal. Appl.* **19**, 67–77.

Payne, L.E. & Sather, D. 1967c. "On some improperly posed problems for quasilinear equations of mixed type." *Trans. Amer. Math. Soc.* **128**, 135–141.

Payne, L.E., Song, J.C. & Straughan, B. 1988. "Double diffusive porous penetrative convection; thawing subsea permafrost." *Int. J. Engng. Sci.* **26**, 797–809.

Payne, L.E. & Straughan, B. 1989a. "Order of convergence estimates on the interaction term for a micropolar fluid." *Int. J. Engng. Sci.* **27**, 837–846.

Payne, L.E. & Straughan, B. 1989b. "Comparison of viscous flows backward in time with small data." *Int. J. Nonlinear Mech.* **24**, 209–214.

Payne, L.E. & Straughan, B. 1989c. "Critical Rayleigh numbers for oscillatory and nonlinear convection in an isotropic thermomicropolar fluid." *Int. J. Engng. Sci.* **27**, 827–836.

Payne, L.E. & Straughan, B. 1990a. "Improperly posed and non-standard problems for parabolic partial differential equations." *Elasticity, Mathematical Methods and Applications: The Ian Sneddon 70th Birthday Volume.* G.Eason & R.W.Ogden (eds.) pp. 273–299. Ellis-Horwood Pub.

Payne, L.E. & Straughan, B. 1990b. "Error estimates for the temperature of a piece of cold ice, given data on only part of the boundary." *Nonlinear Analysis, Theory, Meths., Applicns.* **14**, 443–452.

Payne, L.E. & Straughan, B. 1990c. "Effects of errors in the initial-time geometry on the solution of the heat equation in an exterior domain." *Q. Jl. Mech. Appl. Math.* **43**, 75–86.

Payne, L.E. & Straughan, B. 1996a. "Stability in the initial - time geometry problem for the Brinkman and Darcy equations of flow in porous media." *J. Math. pures et appl.* **75**, 225–271.

Payne, L.E. & Straughan, B. 1996b. "Continuous dependence on geometry for the backward heat equation in an exterior domain." To appear.

Payne, L.E. & Weinberger, H.F. 1958. "New bounds for solutions of second order elliptic partial differential equations." *Pacific J. Math.* **8**, 551–573.

Persens, J. 1986. *On Stabilizing Ill-Posed Problems for Partial Differential Equations Under Perturbations of the Geometry of the Domain.* Ph.D. Thesis, Cornell University.

Plis, A. 1960. "Nonuniqueness of Cauchy's problem for differential equations of elliptic type." *J. Math. Mech.* **9**, 557–562.

Plis, A. 1963. "On nonuniqueness in the Cauchy problem for an elliptic second order equation." *Bull. Acad. Polon. Sci.* **11**, 95-100.

Protter, M.H. 1953. "Uniqueness theorems for the Tricomi equation." *J. Rational Mech. Anal. 2*, 107–114.

Protter, M.H. 1954. "New boundary value problems for the wave equation and equations of mixed type." *J. Rational Mech. Anal. 3*, 435–446.

Protter, M.H. 1960. "Unique continuation for elliptic equations." *Trans. Amer. Math. Soc. 95*, 81–91.

Protter, M.H. 1961. "Properties of solutions of parabolic equations and inequalities." *Canadian J. Math. 13*, 331–345.

Protter, M.H. 1962. "Asymptotic behavior of solutions of hyperbolic inequalities." *Bull. Amer. Math. Soc. 68*, 523–525.

Protter, M.H. 1963. "Asymptotic behavior and uniqueness theorems for hyperbolic operators." In *Proc. Symp. Part. Diff. Equations*, Novosibirsk, pp. 348–353. Acad. Sci., USSR Siberian Branch, Moscow.

Protter, M.H. 1967. "Difference Methods and Soft Solutions." In *Nonlinear Partial Differential Equations*, Academic Press, New York & London.

Protter, M.H. 1974. "Asymptotic decay for ultrahyperbolic operators." In *Contributions to Analysis*, Academic Press, New York & London.

Protter, M.H. & Weinberger, H.F. 1967. *Maximum Principles in Differential Equations.* Prenctice-Hall, Inc. Englewood Cliffs, New Jersey.

Pucci, C. 1955. "Sui problemi di Cauchy non "ben posti"." *Rend. Accad. Naz. Lincei* **18**, 473–477.

Pucci, C. 1961. "Su un problema esterno per la equazione delle onde." *Ann. Mat. Pura Appl.* **56**, 69–78.

Ramm, A.G. 1995. "Examples of nonuniqueness for an inverse problem of geophysics." *Appl. Math. Lett.* **8**, 87–89.

Rellich, F. 1933. "Zur ersten Randwertaufgabe bei Monge-Ampèreschen Differentialgleichungen vom elliptischen Typus; differential geometrische Anwendungen." *Math. Annalen* **107**, 505–513.

Richardson, L.L. 1993. *Nonlinear Analyses for Variable Viscosity and Compressible Convection Problems.* Ph.D. Thesis, Univ. Glasgow.

Richardson, L.L. & Straughan, B. 1993. "Convection with temperature dependent viscosity in a porous medium: nonlinear stability and the Brink - man effect." *Atti Accad. Naz. Lincei* (Ser. IX) **4**, 223–230.

Rionero, S. & Galdi, G.P. 1976. "On the uniqueness of viscous fluid motions." *Arch. Rational Mech. Anal.* **62**, 295–301.

Rionero, S. & Chirita, S. 1987. "The Lagrange identity method in linear thermoelasticity." *Int. J. Engng. Sci.* **25**, 935–947.

Rionero, S. & Chirita, S. 1989. "New reciprocal and continuous dependence theorems in the linear theory of viscoelasticity." *Int. J. Engng. Sci.* **27**, 1023–1036.

Rogers, C. & Ames, W.F. 1989. *Nonlinear Boundary Value Problems in Science and Engineering.* Math. in Sci. & Engng. Ser., **183**, Academic Press.

Russo, R. 1987. "Uniqueness theorems for a partial differential system of parabolic type in unbounded domains." *Commun. Part. Diff. Equations* **12**, 883–902.

Russo, R. 1991. "On the hyperbolicity condition in linear elasticity." *Le Matematiche* **46**, 393–402.

Schaefer, P.W. 1965. "On the Cauchy problem for elliptic equations." *Arch.*

Rational Mech. Anal. **20**, 391–412.

Schaefer, P.W. 1967. "Pointwise bounds in the Cauchy problem for an elliptic system." *SIAM J. Appl. Math.* **15**, 665–677.

Seidman, T.I. & Elden, L. 1990. "An optimal filtering method for the sideways heat equation." *Inverse Problems* **6**, 681–696.

Serrin, J. 1963. "The initial value problem for the Navier-Stokes equations." In *Nonlinear Problems.* Univ. Wisconsin Press, Madison.

Showalter, R.E. 1974. "The final value problem for evolution equations." *J. Math. Anal. Appl.* **47**, 563–572.

Showalter, R.E. 1975. "Quasi-Reversibility of First and Second Order Parab -olic Evolution Equations." In *Pitman Res. Notes. Math.* **1**, 76–84.

Sigillito, V.G. 1977. *Explicit a Priori Inequalities with Applications to Boundary Value Problems.* Pitman Res. Notes Math., **13**, Pitman Press, London.

Song, J.C. 1988. *Some Stability Criteria in Fluid and Solid Mechanics.* Ph.D. thesis, Cornell University.

Straughan, B. 1975. "Uniqueness and stability for the conduction-diffusion solution to the Boussinesq equations backward in time." *Proc. Roy. Soc. London A* **347**, 435–446.

Straughan, B. 1976. "Growth and uniqueness theorems for an abstract nonstandard wave equation." *SIAM J. Math. Anal.* **7**, 519–528.

Straughan, B. 1977. "Growth and instability theorems for wave equations with dissipation, with applications in contemporary continuum mechanics." *J. Math. Anal. Appl.* **61**, 303–330.

Straughan, B. 1982. *Instability, Nonexistence and Weighted Energy Methods in Fluid Dynamics and Related Theories.* Pitman Res. Notes Math., vol. **74**, London.

Straughan, B. 1983. "Backward uniqueness and unique continuation for solutions to the Navier-Stokes equations on an exterior domain." *J. Math. pures et appl.* **62**, 49–62.

Straughan, B. 1985. "The impossibility of complete decay in a compressible, symmetric, toroidal dynamo." *ZAMP* **36**, 179–183.

Straughan, B. 1992. *The Energy Method, Stability, and Nonlinear Convection.* Springer-Verlag: Ser. in Appl. Math. Sci., vol. **91**.

Straughan, B. 1993. *Mathematical Aspects of Penetrative Convection.* Pitman Res. Notes Math., vol. **288**, Longman, Harlow.

Talenti, G. 1978. "Sui problemi mal posti." *Boll. Unione Matem. Ital.* **15A**, 1–29.

Talenti, G. & Vessella, S. 1982. "A note on an ill-posed problem for the heat equation." *J. Austral. Math. Soc. A* **32**, 358–368.

Thess, A. & Bestehorn, M. 1995. "Planform selection in Bénard-Marangoni convection: *l* hexagons versus *g* hexagons." *Phys. Rev. E* **52**, 6358–6367.

Tikhonov, A.N. 1963. "On the solution of ill-posed problems and the method of regularization." *Dokl. Akad. Nauk SSSR* **151**, 501–504.

Tikhonov, A.N. & Arsenin, V.V. 1977. *Solutions of Ill Posed Problems.* V.H. Winston - Wiley, New York.

Trytten, G.N. 1963. "Pointwise bounds for solutions in the Cauchy problem for elliptic equations." *Arch. Rational Mech. Anal.* **13**, 222–244.

Weinstein, A. 1954. "On the wave equation and the equation of Euler - Poisson." Fifth Symp. Appl. Math., pp. 137–147. McGraw-Hill, New York.

Index

M

N

O

Q

S

T

U

W

Y

Mathematics in Science and Engineering
Edited by William F. Ames, Georgia Institute of Technology

Recent titles

T.A. Burton, *Volterra Integral and Differential Equations*

Ram P. Kanwal, *Generalized Functions: Theory and Technique*

Marc Mangel, *Decision and Control in Uncertain Resource Systems*

John L. Casti, *Nonlinear System Theory*

Yoshikazu Sawaragi, Hirotaka Nakayama, and Tetsuzo Tanino, *Theory of Multi-objective Optimization*

Edward J. Haug, Kyung K. Choi, and Vadim Komkov, *Design Sensitivity Analysis of Structural Systems*

Yaakov Bar-Shalom and Thomas E. Fortmann, *Tracking and Data Association*

V.B. Kolmanovskii and V.R. Nosov, *Stability of Functional Differential Equations*

V. Lakshmikantham and D. Trigiante, *Theory of Difference Equations: Applications to Numerical Analysis*

B.D. Vujanovic and S.E. Jones, *Variational Methods in Nonconservative Phenomena*

C. Rogers and W.F. Ames, *Nonlinear Boundary Value Problems in Science and Engineering*

W.F. Ames and C. Rogers, *Nonlinear Equations in the Applied Sciences*

Josip E. Pecaric, Frank Proschan, and Y.L. Tong, *Convex Functions, Partial Orderings, and Statistical Applications*

E.N. Chukwu, *Stability and Time-Optimal Control of Hereditary Systems*

Viorel Barbu, *Analysis and Control of Nonlinear Infinite Dimensional Systems*

Yang Kuang, *Delay Differential Equations: With Applications in Population Dynamics*

K.A. Ames and B. Straughan, *Non-Standard and Improperly Posed Problems*

Z. Gajic and M.T.J. Qureshi, *The Lyapunov Matrix Equation in System Stability and Control*

D. Straub, *Alternative Mathematical Theory of Non-equilibrium Phenomena*